# Teaching Einsteinian Physics in Schools



In our world today, scientists and technologists speak one language of reality. Everyone else, whether they be prime ministers, lawyers, or primary school teachers speak an outdated Newtonian language of reality.

While Newton saw time and space as rigid and absolute, Einstein showed that time is relative – it depends on height and velocity – and that space can stretch and distort. The modern Einsteinian perspective represents a significant paradigm shift compared with the Newtonian paradigm that underpins most of the school education today. Research has shown that young learners quickly access and accept Einsteinian concepts and the modern language of reality. Students enjoy learning about curved space, photons, gravitational waves, and time dilation; often, they ask for more!

A consistent education within the Einsteinian paradigm requires rethinking of science education across the entire school curriculum, and this is now attracting attention around the world. This book brings together a coherent set of chapters written by leading experts in the field of Einsteinian physics education. The book begins by exploring the fundamental concepts of space, time, light, and gravity and how teachers can introduce these topics at an early age. A radical change in the curriculum requires new learning instruments and innovative instructional approaches. Throughout the book, the authors emphasise and discuss evidence-based approaches to Einsteinian concepts, including computer-based tools, geometrical methods, models and analogies, and simplified mathematical treatments.

*Teaching Einsteinian Physics in Schools* is designed as a resource for teacher education students, primary and secondary science teachers, and for anyone interested in a scientifically accurate description of physical reality at a level appropriate for school education.

**Magdalena Kersting** is an educational researcher, physics educator, and science communicator with a keen interest in getting students excited about Einsteinian physics. Based at the University of Oslo in Norway, Magdalena collaborates with scientists, educators, and teachers around the world to bring great science education to as many students as possible. She believes that Einsteinian physics education can help students think more clearly about complex ideas and, eventually, enable them to build a better future.

**David Blair** is a pioneer in gravitational wave research with a passion for education. This led him to found both the Australian International Gravitational Research Centre and the Gravity Discovery Centre. The research centre played a significant role in the discovery of gravitational waves, and is now part of the ARC Centre of Excellence for Gravitational Wave Discovery, OzGrav. The Gravity Discovery Centre was set up as an education centre to bring Einsteinian physics to schools and the public. He leads the Einstein-First Project, which is developing a complete Einsteinian physics science curriculum from primary to high school.

# Teaching Einsteinian Physics in Schools

## An Essential Guide for Teachers in Training and Practice

**Edited by Magdalena Kersting and David Blair**

Routledge
Taylor & Francis Group
LONDON AND NEW YORK

First published 2022
by Routledge
2 Park Square, Milton Park, Abingdon, Oxon OX14 4RN

and by Routledge
52 Vanderbilt Avenue, New York, NY 10017

Routledge is an imprint of the Taylor & Francis Group, an informa business



British Library Cataloguing-in-Publication Data
A catalogue record for this book is available from the British Library

Library of Congress Cataloging-in-Publication Data
A catalog record has been requested for this book

ISBN: 9780367752590 (hbk)
ISBN: 9781760877712 (pbk)
ISBN: 9781003161721 (ebk)

Typeset in Univers LT Std
by KnowledgeWorks Global Ltd.

# Contents

# List of figures

# List of tables

# List of contributors

**David Blair** is Emeritus Professor at the University of Western Australia and a pioneer in gravitational wave research in Australia. Working on methods for their detection for more than 40 years, Professor Blair also founded the Australian International Gravitational Research Centre. This centre became a Western Australian node of the ARC Centre of Excellence for Gravitational Wave Discovery, OzGrav. In 2014, Professor Blair began leading the Einstein-First Project, aimed at introducing Einsteinian Physics to children. The project's partners include Curtin University, Edith Cowan University, the Graham (Polly) Farmer Foundation, the US Air Force Academy, and the Gravity Discovery Centre.

**Jackie Bondell** is Education and Public Outreach Coordinator for OzGrav, the ARC Centre of Excellence for Gravitational Wave Discovery, developing educational content for public outreach events and curriculum for school incursion programmes, focussing on incorporating innovative technology. She designs curriculum-driven science lessons that effectively incorporate VR into student-centred activities while aligning with curriculum standards and learning outcomes. With a passion for promoting STEM to traditionally underrepresented populations, she aspires to reach more female students as well as those in rural and regional areas of Australia. Prior to joining OzGrav, Jackie spent 15 years as a physics instructor in the US. She holds a Master's Degree in Astrophysics, and is a National (US) Board Certified Teacher of Secondary Physical Science.

**Rahul Choudhary** is a PhD researcher at the University of Western Australia. His research interest includes teaching Einsteinian physics and optics. He has previously completed his Master's degree in Physics from India. He is also a member of OzGrav, the ARC Centre of Excellence for Gravitational Wave Discovery. He is active in science outreach and science communication.

**Stanley Delhaye** is a postdoctoral researcher in physics education at the Eindhoven School of Education at Eindhoven University of Technology. His research focusses on teaching general relativity to upper-secondary students and introducing general relativity to the Dutch physics curriculum.

**Lesley de Putter** is assistant professor in science education at Eindhoven University of Technology. Her research focusses on physics and chemistry teaching with modern technology and modern teaching strategies such as VR in the classroom, context-based education, inquiry learning, and research-based practical assignments.

**Stuart Farmer** is the Education Manager at the Institute of Physics (IOP) in Scotland. Prior to this, he taught physics in Scotland for over 30 years. Throughout this time, he has been active in curriculum and assessment development and providing professional learning for teachers. He was Chair of the Scottish Qualifications Authority (SQA) Physics Panel overseeing the Scottish physics curriculum and its assessment for seven years. He has also been Chair of the Association for Science Education (ASE) Trustees, ASE Scotland, and IOP Scotland. Since 2014, he has worked in partnership with the Perimeter Institute in Canada to develop and deliver professional learning for teachers.

**Joris van Heijningen** is a Research Logistician at UCLouvain. He is an experienced instrumental researcher and lecturer who has worked at different large experiments such as CERN, SLAC, and the Virgo and KAGRA gravitational wave observatories. He is active in science outreach and science communication.

**Ellen Karoline Henriksen** is professor of physics education and head of the physics education research group at the University of Oslo, Norway. Her research is related to teaching and learning physics in school and higher education, science teacher education and competence, as well as to students' motivation and educational choices. Her teaching is within science teacher education (pre-service and in-service).

**Friedrich Herrmann** is a retired professor of physics didactics at the Karlsruhe Institute of Technology. He was director of the Department of Physics Didactics where he and his research group developed the Karlsruhe Physics Course (KPK). Throughout his career, he also worked as a physics teacher at a secondary school. There, the concepts of the KPK were tested under real classroom conditions. His work as a university teacher was honoured with the Baden-Württemberg State Teaching Award. At the moment, he is engaged in the dissemination of the KPK in China.

**Floor Kamphorst** is a physics teacher at a secondary school, and a PhD researcher in physics education from the Freudenthal Institute for Science and Mathematics Education, Utrecht University. Her research is on teaching and learning special relativity in upper secondary school.

**Tejinder (Jyoti) Kaur** is a research fellow of Physics Education at the University of Western Australia. She has a PhD in teaching Einsteinian physics to high-school students. She is now continuing her research in the same field, focussing on creating and testing Einsteinian curriculum at primary and secondary school levels. She loves to read scientific articles before her five-year-old son and feels extremely happy to answer his curious questions.

**Magdalena Kersting** is an educational researcher, physics educator, and science communicator working to bring great science education to as many people as possible. She works as a postdoctoral researcher at the University of Oslo, Norway, and studies how students experience abstract knowledge in digital and virtual learning environments in formal and informal learning contexts. As part of her PhD research, Magdalena developed an online learning environment about Albert Einstein's theory of general relativity, and she continues to have a keen interest in getting students excited about Einsteinian physics.

**Hongbin Kim** is a postdoctoral researcher in the physics education department at Seoul National University. His research focusses on general relativity education for physics teachers. A historical and cultural approach by exploring the original science literature in physics is the key feature of his study.

**Ute Kraus** is professor of physics and physics education at Hildesheim University, Germany. She works on new approaches for teaching relativity at different levels: secondary school, physics teacher education, and undergraduate physics. Her other research interests are relativistic visualisation and the theory of binary X-ray pulsars. She is maintaining Open Educational Resources (spacetimetravel.org) that make her research outcomes available for teaching (joint work with Corvin Zahn).

**Gyoungho Lee** is professor and head of the department of physics education at Seoul National University, South Korea. He is interested in the establishment of new science education based on the meaningful understanding of science. He is teaching several courses, "Foundation of Physics Education", "Theory and Practices in Physics Teaching", and "Historical Development of Physics Concepts", etc.

**Karl-Heinz Lotze** is a theoretical physicist who, in the 1990s, started a second career in physics and astronomy education at the Faculty of Physics and Astronomy of the Friedrich Schiller University Jena, Germany. His main interest is teaching relativity theories, relativistic astrophysics, and cosmology to various audiences. He is now a senior professor at the Wilhelm and Else Heraeus Foundation.

**Carolyn Maxwell** is a PhD researcher with the Einstein-First Project at The University of Western Australia and a member of OzGrav, the ARC Centre of Excellence for Gravitational Wave Discovery. She is a qualified science teacher and holds a Master's degree in Education from The University of Western Australia; she was awarded the Australian College of Educators'

Prize for her Master's work. Her current research includes investigating students' learning progressions in the Einstein-First project's modernised physics curriculum and evaluating whether their conceptual understanding of Einsteinian physics is sustained over time. Carolyn is passionate about education and increasing the uptake of STEM subjects in schools by students from all backgrounds.

**Mark Myers** has an undergraduate degree in Computer Science and a Masters in Astronomy and Astrophysics. He currently designs and builds animations and interactive experiences for the Australian Research Council Centre of Excellence for Gravitational Wave Discovery (OzGrav) as a part of the outreach team. He has worked as a software developer and consultant, and has also worked professionally as an actor, circus performer, pilates teacher, acting and stage combat instructor and tall ship sailor. He has a passion for astrophysics and science in general, and the human experience within it and aims to bring the joy of science for the general public in an accessible form.

**Birgit Pepin** is professor of STEM education at Eindhoven University of Technology. She supervises PhD students and post-docs investigating the teaching and learning of mathematics and science, amongst them those who research quantum mechanics and general relativity education. In particular, her research is concerned with (digital) curriculum resources designed/used for innovative teaching in secondary and university education.

**Michael Pohlig** is a retired high school physics teacher. Since 2007, he has had a teaching position at the Karlsruhe Institute of Technology (KIT), and is based at Didactics of Physics. He participated in the development of a new physics curriculum and specialised in the didactics of relativity. He is involved in the dissemination of the Karlsruhe Physics Course (KPK) in China.

**Markus Pössel** is a theoretical physicist turned astronomical outreach scientist. He is the managing scientist at the Centre for Astronomy Education and Outreach Haus der Astronomie in Heidelberg, Germany, and the director of the International Astronomical Union's Office of Astronomy for Education.

**Matej Ryston** is a physics teacher and PhD researcher in the Department of Physics Education at Charles University in Prague. In his research, he develops new ways of introducing general relativity in secondary education with practical equipment as well as visualisations of relativistic phenomena, and runs special school programmes popularising special and general relativity.

**Roberto B. Salgado** is a visiting assistant professor of physics at Colgate University. His research interests are in relativistic physics (spacetime structure and relativistic pedagogy), applied mathematics, physics pedagogy, computational physics, and scientific visualisation. In particular, he develops new visualisations of relativity and new ideas for teaching relativity, which can be viewed at visualrelativity.com.

**Elwin Savelsbergh** is associate professor at the Freudenthal Institute for Science and Mathematics Education, Utrecht University, and professor of Science and Technology Education at the University of Applied Sciences Utrecht. His research focusses on curriculum design, science teacher education, and the development of interest and attitude in science education.

**Bernard Schutz, FRS** holds a half-time professorship at Cardiff University, and is a member of the Gravity Exploration Institute in the School of Physics and Astronomy. His work is primarily on gravitational wave detection and astronomy; he is a senior member of the LIGO Scientific Collaboration. He is also an Emeritus Director of the Max Planck Institute for Gravitational Physics (Albert Einstein Institute) in Potsdam, Germany, where he was a director since its foundation in 1995 until his retirement in 2014. While working there, he established in 1998 the open-access journal Living Reviews in Relativity, which has become the highest-impact open-access journal in the world. He holds the position of Adjunct Professor of Physics at Georgia Institute of Technology in the USA. He is a member of the US National Academy of Sciences. His work on gravitational waves has recently been recognised with the Eddington Medal of the Royal Astronomical Society and the Isaacson Award of the American Physical Society.

**Silvia Simionato** is an Italian PhD Researcher in Astrophysics Education at the Jena University (Germany). She studied Astronomy at the Padua University (Italy), writing her Master Thesis "Development of Astronomy Educational Resources: the Case of Universe Awareness" in collaboration with the Leiden University and the international foundation "Universe Awareness – UNAWE" (The Netherlands). Her research interests concern all educational aspects and the elementarisation of astrophysical content. The aim is to better understand and teach physics with the help of fascinating exercises and examples from cosmology, astronomy, and astrophysics. One of her focusses is on gravitational lenses, where she combines general relativity and optics to visualise light on curved paths.

**David F. Treagust** is John Curtin Distinguished Professor of Science Education in the School of Education at Curtin University in Perth, Western Australia. He taught secondary school science for ten years prior to working in universities in the USA and Australia. His research interests are related to understanding students' ideas about science concepts, especially chemistry concepts, how they contribute to conceptual change and can be used to enhance the design of curricula and teachers' classroom practice.

**Wouter van Joolingen** is professor of Science and Mathematics Education at the Freudenthal Institute for Science and Mathematics Education, Utrecht University. His research interests include the use of modern technology in education, such as simulations, modelling tools, creation tools in order to foster the development of scientific literacy.

**Marjolein Vollebregt** is a physics teacher at a secondary school, and a teacher educator at the Freudenthal Institute for Science and Mathematics Education at Utrecht University. Her PhD focussed on the teaching and learning of an initial particle model.

**Sven Weissenborn** is a PhD-researcher in physics education at the University of Hildesheim, Germany. In his research, he develops and evaluates digital learning environments to teach relativistic phenomena. His web application, called ViSeMo, is based on the Sector model from U. Kraus and C. Zahn.

**Marjan Zadnik** was the inaugural Dean of Teaching and Learning in Science and Engineering at Curtin University and is now Adjunct Associate Professor at the University of Western Australia. He previously worked at the Enrico Fermi Institute at the University of Chicago, and at the Max Planck Institute in Mainz, Germany. He has won many awards for excellence in teaching, including the inaugural Education Medal from the Australian Institute of Physics and a National Teaching Fellowship. He has also won many competitive research and teaching grants. Current educational research projects include measuring the effectiveness of teaching Einsteinian Physics at schools, as well as investigating how students learn in science and engineering laboratory environments.

**Corvin Zahn** is a senior researcher in physics at Hildesheim University, Germany. Since his doctoral research on the visualisation of relativity, he has been working in this field. He also does research in the development of physical and digital models and their use in teaching relativity at school and university. He has developed Open Educational Resources (spacetimetravel.org) that make his research outcomes available for teaching (joint work with Ute Kraus).

# Preface

Physicists and educators around the world aim to inspire and educate future generations by teaching students our best understanding of the universe. Einsteinian physics currently comprises this best understanding. It is based on two major theories of physics: the theory of special and general relativity that describes space, time, and gravity at the cosmic scale, and quantum physics that describes the interactions of matter and radiation at scales down to the smallest subatomic particles. The term "Einsteinian physics" is a semantic convenience, inspired by Albert Einstein's fundamental role in the development of both theories.

The modern Einsteinian perspective represents a significant paradigm shift compared with the Newtonian world view that underpins most of the school science education today. It is quite clear to us that the traditional approach of learning science, in which only those who advance to specialist courses in physics at senior high school and university get to learn the Einsteinian paradigm, is most inefficient. The students are burdened with a prior scaffold of Newtonian concepts that must be put aside as they learn the Einsteinian language of reality.

A consistent education within the Einsteinian paradigm, thus, requires radical rethinking of science education across the entire school curriculum internationally. With this book, we wish to contribute to the ongoing transformation of the educational landscape to make humanity's best understanding of reality available to all. The book is an outcome of a shared vision that has been developed within the international Einsteinian Physics Education Research (EPER) collaboration. Members of the EPER collaboration have been developing innovative instructional approaches and learning resources while finding new ways of conducting physics education research on a global scale.

Reflecting the diversity of our international efforts to firmly establish Einsteinian physics within the school curriculum, this book brings together a coherent set of chapters written by leading experts in the field of Einsteinian physics education. The book is designed as a resource for science teachers and trainee teachers, textbook writers and developers of learning resources as well as for those who formulate national curricula and conduct physics education research. If readers want to read the book from the beginning to its end, we hope they will find insight and pleasure in this order. But this order is not meant to limit readers, who are, of course, free to choose topics of interest and dip in and out as they wish.

The first section motivates the need to teach Einsteinian physics and explores the fundamental concepts of relativity and quantum physics. Our authors provide a set of foundational concepts that teachers can use to give school students the language with which to discuss frontiers of knowledge while emphasising that science is an incomplete discipline in which there are still many problems left to explore.

The second section of the book continues to present instructional approaches to Einsteinian physics that emphasise and discuss hands-on activities, geometrical approaches, evidence-based models and analogies, simplified mathematical treatments, and computer-supported learning resources. These instructional approaches combine student-centred classroom practices with a focus on learning for meaningful understanding. Our authors identify the most challenging concepts and weaknesses associated with various instructional approaches, such as critiques of the models and analogies which, while essential to conveying abstract concepts, can also lead to misconceptions. Findings of science education research are presented throughout the chapters to highlight benefits in terms of student attitudes, conceptual understanding, and gender inclusiveness. A very important part of the book is the discussion of classical approximations of modern concepts and how to introduce the Newtonian theory within the context of Einsteinian physics. All educators recognise the value of Newtonian physics, so a key aspect to learning the Einsteinian paradigm is to connect it back to Newtonian gravity and electromagnetic waves.

The last section of the book gives an overview of the current status of Einsteinian physics education research from around the world to show how different research groups have pioneered different approaches to and understandings of Einsteinian physics education. Many of the authors have encountered similar obstacles in teaching and learning of Einsteinian physics in schools and in implementing these topics in national curricula. Bringing together diverse perspectives allows utilising our complementary approaches so that we can learn from each other and move together towards a great future of modern physics education.

Paradigmatic changes do not happen overnight. Our responsibility as physics educators is to be more than responsive to change. We have to think about the world as it can be and envision opportunities and possibilities in science classrooms to actively move towards our vision of what is better. We hope that our vision presented in this book will entice readers, establish understanding, instil appreciation, and build confidence and motivation in the emerging learning domain of Einsteinian physics. Above all, we want everyone to be able to share the astonishing story of our existence, a story that can only be told in the language of Einsteinian physics.

***An Einsteinian language of reality***

It takes a long time for new thinking to be incorporated into widespread common understanding. Einsteinian physics was developed in the first three decades of the 20th century. It imposes a radically new way of describing our reality and has propelled physics into the modern arena of scientific inquiry and invention. From the Big Bang to black holes, from quantum cryptography to gravitational wave astronomy, Einsteinian concepts define the way we look at the universe. Although the Einsteinian paradigm has entirely replaced the Newtonian paradigm as our best understanding of the universe, the Newtonian conception of the world is still learnt implicitly in schools, because it is deeply embedded in our language, our approach to mathematics, and our way of thinking. This way of thinking is difficult to change because everyone, including teachers, has been brought up with it.

Language plays a key role in the process of conceptualising the world around us. Indeed, research in psychology, cognitive science, and linguistics suggests that language is more than a mere system of communication. Vocabulary itself is part of the conceptual framework of reality. As such a well-chosen vocabulary helps us develop an understanding of scientific concepts, which, in turn, is essential for being able to engage in scientific conversations. Many scientific discoveries can only be discussed in the language of Einsteinian physics: those without that language are disenfranchised from some of humanity's greatest achievements.

Just as with learning a foreign language, fluency in physics requires practice at speaking, and not just listening. Acknowledging the close links between scientific thought and the adequate use of scientific vocabulary, we have compiled a glossary of the language of Einsteinian physics. We hope that a concise overview of the Einsteinian vocabulary of space, time, light, and matter will help readers ease into the subject matter as well as support teachers in facilitating productive discussions when introducing Einsteinian concepts in their classrooms.

Seeking to change the emphasis of physics education around the world, we have to give reasons, as well as suggestions and resources for change. We hope that this book provides these reasons, suggestions, and resources so that children will grow up with a new common sense. It is our wish that students will be able to better appreciate the wonders of physical reality, be better equipped to live in the modern world, and are prepared to take the next steps in the human journey of understanding and discovery.

**Magdalena Kersting and David Blair**

# Acknowledgements

*Teaching Einsteinian Physics in Schools* is the outcome of joint efforts and commitments of many research groups and dedicated individuals around the world. We owe much to these physicists, educators, and teachers who all have pioneered innovative approaches for researching, teaching, and learning Einsteinian physics. We wish to acknowledge the great efforts of Einstein-First in Australia, ReleQuant in Norway, and Spacetime-Travel in Germany. All three projects have organised workshops and seminars to bring the Einsteinian physics community together to modernise physics education. Special thanks to Ute Kraus and Corvin Zahn for having arranged the Heraeus seminar "General Relativity as a Challenge for Physics Education". It was at this seminar that the idea of this book was born.

Our editorial duties have been personally and professionally rewarding, and we have learned much by working with 30 authors with diverse backgrounds coming from nine countries. We are grateful to all of them for having joined us in this project, and for having reviewed, commented, and revised drafts of the chapters diligently.

# Section 1

# Motivations and needs to teach Einsteinian physics

# 1 Intuition in Einsteinian physics

*Bernard Schutz, FRS*

**WHAT?**

This chapter sets the stage for the rest of the book by exploring the role of intuition as a tool to deepen the understanding of Einsteinian physics. Drawing on examples from the history of general relativity, Bernard Schutz argues that the development of physical intuition is a crucial goal in physics education in parallel with any mathematical development of a physics subject.

**FOR WHOM?**

This chapter is intended for readers who wish to learn how expert physicists think conceptually about their subjects to understand them, and also, for those who wish to see how we can introduce Einsteinian physics to students by developing their intuition as well as teaching them mathematics.

## INTRODUCTION

If you say the word "Einstein" to an average person, the first thing likely to come to their mind is an image of Einstein's iconic wild hair. The next thought will probably be that his science, however important, is far beyond the understanding of the average person. Both of these things place Einstein outside everyday society, a curious phenomenon to be wondered at. The teaching of physics at schools has until recently reinforced this view, mostly leaving Einsteinian physics to specialist university courses. The mathematical framework for this physics is challenging, and that has seemed a good reason to teach it only to those advanced university students who have mastered enough mathematics.

But it can't be left to the boffins any more. Everyone has heard of black holes; even small children want to know what they are. Recently, gravitational waves have also entered our vocabulary: their first detection in 2015 caused a media sensation, coming as they did from *two* black holes *merging together*! Other terms in common use that could come straight out of a general relativity textbook include the big bang, wormholes, and even the warp drive. And then, there is the quantum side of Einsteinian physics. Our lives would come to a halt without the Internet, computers, mobile phones, microwave ovens, and other devices that only work because of quantum principles. I would guess that well more than half of the world's economic activity today is enabled by physics that was only developed in the last 100 years and which is still thought to be too difficult to teach at the school level.

Does this leave us in a tough spot? Is the world to be divided between "users" who have no idea what is inside the box and "experts" whose work is too arcane to explain to the rest of the world? That would be dangerous for society. If we don't want that, can we open up the world of physics concepts to the people who don't have the mathematics to deal with the equations in which the physics is framed?

I believe that we can do it, and that we can, at the same time, improve the way we teach Einsteinian physics even at the advanced level. The way to do it starts with realising that even the experts think conceptually about their subjects in order to understand them; they don't go around just solving complex equations any time they want to answer a question. They use what we call *physical intuition*: a basic understanding of how things work and fit together that requires little or no mathematics. We should be introducing Einsteinian physics to students by developing their intuition as well as teaching them the mathematics.

Without intuition, science just would not work. Scientists today are specialists, and communication between different specialties requires a shared physical intuition. For example, if an astronomer wants to understand how a black hole might affect a star that is near it because he might want to observe the star in his telescope, he does not need to be able to solve Einstein's equations himself. Rather, he can approach a specialist in general relativity for insight, and she will explain to him how the gas gets pulled off the star by the black hole's gravity, how it will then get hot and therefore bright as it gets near the hole, and finally how it will vanish from view after being swallowed by the hole. None of their conversations will use arcane mathematics. Instead, they will rely on shared intuitive concepts about fluids, heat, and gravity. And these are concepts that are taught in school.

What is not always taught, either at school or at university, is an appreciation of the important role played by intuition in science. Some physicists I know don't display much intuition; they work mostly with the mathematics of their specialty, and they consequently don't have much interaction with other specialties. However, other physicists have been famous for their intuition. One example was Richard Feynman, who won the Nobel Prize for helping develop the theory of quantum electrodynamics. He was a member of the scientific panel that investigated the deadly 1986 Challenger disaster, when a space-shuttle launch failed because its booster rockets exploded. Feynman famously demonstrated on television that the fault lay with the decision to launch in cold weather; he dipped a piece of rubber used in the booster assembly into ice water, and then simply snapped it in half. End of discussion. There are plenty of other stories showing how Feynman was able to cut through confusions in apparently complex physics problems with a simple physical argument that even a high-school student could understand.

It is important to be clear about what intuition is. I would distinguish between the kind of intuitive thinking Feynman demonstrated and "popular science" presentations, although there is an obvious overlap. Popularising science is important, but such talks often aim at entertaining the audience. Here, we aim at teaching young scientists to use intuition in order to *do* science more effectively.

I believe that intuition plays a key role in scientific thinking, even though in casual discourse, it is often used as a synonym for sloppy thinking. Intuition is not the opponent of logic; rather, it complements logic. A brain governed entirely by logic would be a brain in paralysis, unable to generate an original thought. We educators need to guide students toward a fruitful interplay between the creativity and model-building of the intuitive process and the analytical and critical facilities of logic.

I will, therefore, focus in this chapter on *developing* physical intuition as part of learning physics, and then using intuition as a tool for deepening understanding, one that works in parallel with any mathematical development of a physics subject. The teacher can then grab hold of those intuitive concepts in students' minds to link up with other parts of physics, even parts where the mathematics is more complex. This should be done at school (and it often is) and university (often not). Doing this is important because, as I will argue later, intuition is *central* to how our brains work and how we make sense of the real world. We are not computers, and teaching us physics is not like programming a computer. So, my aim in this chapter is to explore how we can use the way intuition works in physics and in the physicist's brain to improve the learning of science.

## INTUITION IN GENERAL RELATIVITY: EXAMPLES

First, I will try to make the notion of intuition more concrete by giving three examples from the most mathematically complex work by Einstein: general relativity. The first example is from the history of the subject. The equations were more or less complete by late 1915 (the cosmological constant came later), the fundamental solution for what we now call a black hole was discovered by Schwarzschild in 1916. Einstein discussed gravitational waves already in 1916 and arrived at what we now call the "quadrupole formula" showing how systems emit gravitational waves in 1918. In the 1920s, the theory was applied to cosmology, ready for Hubble's discovery of the expansion of the universe in 1929. So, one might have expected rapid progress following on from these major achievements, placing general relativity on a firm physical basis. But instead, we find that even in the 1950s, physicists could not agree on whether Schwarzschild's solution described real objects, nor on whether gravitational waves were physically real or just an artefact of the complicated mathematics. Einstein himself changed his mind on these questions several times.

In my view (Schutz 2012), the root problem was that, in the period 1930–1950, people who worked on relativity approached it mainly mathematically. Naturally, we don't come into the world equipped with a relativity intuition: we have to gain it through "experiences" in the subject. Scientists commonly do this when entering unfamiliar territory, using an intuitive tool called heuristics, in which complex systems are approximated by a few simple characteristics that might be found by studying particularly simple examples. The mathematics of relativity gets quite complex for realistic problems, and occasionally people (including Einstein) wrote papers with fundamental errors that simply clouded the subject. But in the early period, few of the mathematicians working on relativity tried to work on heuristics; instead, they trusted only what they could get mathematically from the full Einstein equations.

In the 1950s real progress in relativity began because some brilliant physicists who were already skilled in heuristic thinking began working in the field. A famous example is Feynman himself, who at a key conference in 1957 (DeWitt & Rickles 2017) gave a simple physical argument showing that gravitational waves had to be real and had to carry energy (up until then hotly questioned). Feynman also pointedly remarked at this meeting that progress in physics did not always come from doing the mathematics better, but could come instead by working on simple situations or approximations, and then by doing experiments. What he was saying that relativists needed better physical intuition.

Physicists spent the next several decades developing this intuition: working on simple examples, using approximations to bridge between the simple examples, asking physical questions about what could be measured by experiment and observation, and doing experiments and observations. The 1957 conference kicked off discussions aimed at intuition-building among a growing number of scientists working in this field. The more scientists joined in, the more important it was to develop a language, a body of heuristics, an intuition that could be shared, that would help them communicate with one another.

By around 1990, there was a well-developed corpus of heuristics about black holes, gravitational waves, gravitational collapse, and cosmology, the study of which led relativists to a deeper intuitive grasp of the theory. In parallel, astronomers had been opening up a fascinating universe, beginning to gather evidence for black holes, gravitational waves, and the Big Bang. This input of data nicely validated the heuristic concepts that the theorists were developing. At this point, 75 years after its auspicious beginning, general relativity was at last ready to become a working part of astrophysics. Mathematics alone had not been enough to get there.

My second example follows one part of the above story. It is how physicists finally came to terms with black holes. We now understand that the Schwarzschild solution represents a black hole, deep inside of which is a "singularity", a place where gravitational forces become infinitely large, but that surrounding the singularity is a surface we call the horizon, which is a one-way membrane: things from outside can cross to inside, but not the other way around. We are troubled by the singularity because the laws of physics, as we currently understand them, don't tell us what happens when an infalling body reaches it, nor what happens next. But this is something we can live with because it is not even in principle observable: no information can get out to us through the horizon. Crucially, therefore, it does not disturb our ability to use physics to describe things outside the black hole. Notice that the description I have just given is intuitive: without mathematics, you can still make sense of it because I have linked it with your understanding of causality in the physical world.

But it took a very long time for physicists to get to this intuitive picture. One reason is that it seems, from Schwarzschild's equations, that on the horizon time stands still. This was not consistent with the notion of causality that physicists had in the early 20th century. But early work by Painlevé, Gullstrand, and Lemaître had shown by 1935 that if you describe the Schwarzschild solution in what might be called "co-moving" coordinates – coordinates that are dragged along with infalling particles – then time does not stop at the horizon; space is smooth and regular there; gravity is not infinite; and the particles just keep on going inwards. So, the horizon was not a singular place, and the apparent problem with time was caused by the coordinate system Schwarzschild had used. Moreover, only a few years later Oppenheimer and Snyder (1939) had argued (using part intuition, part heuristics, and part mathematics) that a sufficiently massive collapsing star goes all the way down to its "gravitational radius", at which point — as we now understand — it turns into a black hole, leaving the horizon at that location. Though these results seem to be convincing to us today, they did not convince other people working in this field at that time, including Einstein, to take black holes seriously.

The final step wasn't taken until almost 20 years later, in 1957. John Wheeler, a distinguished quantum field theorist (he had been Feynman's PhD supervisor), felt it was time for relativity to be turned into a part of physics. To him, the most formidable problem was whether the Schwarzschild solution described real objects. A key issue that nobody had yet addressed was that the solution was perfectly spherically symmetric, and real bodies simply aren't. This symmetry was an idealization that had enabled Schwarzschild to simplify the Einstein equations and get an exact solution in the first place. Schwarzschild's solution was, therefore, a heuristic, a simplified model of the real thing. So Wheeler asked, how faithful is this model to the real world? If a real star collapsed, would it really become like the Schwarzschild solution, the way Oppenheimer and Snyder had argued?

It is said that Wheeler hoped that the answer would be no, which means that any small initial asymmetries of the star would get larger and larger as it collapsed, so that by the time it reached its gravitational radius, it would fragment or do something else that would prevent it getting any denser, and prevent the formation of the horizon. So, he set to work with Tullio Regge to prove this in the simplest way: study mathematically what happens to very small non-spherical changes in Schwarzschild's gravitational field. Wheeler hoped to prove that these small perturbations would grow exponentially with time, the sign of an *instability*. If the Schwarzschild solution were unstable, nothing like it would exist in the real world.

But in fact, Regge and Wheeler (1957) found the opposite: all perturbations either died away exponentially or became constant in time. This meant that a non-spherical star could indeed collapse through its gravitational radius, and the horizon would afterwards simply settle down exponentially rapidly into something very similar to the Schwarzschild solution. The Schwarzschild model did indeed resemble objects that could exist in the real world!

Regge and Wheeler's work was still, of course, a simplified case, a heuristic itself that validated the original black hole heuristic. It was only an approximate solution of Einstein's equations, valid only for tiny perturbations. A full exact solution was beyond anyone's ability until we became able to do fully numerical simulations on supercomputers. But it was exactly the kind of step that was needed according to Feynman's assertion (in the same year): do the simple calculations, and use them to fit the theory (in this case, black holes) into the real world. It convinced Wheeler and he became the field's biggest proponent of the reality of black holes in astronomy. Heuristics and intuition are powerful drivers of scientific investigation.

My third example regarding intuition is drawn from my own teaching. Many years ago, I wrote a textbook called *A First Course in General Relativity*, which is now in its third edition (Schutz 2022). It introduces the full mathematics needed for Einstein's equations, aiming to get students to the point where they can solve those equations and understand the physical meaning of the solutions. But, a one-year university course has only a certain number of hours, and so, if you do mathematics, you are limited in the amount of heuristic/intuitive development you can include. I used exercises in each chapter to help with this and stressed intuitive ideas in my lectures. But, it raised a challenge: how far into general relativity can one go without much mathematics, relying much more on intuition and heuristics?

I worked for almost 20 years on this challenge, and by doing so, greatly deepened my own intuition about the theory. The result was the book *Gravity from the Ground up* (Schutz 2003), and it illustrates the point of this chapter that one can genuinely *teach* students to understand a subject like gravity – starting with Galileo and going all the way to general relativity – using only pre-calculus mathematics. In fact, most of the mathematics in this book is reserved for "boxes" that interested students can attempt, or they can just accept the results, incorporate them into their thinking, and go on to what comes next. The pedagogical aim of writing this book was to find the right way to link new concepts with the ideas that students had already learned at school. The fact that this was possible taught me how the complexity of the mathematics of general relativity tends to hide the continuity of physical concepts from Galileo, through Newton, to Einstein. Teachers of relativity would do well to bring this intuitive continuity to the front as often as possible.

In view of this, I threw away the usual (chronological) order of presentation of gravity theory, where one first discusses all about Newtonian gravity and only then goes on to Einstein. Instead, it made much more sense to me to treat Newtonian gravity from the start as a special case of Einsteinian gravity, and to immediately employ Newtonian concepts, as soon as they are developed, as a way to introduce their Einsteinian relatives.

This approach starts right away in the first two chapters (there are 27 in all). Galileo's experiment of dropping balls from the Leaning Tower of Pisa is described in the first chapter, and Newton's extension of Earth-bound gravity to the whole Solar System is treated in the second. Now, Galileo's experiment is historically the first statement of what we call the equivalence principle, which in turn is a foundation stone of general relativity. So, Chapter 1 already explicitly lays that part of the foundation. And in Chapter 2, we re-express the equivalence principle in Einstein's form: an observer who is falling freely in a gravitational field measures no local effects of gravity at all; she is weightless. This allows us to prove in this chapter that light is redshifted as it climbs up out of the Earth's gravitational field. Einstein regarded this gravitational redshift as the most important prediction of general relativity. So here, we have Galileo, Newton, and Einstein arm in arm. Any intuition that students may already have about Newtonian gravity begins building their relativistic intuition.

To me, this approach makes a lot of sense. The "weird" concepts of general relativity like black holes are more approachable if introduced alongside relevant parts of Newtonian gravity. Black holes themselves make their first appearance as early as Chapter 4, where I describe the 18th century discussion of stars that are so massive and compact that they can in principle trap light, something that follows from Newtonian gravity and the equivalence principle. Students can take away from this that, when they later learn about black holes, their trapping of light is not something new with Einstein; it is a concept they can comfortably carry over from Newton. The dark Newtonian star is a useful heuristic on the road to Einsteinian black holes.

Intuition and heuristics not only help students learn but also guide good physicists in opening up new ideas. The 18th century discussion of hypothetical dark stars is an example. Another remarkable example is the investigation in the 19th century by the famous mathematician Laplace of the possibility that gravity would have a finite speed of transmission (not allowed by Newton). Laplace realized that this would lead planets to very gradually spiral in toward the Sun (Laplace 1839). We know from Einstein that this is indeed a consequence of the emission of gravitational waves, which themselves are a consequence of the fact that relativistic gravity has a finite speed (that of light). Although the in-spiral effect in the Solar System is unmeasurable, it is today being observed in every gravitational wave detection. In fact, these detections beautifully combine these two Newtonian and pre-Einsteinian concepts: the waves come from black holes spiralling together. So, while these early speculations about dark stars and inspiralling planets did not directly lead anywhere, they show that Einstein's general relativity emerged organically from some of the physics we teach at school. So, it stands to reason that students at school who are developing their physical intuition should also find much of Einstein's physics natural.

## WHAT IS INTUITION, AND HOW DOES IT RELATE TO MATHEMATICAL LOGIC?

Physics is often presented primarily as a framework of laws, which are expressed in the language of mathematics. If laws make experimentally verified predictions in their respective domains of validity (and we don't yet have a "theory of everything"), then in this picture, we "understand" that domain. But is this really what understanding a domain means? It is true that if we can calculate the mathematical consequences of a theory, then we can predict experimental and observational outcomes. When Laser Interferometer Gravitational-wave Observatory (LIGO) made its first detection of gravitational waves (Abbott, et al. 2016), the observed waveform fit very well with prior predictions, obtained using elaborate analytical calculations and supercomputer simulations, of waves from black hole inspiral and merger. This gave us confidence that the source really was a black hole binary system. It validated our understanding of the way two such bodies orbit and merge, for example. But does our *understanding* consist just of the outputs of mathematical and computer calculations?

Physicists work with what we call models. Models can be elementary, even oversimplified. There is a common joke about the theorist who starts out saying "Let's consider a spherical cow"; it captures the danger of too much simplification. But, even our deepest theories are themselves models, since we know they don't describe everything out there. And even if they did, our brains could not take it all in.

Models match how our brains work. Animals use their brains to model the outside world. The models work because they include what the animal needs for survival, but they exclude irrelevant stuff. And the animal's model of the outside world itself includes and fits together a number of sub-models, which, for example, may handle the information from different sensory systems. In much the same way, our model of a black hole binary system contains sub-models of the two individual black holes. These models exclude irrelevant information such as how and when the black holes were formed. When the binary separation is large enough, we can compute the orbit from Einstein's equations without knowing anything about the holes except their positions, masses, and spins. Our model of a single black hole tells us that ignoring its detailed structure at this point is fine. When the holes get closer to one another, we have to replace this simple quasi-Newtonian model (technically called a post-Newtonian model) with a model based on a full numerical simulation that uses Einstein's equations to compute how holes interact with each other, deform, and merge. Both models tell us what the gravitational waves emitted by the systems will look like. Based on them, we have another model that computes how these waves travel through the universe to our detector. That model excludes irrelevant things like what happened to the binary system itself. And then, we need another model for our detectors' response to the incoming gravitational waves. This model ignores where the waves came from or how they got here. The detector physics model is itself, of course, a hugely complex assemblage of models of the different parts of the detector. The models for different parts may be using the laws of physics of different domains like surface physics or the quantum theory of light.

I understand some of these models, but as a theorist of general relativity, I don't understand many of sub-models of the detector model. Yet, I will assert that I understand detectors and how they work. This is because I have a greatly simplified model of the detector in my head, but one which is faithful enough for my purposes. And that is the key – my purpose in understanding the detector is to know that I can rely on it to tell me about the gravitational waves, which are what I am really interested in. My colleagues who are detector physicists have a rather different set of objectives, which require a much more sophisticated detector model than mine. But for each of us, the model we have in our heads captures much more closely the notion of *understanding* than does simply our ability to write down the laws of physics that support and validate these models.

I believe that these models form the basis of our physical intuition. When a group of detector physicists want to understand what is going wrong with a detector, they do not immediately go back to the first principles and begin solving a lot of fundamental physics equations. They first use their sub-models of different parts of the detector to try to pin down which sub-model might capture the domain where the problem originates. Then that sub-model is taken apart, into its constituent sub-sub-models, and so on. The problem would be found in one such model or perhaps in the way that their mutual interactions have been (incompletely) modelled. At some point, some difficult calculations might be needed, or computer simulations, but rarely will these involve going right back to the fundamental laws of physics. Learning physics, therefore, is much more than learning the mathematical logic of the fundamental laws. It is learning to keep models in one's head and how to knit them together into a representation of the real world. An intuitive physicist can do this really well. Therefore, a student who is learning physics needs not only the equations but also the key heuristic models that can be put together to make the whole domain understandable. These key heuristics were lacking for the physicists who, in the period 1930–1950, were struggling to make sense of general relativity.

Not coincidentally, this kind of structuring of physical understanding reflects what happens in our brains when we learn and think. Neuroscientists, cognitive scientists, and philosophers have in recent years been learning a lot about how the brain "knows" things. Much still remains to be discovered, but what is understood today is germane to our discussion. Decision-making in our brains is done via processes that resemble the working of neural networks in our computers, where information (both sensory and memory) is filtered through many layers until the "best" response is triggered. The process resembles what scientists call "Bayesian statistics", where one's confidence in the value returned by some measurement is computed not just from the known uncertainties in the measuring procedure but also from uncertainties one assigns to the assumptions one makes. These assumptions, called priors, could be things like environmental factors that might have affected the measuring apparatus, or they could include information from previous measurements of the same thing. In astronomy, for example, they might include one's confidence in what astronomers already believe they know about the system being observed.

The brain makes decisions in what I would call a quasi-Bayesian way, relying on memory and hard-wired "instincts" for relevant priors and their uncertainties, and relying on senses or other mental events to trigger the need for a decision and to input new information that is relevant. The different neural network layers apply tests of what is relevant and what is important for the decision. But, where Bayesian theory would lead to a range of possible measured values with their probabilities, the brain is constructed to arrive typically at a single decision, which hopefully is the best for the animal in the circumstances. The way the brain makes decisions, therefore, is not deductive. It is fundamentally probabilistic, and it is only as good as the information that is used for it, including the stored experience from relevant previous situations.

It follows that when we are presented with a problem, ideas about how to address it percolate up from below, and this percolation is unconscious. At some point, ideas become conscious, and there is considerable debate among neuroscientists today about what exactly consciousness means. But for our purposes here, it is enough to know that we rely on our "gut" level for input to our consciousness. In other words, much of what we "think" is arrived at statistically, not deductively. This captures our intuitive way of understanding the world, and its statistical nature is not a reason to distrust it: decision-making has been well-adapted by evolution to deal well with situations where a lot is known, and equally well with situations where little information is available to guide decisions. This point has been strongly argued, for example, in the book *Gut Feelings: The Intelligence of the Unconscious* (Gigerenzer 2007).

Even more challenging to the primacy of logic in human thinking is the recent book *The Enigma of Reason* (Mercier & Sperber 2019). They argue that logic and reasoning are useless to animals that cannot communicate about "how" they arrived at decisions: such animals simply follow their gut feelings, arrived at in this quasi-Bayesian way with inputs from built-in instincts, learning, and senses. If the animals are part of a social group, then hierarchy and other externals determine which animal's decisions are followed by the group. But, when language arose among humans, it became necessary to be able to persuade other people of one's own point of view in order to preserve the working of the social group. So, giving reasons became a way to influence others. Of course, then one had also to listen to other individuals' reasons, to respond to other individuals' challenges. A way had to be found to choose among conflicting decisions; the language of logic originated here. We have acquired the notion of logic through our cultural evolution. It is not biologically wired into our brains.

Simple as this line of reasoning about the evolution of logic may be, it has profound implications. It suggests that people normally do not arrive at decisions by exercising their faculty of reason, although of course, they can force themselves to do if they are dealing with mainly objective facts (something that doesn't often occur in everyday life!). But even if they have taken a decision at gut level, believing they "know" it is right, people will invent reasons post-hoc if challenged. If they are part of a social group that all are inclined to the same choice (as much of politics seems to be in the age of the internet), then their reasons do not have to be particularly logical, since they won't be seriously challenged. Aristotle may have felt that the human brain is unique in being logical, but in the Mercier-Sperber view, the human brain is like that of any other animal, except it is good at crafting justifications when it has to!

This allows us to reconsider how animals' brains use models of the real world around them. The purpose of such a model is to condition the way the animal reacts to new situations, new challenges, and help it to survive. We should think of these models, not as map-like representations of the world, but as algorithms for dealing with them. Intuitions about physics, in this view, can be seen as prescriptions for fitting different parts of physics together in the mind, prescriptions that help a physicist decide what to expect from a new experiment, decide whether to be comfortable or uncomfortable with the suggested explanation of a new phenomenon. They are particularly important in the creative side of physics, where new ideas are needed. Our intuitions are algorithms that guide the creation of these ideas, ensuring that we don't waste time on totally random junk. All of this works mainly in the unconscious mind, feeding decisions upwards to our conscious selves.

## CONCLUSION: TEACHING PHYSICS – THE INTERPLAY OF INTUITION AND LOGICAL REASONING

How do all these recent ideas about how our brains work affect our view of intuition and mathematics in physics? I have – and this is, of course, necessarily a simplification – identified intuition with the concepts/models that win the unconscious Bayesian competition among alternatives in our brains when we seek solutions to problems in physics. I would correspondingly identify mathematics as the language, and logic as the method, which we use post-hoc to justify the correctness of our intuition. Intuition is how we "know" science, and logical testing of intuition against experiment/observation mediated by the language of mathematics is how we make sure that what we "know" is "right".

As a social group, scientists put up a much stronger set of challenges to intuitions than one typically experiences in normal social activity. Competition among scientists, the insistence on validation by experiment or observation or mathematical calculation, and the tradition of refereeing research papers all ensure that ideas are challenged, that the proposer should not be able to get away with justifications that don't conform to high standards of logic. In particular, the ultimate appeal to objective experimental validation enables the whole community to agree on the way ideas are challenged and encourages scientists themselves to be more self-critical of their gut instincts or their intuition than is common in normal life. Politics and business, for example, operate on quite different rules. That is, I think, one reason why science is by-and-large successful in its own terms, why it normally advances its frontiers and rarely has to retreat and abandon territory it had claimed to understand. But, its success does not come from relying purely on logic. It cultivates an especially self-critical form of intuition. Nevertheless, the interplay between intuition and logic is as much a part of science as of any other human activity.

So, when physics is taught to students, the development of physical intuition must be a key goal. It needs to be cultivated. New concepts should be introduced using both intuition and mathematics at an appropriate level to the students. Explicit attention should be paid to "meaning". Difficult subjects, like Einsteinian physics, can be introduced successfully using intuitive concepts, and readers will find creative and exciting suggestions in later chapters of this book about how this can be done. Many of these intuitions will link up with the ones that students already have, leading to a grasp of much of the physics even without elaborate mathematics. At school level, most of the physics in these advanced subjects is discussed at the intuitive level. But even in a university course, teachers should explicitly try to cultivate intuition and to use it to make mathematics more understandable.

We are talking about scientific intuition here, which means it has to conform to the scientific method and has to be ready to withstand strong challenges from other scientists. Therefore, the instruction needs to pay attention to the soundness of the intuitions that students are developing. Intuitions act below the level of consciousness, so it can be difficult for a teacher to know whether students are picking up useful or faulty intuitions. Teachers, therefore, need to create the kinds of challenges to intuition that scientific progress requires.

To ensure that students' intuitions are sound, students need to be self-critical about them, and they need to be able to defend those intuitions against challenges from others. Without introducing this kind of "quality control", for example, using group discussions or individual student presentations, teachers risk leaving students with faulty concepts. We are good at assessing students' mathematical facility with a physics subject: we just give them written tests. But, if we are going to guide the development of intuition, we need to enable students to do what humans have done for millennia: justify their intuitions by explaining them to others, by giving convincing reasons.

Developing students' physical intuition can serve another important purpose, particularly because most of our school students won't become scientists.

We want students to find science interesting, but we also want them to respect it. We want them to come away understanding how it deals with reality and how it makes progress. In a world full of fake news and reality-deniers, students need to develop their own scientific intuitions and experience scientific debate in order to understand how working scientists "know" about the world. The acquisition of scientific expertise must not be seen as an arcane ritual, opaque to and hence untrusted by the general public. Paying attention to intuition gives us the opportunity to teach physics in an interesting way and to show all students that anyone can learn to think like a scientist.

If we want to accomplish this, then we should recognise that words and pictures and experiments matter just as much as equations, and that argument and discussion matter just as much as written exams.

## ACKNOWLEDGEMENTS

I would like to acknowledge stimulating discussions with many of the contributors to this book, and especially to the editors, M. Kersting and D. Blair. I am also grateful for insightful comments from C. Gorman.

## REFERENCES

Abbott, B. P., et al. (2016). "Observation of Gravitational Waves from a Binary Black Hole Merger." *Phys Rev Lett*, 116, 061102.

DeWitt, C. M. & Rickles, D. (Eds.). (2017). *The Role of Gravitation in Physics: Report from the 1957 Chapel Hill Conference*, Edition Open Access, Max Planck Institute for the History of Science.

Gigerenzer, G. (2007). *Gut Feelings: The Intelligence of the Unconscious*. Viking Press.

Laplace, P. S. Marquis de (1839). *Celestial Mechanics Vol. 4*, transl. N. Bowditch. Little & Brown, Boston (Book 10, Chapter 7, Section 22). Translation from the French 2nd edition 1829.

Mercier, H. & Sperber, D. (2019). *The Enigma of Reason*. Harvard University Press.

Oppenheimer, J. R. & Snyder, H. (1939). "On Continued Gravitational Contraction." *Phys Rev*, 56, 455–459.

Regge, T. & Wheeler, J. A. (1957). "Stability of a Schwarzschild Singularity." *Phys Rev*, 108, 1063–1069.

Schutz, B. F. (2003). *Gravity from the Ground Up*. Cambridge University Press.

Schutz, B. F. (2012). "Thoughts about a Conceptual Framework for Relativistic Gravity." In *Einstein and the Changing Worldviews of Physics*, Lehner, C., Renn, J., & Schemmel, M., (Eds.); Birkhaueser; pp. 259–269.

Schutz, B. F. (2022). *A First Course in General Relativity*. Cambridge University Press, 3rd ed.

# 2 Time for changing paradigms in science and education

*David F. Treagust*

**WHAT?**

This chapter provides a broad picture of science and education as two human endeavours. Tracing parallels between the historical developments in both disciplines, David Treagust shows how new ways of thinking enable us to make progress in science and education.

**FOR WHOM?**

This chapter is intended for readers who wish to obtain historical and philosophical background knowledge of Einsteinian physics education.

## INTRODUCTION

Perhaps one of the major influences on our understanding of how scientific research and scientific knowledge evolves and develops was the publication of Thomas Kuhn's (1962) *The structure of scientific revolutions*. This small book really changed the way we look at the enterprise that is science. Essentially, Kuhn's work challenged the then-current conception of science as being a steady progression of the accumulation of new ideas. Rather, Kuhn argued that science mostly advances by occasional revolutionary explosions of new knowledge with the introduction of new ways of thinking, which Kuhn called paradigms. This term paradigm is now universally used to recognise the framework that provides model problems and solutions to the achievements of a community of scientists. Essentially, paradigms are the underlying frameworks that underlie a way of thinking about a phenomenon such as falling bodies.

Consequently, when the paradigm changes, the way of thinking about the phenomenon of falling bodies changes. Conducting research in science and conducting research in education are both human endeavours, that is, scientists investigate the world around us, and educational researchers investigate how individuals learn and understand the world around us. This chapter considers the question, do the same processes of knowledge creation and theory generation in science also apply to education?

The parallels between paradigm changes in science and education are indeed intriguing, having essentially the same elements – a dominant way of thinking about a particular phenomenon and conducting research, new evidence and ways of thinking, resistance to change and obstacles to overcome to move forward with new ideas and thinking. Philosophers of science have provided fruitful perspectives on the development of science, and the notion of paradigm is well-established. When educators introduce a new school curriculum such as Einsteinian physics, the topic of this book, it is important to take a broad perspective and examine new theories and ideas about teaching and learning that have been introduced in education and the ways to measure the success or failure of this introduction using educational research.

In this chapter, I describe several paradigmatic changes in science where the ideas-models-theories within a new paradigm became the new norm, and also provide examples where the ideas-models-theories did not progress to a new paradigm. Some ideas are ahead of the Zeitgeist – the predominant ideas and ways of thinking at that time – and are initially ignored, sometimes for a very long time before the ideas are seen as being the new norm.

I then provide a comparison of paradigmatic changes in science with paradigmatic changes in education, particularly as they apply to teaching and learning of the modern concepts of science and especially physics. I illustrate where the ideas-models-theories become the new norm in education to help guide the improvement of teaching and learning of science, and also cases where these ideas-models-theories did not progress.

In science, there is often a radical difference between two paradigms – so much so that it is almost impossible to even talk about the different theories using the same language because there is an incommensurability between the two different paradigms. In education, the differences are less dramatic and the two paradigms are largely incompatible rather than incommensurable. Education is much less well-developed theoretically than science, and strictly speaking, education may be considered to be in a pre-paradigmatic state as there is no universal agreement on any underlying theory. However, educational researchers use the term paradigm to distinguish contrasting approaches to teaching and learning that serve particular purposes (Shuell, 1988; Watson & Reigeluth, 2018). These different educational ideas have gained more credence over time; so essentially, the new more usable ideas behave like paradigms.

The parallel structure of science and education as two human endeavours that are continually developing further prompts the question of what we can learn from paradigm changes in science and how we can translate insights from the natural sciences to the educational and social sciences. Redish (2014) addressed this issue in detail in his Oersted Lecture 2013 by convincingly arguing that teaching and learning should be based on educational research and be theoretically informed much as the same way that physicists apply research findings in their own work. What can we learn from the practice of research in science and physics for guiding the implementation of a new norm of teaching and learning Einsteinian physics in schools?

## CHANGING PARADIGMS IN SCIENCE

As research progresses, scientists notice anomalies in scientific research findings which cannot be explained by the existing ways of thinking, that is, by the ways of thinking within the existing paradigm. Over time, the model problems and solutions change, and some scientists begin to work within a different framework that provides new model problems and solutions. This is the notion of paradigm shifts. However, these paradigmatic shifts or changes do not happen overnight. New ideas-models-theories are only accepted by the majority of scientific researchers when they are based on reliable, valid, and repeatable empirical data. When these new ideas-models-theories can better explain phenomena and better predict new and future phenomena, they are accepted by the majority of scientists.

Not all scientists conduct research within this framework of new ideas-models-theories. Some scientists persist with "former" ideas-models-theories, but the number of these scientists lessens as the time goes by, and eventually, the "former" ideas-models-theories become no longer acceptable by the majority of the scientists. Nevertheless, there are specific situations where ideas-models-theories from the former paradigms still function in everyday explanation (such as Newton's law of falling bodies works perfectly well on the Earth), but no longer provide a complete explanation of the phenomenon (such as gravitational bending of light around stars or black holes).

### *Changes in thinking and researching in science that become the new norm*

Some examples are needed to make the idea of changing paradigms more comprehensible. Examples of paradigm changes include:

1. From Ptolemaic Cosmology (where the Earth was the centre of the solar system) to Copernican Cosmology (where the Sun was the centre of the solar system);
2. From Aristotelian mechanics (where for example, objects rise or fall according to the ratio of the four elements of which they are composed, namely, Earth, air, fire, and water) to classical mechanics (where the movement of bodies is understood by mathematical laws and principles such as Newton's laws of motion);
3. From the theory of spontaneous generation (where living organisms are hypothesised to develop from non-living matter, later disproved by Francesco Redi) to the theory of biogenesis (where complex living organisms come only from other living organisms by reproduction, attributed to Louis Pasteur); and

4. From Pangenesis (where the body emitted its own tiny particles called gemmules that migrate to the gonads and the reproductive cells, attributed to Charles Darwin) to Mendelian inheritance (where features of the parents are passed on to offspring through genes).

These four examples and others[1] of paradigm changes illustrate how specific ideas-models-theories developed by their protagonists were displaced and discarded once new ideas-models-theories could better explain the existing observations.

***Changes in thinking and researching in physics that become the new norm***

The examples in this section are more specific to physics and include ideas-models-theories that can still explain phenomena but are not able to explain newer phenomena, and hence, are seen to be limited in terms of providing a broader understanding of the phenomena.

1. From Newtonian physics (where movement of bodies is understood in terms of Newton's laws of motion that unfold against the backdrop of absolute space and time) to an Einsteinian relativistic worldview (where space and time are dynamic entities that shape the laws of physics). An example often discussed in the Einsteinian physics context is the notion of gravity as a manifestation of spacetime caused by objects with mass compared to gravity as a force in Newtonian physics.
2. From classic mechanics (where events are deemed to be strictly determined and can be explained at a macro level, but not at a micro level) to quantum mechanics (where events are seen as probabilistic and statistical and can be explained at the micro level).

A well-illustrated case of a period of changing paradigms in physics was evident in Einstein's *Annus Mirabilis* – the year of miracles – in 1905 when he published four papers on the particulate nature of light – "quanta", Brownian motion, electrodynamics of moving bodies, and mass – energy equivalence.[2] According to Renn and Hoffmann (2005), these four papers today are recognised as tremendous achievements, but at the time, they were not noticed by most physicists as being important, and many scientists who did know about the papers rejected the findings outright. This is not an uncommon response by those working in a currently established and existing paradigm who observe findings that are incompatible and incommensurate with their own work. Indeed, the findings expressed in these four papers did not fit the then-currently existing framework of ideas-models-theories for conducting research in physics. Some of this work – such as the theory of light quanta – remained controversial for years, that is, a substantial number of physicists remained committed to the previous paradigm of light travelling as rays.

### *Some potential new ideas did not progress to be the new norm*

Science progresses when a new knowledge comes to the fore and is appraised by fellow scientists who decide to conduct further investigations into the ideas-models-theories. As noted above, at first, Einstein's new ideas-models-theories were dismissed or ignored, but over time, they have been accepted by most scientists. However, some new ideas based on research findings did not become the new norm with a different framework of ideas-models-theories for conducting research. Two examples from the recent history of science are Lysenkoism and cold fusion.

*Environmentally acquired inheritance – Lysenkoism (1940s–1950s)*

Lysenko's ideas that animals and plants could have environmentally acquired inheritance gained a lot of support in the Soviet Union. Lysenkoism was based on Marxist-Leninist ideology and was supported by Stalin. Lysenko campaigned against Mendelian genetics by considering the heritability of variation in population by mechanisms involving the environment other than changes in the DNA sequences – he argued that winter varieties of wheat could be transformed into spring varieties. Repeated experiments conducted at the University of Cambridge could not replicate Lysenko's results. Therefore, this new and different framework of ideas-models-theories for conducting research in genetics gained no supporters or followers – at least outside the Soviet Union. The strong support in the Soviet Union, where the ideas matched the existing ideology, had disastrous consequences for wheat yield during Stalin's time. Lysenkoism negatively affected Soviet genetics, evolutionary and developmental biology, and Soviet research took a long time to recover. Strangely, there is an increasing interest in Russia to rehabilitate Lysenko's image (Kolchinsky et al., 2017).

*Cold fusion – Fleischmann and Pons (1989)*

Fleischmann and Pons, both at the University of Utah, made headline news around the world (I remember this announcement in 1989 very clearly) with their claim to have achieved fusion in a simple tabletop apparatus – a test tube – working at room temperature. If found to be an accurate claim, this result would be revolutionary in science and bring about a new framework for thinking about fusion. Fleishmann was an eminent electrochemist, so this research, though considered with some scepticism, also had some level of credibility. However, other experimenters failed to replicate their work, so the original research became suspect. Soon, most of the scientists no longer considered cold fusion to be a real phenomenon. Nevertheless, research continued for many years, especially in Japan, where a small minority of scientists was privately funded. The potential financial benefits would be enormous if cold fusion did work. Indeed, when some experiments showed a small fraction of excess heat in reactions, checking for experimental error was the next task, and these errors tended to be where any heat change lay (Pollack, 1992). Even today, research continues with new theories in the desire to find the elusive heat and commercial opportunities (Ritter, 2016).

### *Ideas ahead of the Zeitgeist*

The term Zeitgeist refers to the spirit or mood of a particular period of history as shown by the ideas and beliefs of the time. In the context of science, the time of the Zeitgeist may be considered as a period when scientists work within a paradigm, an accepted framework of ideas-models-theories for conducting research. As discussed above, initially, Einstein's ideas were not considered important and were not accepted by many scientists because they did not fit the existing paradigm – his ideas were ahead of the Zeitgeist. There are other examples in science worth noting that were ignored or not considered worthy of further attention, but subsequently have been shown to be correct and more productive in explaining the observed phenomena.

#### *Inherited characteristics – Gregor Mendel (1866)*

Mendel published his laws of heredity in the German language in 1866 as "Experiments on Plant Hybridization" in the Proceedings of the Nature Research Society of Brünn (now Brno, in the Czech Republic). Mendel chose to circulate and communicate his work and sent a printed copy to 40 leading biologists in Europe. History shows that most of these copies were unread. At the time, most natural science was descriptive, and one biologist who did reply stated that he could not understand the mathematics and did not see the value of this quantitative approach in biology. Mendel's paper was cited only three times in 35 years. The German-speaking Englishman, William Bateson translated Mendel's paper, defended Mendel's work and brought it to the attention of the scientific community, which by that time was able to understand the significance of Mendel's work (for details see Darden, 1977 and Henig, 2001).

#### *Gravitational field acting on light rays – Johan Georg Von Soldner (1804)*

Soldner, a German physicist, mathematician and astronomer, calculated the bending of light rays grazing the Sun's disc and thereby recognised that light rays could be considered as heavy objects being acted upon by gravity (Jaki, 1978). He also identified that light rays do not travel in a straight line, but in a hyperbola. Soldner's numerical calculation of the bending of light around the Sun in 1801 was almost the same quantity as that derived by Einstein (Jaki, 1978, p. 928). Soldner's ideas were too early to be understood and were ignored. This initially suffered the same fate as Einstein's ideas.

#### *Continental drift – Alfred Wegener (1912)*

Wagener, a German polar researcher, geophysicist, and meteorologist, argued that all the continents were once joined together in a single landmass, called Pangaea, and had since drifted apart. However, with no mechanism to explain the phenomenon, the idea did not have many supporters, and by 1930, his theory had been rejected by most geologists. His idea sank into obscurity for the next few decades, only to be resurrected as part of the theory of plate tectonics during the 1960s (Bryson, 2003).

What is common in these three examples is that the scientists' research findings were incompatible and also incommensurable with the existing ideas-models-theories to explain these phenomena at the time and so, were ignored. Next, we look at similar developments in education.

## CHANGING PARADIGMS IN EDUCATION

In their work, educational researchers use different ideas-models-theories that have developed over time to study human behaviour and human learning. Three commonly accepted major schools of thought (there are others) which can be traced back to different philosophies are humanism (based on the needs and interests of individual learners), behaviourism (based on the external and environmental stimuli to cause observable learning), and cognitivism (based on how students think). The most consequential development in educational practice over the past 50 years is the change of focus from classrooms being teacher-centred to being student-centred. Fundamentally, classrooms changed from students learning to remember and repeat back in tests to learning for meaningful understanding. In the former based on behaviourism, the focus is on ways to improve the delivery of information and ideas to the students; in the latter based on a cognitive approach, the focus is on the learner and how to improve opportunities for students to learn in meaningful ways.

In the same manner that science research changes when the majority of scientists accept the new paradigm, the same situation and actions occur in education when the majority of educators conduct their research within the paradigm, and this influences their classroom practice. However, some educators continue to work and think within the existing paradigm focused on behaviourism because it meets their needs to prepare and present lectures to deliver information, even to small classes. In a similar way to paradigmatic changes in science, paradigmatic changes in education are slow to change.

### *Changes in thinking and researching in education become the new norm: From teacher-centred to student-centred classrooms*

#### *Teacher-centred classrooms*

Learning which tends to be teacher-centred is based on the theory of behaviourism (Skinner, 1953). Behaviourism is an empirically derived model or theory of human learning based on scientific principles derived from data on learning tasks that human behaviour is predictable and controllable. In classrooms that follow a behavioural style of instruction, the primary focus on learning is to remember in order to repeat back in tests. Underlying a behaviourist approach to teaching is the belief that learning is governed by invariant principles, which are independent of conscious control on the part of the learner.

From a behaviourist perspective, a student is presented with specific information (or stimuli), and effective reinforcement by the teacher elicits a positive or negative response. Behaviourist teaching methods rely on skill and drill exercises to provide consistent repetition and rely on reinforcement such as verbal praise, grades, and prizes. Skinner's theory of learning is certainly useful for rote learning of facts and formulae or standard procedures such as multiplication tables, or for compliance with industrial or business standards or processes that are invariant and do not require individual judgement. Indeed, behaviourism has provided the underlying theoretical basis for the development of teaching machines, measurable learning objectives, computer-assisted instruction, and multiple-choice tests. The influence of behaviourism is still strong in corporate and military training as well as in some areas of science, engineering, and medical training, and also, at university level where learning opportunities for "gaining" knowledge are dominated by lectures. In these situations, in a broad sense, the lecturers/teachers are active – talking and explaining, while the students are passive – being presented with information, listening and making sense of the lecture. In recent times, lectures have been changed to incorporate some student-centred learning, providing students with a more active role in their learning as discussed below.

*Student-centred classrooms*

The notions that learning is governed by invariant principles and that these principles are independent of conscious control on the part of the learner were not accepted by many educational theorists who had different interpretations about human learning. Educational theorists like John Dewey, Jean Piaget, Jerome Bruner, Lev Vygotsky, and others had different underlying theoretical explanations for learning. However, all of them shared the idea that in each case, the students actively construct their own ideas and knowledge rather than being dependent on the teacher or lecturer to present these ideas and knowledge to them. In these classrooms, the focus is on learning for meaningful understanding. The argument from these theorists is that students learn and retain more information when they are actively involved in the learning process, and that learning is most meaningful when topics are relevant to students' lives and interests. A related notion is that of situated cognition, where learning takes place in the context where it is applied (Brown, Collins, & Duguid, 1989).

Research has shown that individuals learn at different rates and in different ways, which is not always compatible with the theory of behaviourism. The underlying theory of knowledge behind the theorists like Dewey, Piaget, Bruner, and Vygotsky is constructivism, which emphasises the importance of consciousness, free will, and social influences on learning, which is the opposite of the invariant principles of behaviourism.

The influence of Piaget and Vygotsky, in particular, has led to many changes in primary and secondary classrooms, and many readers are familiar with these two names. These changes in educational thinking may be referred to as a student-centred paradigm involving learning for meaningful understanding in

*Table 2.1* **Comparison of teacher-centred and student-centred paradigms of teaching**

| *Entity* | *Teacher-centred paradigm* | *Student-centred paradigm* |
|---|---|---|
| Knowledge | Transmission | Construction |
| Teaching | Expert provider | Coach and facilitator |
| Teachers | Classify and sort students according to performance | Develop students' competencies and talents |
| Students | Passive | Active |
| Mode of learning | Memorising | Relating |
| Assessment | Monitor and measure learning | Promote and diagnose learning |
| Context | Competitive/Individualistic | Cooperative learning |
| Learning | Individual | Social process |

*Source:* Adapted from Smith and Waller (1997).

contrast to a teacher-centred paradigm dominated by behaviourism involving learning to remember and repeat back in tests. The different features of the two paradigms are exhibited by the descriptors shown in Table 2.1, many of which have already been discussed.

### *Research findings in education that support the paradigm shift in teaching and learning*

In the paradigm of student-centred learning, students take responsibility for their own learning, connect learning to their own lives and thereby increase their engagement and have more opportunity for higher-order thinking. A move away from an emphasis on the use of multiple-choice type tests in a behaviourist paradigm allows for alternative forms of assessment such as portfolios and oral presentations. Referred to as 21st century skills (Trilling & Fadel, 2009), and now a central part of science education and physics education, are critical thinking, creativity, flexibility, collaboration, leadership, communication, social skills as well as technology literacy, information literacy, and media literacy (see, for example, Griffin & Care, 2015). Indeed, insights of these 21st century skills and the student-centred paradigm have informed much of the direction for Einsteinian Physics Education Research. In learning about Einsteinian physics, students are engaged in hands-on activities, creating analogies, and role-plays, which help them take responsibility and become both physically and mentally active in the classroom (see, for example, Kaur et al., 2017). A summary of student-centred teaching/learning and assessment practices includes well-known classroom activities such as concept mapping, role-playing, cooperative learning, problem-based learning, and project-based learning shown in Table 2.2 (see, for example, Daniels, 2017). Implementation of these student-centred activities has been investigated in numerous studies and shown to have successful learning outcomes, including in the teaching of Einsteinian concepts of physics.

***Table 2.2*** **Currently used teaching/learning and student-centred assessment practices**

| | | |
|---|---|---|
| Concept mapping | Concept cartoons | Portfolios |
| Group work | Role playing | Socratic dialogue |
| Model-based learning | Personalised learning | Flipped classroom |
| Cooperative learning | Problem-based learning | Project-based learning |

Just as in fields of science, these paradigmatic changes in education do not happen overnight. New ideas-models-theories are only accepted when based on reliable, valid, and repeatable empirical data. Indeed, these new ideas-models-theories only become accepted when educators, schools, and universities find that these new ideas-models-theories result in increased student engagement and improved student learning and provide a deeper understanding of concepts to be learned. There are opportunities for learning from presentations in lectures, so behaviourism still has its place and can be relevant in some learning situations. Indeed, there are strong challenges to the notion of constructivism from colleagues who favour lectures and consider the approach where students actively create their own knowledge as being too time-consuming and inefficient.

Within society, the expected goals of education and science education, in particular, change over time. As an example, being proficient in reading and writing and having high levels of content knowledge was the main goal of education in the 1960s. As demands for citizens to contribute to society become more complex, the goal of education changes for students to become critical thinkers, demonstrate creativity and have high-level communication skills, as well as to become proficient in reading and writing and have high levels of content knowledge. The evolution of society continues to contribute to the changes of paradigms in education.

At an individual level, as in science, some educators persist with the "former" ideas-models-theories, but their number lessens as the time goes by when the "former" ideas-models-theories are not accepted as the dominant teaching approach by the majority of the educationalists. An example is the change from a typical lecture mode of teaching where the only person speaking is the lecturer to one of a flipped classroom where students are given tasks to do before the lecture/lesson and respond to questions from the instructor during the lecture/lesson. University chemistry education has changed to be student-centred with the introduction of Process Oriented Guided Inquiry Learning (POGIL) classes, where students work in small groups with assigned roles as manager, recorder, presenter, and reflector using highly structured activity sheets to develop conceptual understanding (see Qureshi et al., 2017).

Besides considering the theoretical issues, there are many challenges in implementing any new teaching and learning opportunities and putting these new ideas-models-theories into practice. Teachers may lack confidence in trying to use student-centred active learning approaches, especially when including new content. They fear that the new methods will take too much time so that there will not be enough time available to cover the content, especially with concern about pre-university examinations, and that the students will not know how or be able to take control of their own learning. Finally, some teachers fear losing control of the class and so, the former paradigm of teacher-centred education is preferred.

### *Some potential new ideas did not progress to be the new norm*

#### *Programmed instruction – Skinner (1950–1960)*

Similar to the ideas in science, there were some ideas in education too, which did not become the new norm. Many titles of programmed instruction based on Skinner's initial ideas of reinforcement were produced in the USA and the UK from 1965 to 1969. However, despite the excitement by educators at the time, programmed learning based on Skinner's ideas did not last very long because the immediacy of reinforcement, which was the theoretical underpinning, did not prove to be critical for a great many types of learning tasks beyond drill and practice (Molenda, 2008). Nevertheless, there have been developments in individualised learning, especially with the emergence of soft technologies and incorporating human interventions, which have led to programmed tutoring and direct instruction, and they have gained strong support from their proponents.

#### *Open-enquiry learning (1960s–1970s)*

A response to the influence of behaviourism and the domination of the textbook in the classroom was the development in many countries, particularly the USA, the UK, and Australia, of projects using hands-on approaches for learning science. These approaches ranged from being structured with step-by-step support (in today's parlance, this would be referred to as scaffolding) to unstructured open-ended discovery. Over time, a growing body of research provided a better understanding that students learn where hands-on activities need to be coupled with minds-on activities for effective science achievement. Consequently, projects designed for unstructured open-ended discovery – which left students to find out what they want to know – were much less successful in providing learning outcomes than using enquiry approaches with more teacher guidance (Furtak et al., 2012).

### *Ideas ahead of the Zeitgeist*

Similar to researchers and their scientific discoveries, in science education, several researchers have been ahead of the Zeitgeist in publishing findings on the teaching and learning of Einsteinian physics. Two examples of relativity education are shown in Table 2.3. The research by Haddad and Pella (1972) sought to examine whether there were significant differences in understanding of

***Table 2.3*** **Research on teaching and learning Einsteinian physics in schools that was ahead of the Zeitgeist**

| |
|---|
| Haddad, W. D. & Pella, M. O. (1972). "Relationship between mental maturity and level of understanding of concepts of relativity in Grades 4-8." *The Journal of Experimental Education,* 41(1), 22–32. (**6 citations – only 2 before 2014**) |
| Baldy, E. (2007). "A new educational perspective for teaching gravity." *International Journal of Science Education,* 29(14), 1767–1788 (**28 citations – only 4 citations before 2013**) |

classical, special, and general relativity between grades 4 and 8 students based on 29 identified concepts of relativity. Four pieces of equipment were designed and used in the lessons, including a spacetime continuum model consisting of a heavy round object on foam rubber surface supported by a frame. The authors concluded that "grades 4, 5 and 6 could be taught the conceptual schemes of classical, special and general relativity at the knowledge level. Grades 7 and 8 could be taught classical relativity at the knowledge, comprehension and application levels and special and general relativity at the knowledge, comprehension levels"(p. 31). Despite the relevance of this research, the findings of Haddad and Pella were not known or replicated, and hence. were not put into practice. Maybe like Mendel, Haddad and Pella published their work in a journal which is not often and not likely to be read by science and physics educators.

The purpose of Baldy's study was to describe students' initial conceptions about the phenomenon of falling bodies and to compare how these conceptions evolved during a lesson taught using two different teaching methods – a Newtonian approach where all bodies are attracted to each other, and an Einsteinian approach using a pillow model. The research showed that "Einstein's theory can be understood by ninth-graders and promotes more conceptual change than a method based on Newton's theory of bodies interacting at a distance" (Baldy, 2007, p. 1767). Published in a journal read by science educators and some physics educators, Baldy's work received little attention until 2013. At the time of publication in 2007, there were insufficient like-minded researchers conducting similar research in schools, and so, there was no follow-up and checking of these findings. An interesting point is that Baldy did not cite the work of Haddad and Pella, again indicating that *The Journal of Experimental Education* was not a likely source of papers for review by science educators.

#### *Published work on Einsteinian physics in schools and universities*

Over the past decade, there has been an increasing number of publications reporting research findings of Einsteinian physics being implemented in schools (as evidenced by the chapters in this book) from researchers in Australia, Norway, Germany, Italy, the Netherlands, and Korea among others. As is evident from the

***Table 2.4*** **Recent research on teaching and learning Einsteinian physics in schools (Note: Citations as on 6 March 2020)**

Levrini, O. & Fantini, P. (2013). "Encountering productive forms of complexity in learning modern physics." *Science & Education*, 22(8), 1895–1910. doi 10.1007/s11191-013-9587-4 (**Italy, 48 citations**)

Pitts, M., Venville, G., Blair, D., & Zadnik, M. (2014). "An exploratory study to investigate the impact of an enrichment program on aspects of Einsteinian physics on year 6 students." *Research in Science Education*, 44(3), 363–388. DOI:10.1007/s11165-013-9386-6 (**Australia, 17 citations**)

Zahn, C. & Kraus, U. (2014). "Sector models—A toolkit for teaching general relativity. Part 1: Curved Spaces and Spacetimes." *European Journal of Physics*, 35, 055020. DOI:10.1088/0143-0807/35/5/055020 (**Germany, 23 citations**)

Bungum, B., Bøe, M. V., & Henriksen, E. K. (2018). "Quantum talk: How small-group discussions may enhance students' understanding in quantum physics." *Science Education*, 102(4), 856–877. https://doi.org/10.1002/sce.21447. (**Norway, 8 citations**)

Kaur, T., Blair, D., Stannard, W., Treagust, D., Venville, G., Zadnik, M., … Perks, D. (2018). "Determining the intelligibility of Einsteinian concepts with middle school students." *Research in Science Education*. https://doi.org/https://doi.org/10.1007/s11165-018-9791-y. (**Australia, 3 citations**)

Kersting, M., Henriksen, E. K., Bøe, M. V., & Angell, C. (2018). "General relativity in upper secondary school: design and evaluation of an online learning environment using the model of educational reconstruction." *Physical Review Physics Education Research*, 14(1), 010130-1-010130–18. https://doi.org/10.1103/PhysRevPhysEducRes.14.010130. (**Norway, 17 citations**)

Kim, H. & Lee, G. (2018). "Reflection and outlook on special relativity education from the perspective of Einstein: Focusing on research papers (published in Korean)." *Journal of the Korean Physical Society*, 73(4), 422–435. https://doi.org/10.3938/jkps.73.422. (**Korea, 3 citations**)

Kamphorst, F., Vollebregt, M. J., Savelsbergh, E. R., & van Joolingen, W. (2019). "Students' preinstructional reasoning with the speed of light in relativistic situations." *Physical Review Physics Education Research*, 15, 020123. https://doi.org/10.1177/0013164468028002l6. (**The Netherlands, 1 citation**)

publications reported in Table 2.4, now there would appear to be a critical mass of researchers and research groups who are regularly researching and publishing Einsteinian physics teaching and learning in schools that are being widely cited. There are other recent papers not yet cited that I have excluded from this list of publications.

## CONCLUSION

It is well-known that science has paradigm shifts such as from a Newtonian view to an Einsteinian relativistic view of movement and forces. In this chapter, I have provided examples to emphasise some of these changes at a fundamental and philosophical level. At the outset of these changes, the reason for these different changes may not be well understood by the majority of scientists. Some ideas are ahead of the Zeitgeist and do not become the new norms of a paradigm. However, once a critical mass of scientists begin to accept the new ideas-models-theories, these then become the acceptable norms for conducting research in that field. Less well-known is that education has paradigm shifts, although education as a whole could be argued to be in a pre-paradigmatic state. For example, practices like lectures in the paradigm of teacher-centred learning are still used in the new paradigm of more active student-centred learning.

Indeed, in both science and education, the previous paradigms can remain useful and important in the given situations. For example, Newtonian physics is very useful for understanding the movement of cars or trucks on roads and determining stopping distance of each at different speeds. However, these laws and principles of Newtonian physics do not provide sufficiently accurate measurements at the quantum or astronomical levels, hence the need to consider Einsteinian physics principles. In education, lectures are useful to large audiences to gain new knowledge, but to develop critical thinking, a student-centred approach is needed so that students can discuss ideas and develop new and meaningful ways of thinking.

Knowing about paradigms in science, what effect they have had on transitions to new ways of thinking can be helpful in thinking about different educational models as paradigms and how this perspective enables us to make progress in education. More recently, paradigm changes in education are becoming more evident with more agreement with the outcomes of empirical research supporting student-centred teaching and learning.

As shown in the last part of this chapter, for educational changes such as the teaching and learning of Einsteinian physics in primary and secondary schools to take place, there needs to be a critical mass of researchers and research reports of successful interventions so that this teaching becomes the new norm in physics education. Successful interventions of Einsteinian physics with primary and middle school (grades 4–8) in Lebanon in 1972, and with grade 9 students in France in 2007 were ahead of the Zeitgeist for these fundamental changes in the school science curriculum, and therefore, did not engender much interest. As shown by the recently published research papers in Table 2.4, there is an increasing quantity of research findings to demonstrate the effectiveness of implementing Einsteinian physics in both primary and secondary schools. The challenge is to maintain this momentum of research by colleagues in our international network and continue to present increasingly strong evidence that Einsteinian physics concepts can be learned and understood by the students in primary and secondary schools. This teaching and learning will then become the new norm illustrating that a paradigm shift has taken place in education.

## ACKNOWLEDGEMENTS

Many thanks to Dr Magdalena Kersting (University of Oslo) and Dr Robin Groves (Curtin University-retired) for insightful comments and useful suggestions for improving this chapter.

## NOTES

1. These paradigm changes and many more can be found at https://en.wikipedia.org/wiki/Paradigm_shift#Examples_of_paradigm_shifts_in_the_natural_sciences.
2. For the citations of the four papers, see https://www.loc.gov/rr/scitech/SciRefGuides/einstein.html.

## REFERENCES

Baldy, E. (2007). "A new educational perspective for teaching gravity." *International Journal of Science Education*, *29*(14), 1767–1788.

Brown, J. S., Collins, A., & Duguid, P. (1989). "Situated cognition and the culture of learning." *Educational Researcher*, 18(1), 32–42.

Bryson, B. (2003). *A short history of nearly everything.* Sydney, NSW: Black Swan Books.

Daniels, H. (2017). *The curious classroom: 10 structures for teaching with student-directed inquiry.* Portsmouth. Heinemann.

Darden, L. (1977). "William Bateson and the promise of Mendelism." *Journal of the History of Biology*, 10(1), 87–106.

Furtak, E. M., Seidel, T., Iverson, H., & Briggs, D. C. (2012). "Experimental and quasi-experimental studies of inquiry-based science teaching: A meta-analysis." *Review of Educational Research*, 82(3), 300–329.

Griffin, P. & Care, E. (Eds.). (2015). *Assessment and teaching of 21st century skills: Methods and approach.* Dordrecht. Springer.

Haddad, W. D. & Pella, M. O. (1972). "Relationship between mental maturity and level of understanding of concepts of relativity in grades 4–8." *The Journal of Experimental Education*, 41(1), 22–32

Henig, R. M. (2001). *The monk in the garden: The lost and found genius of Gregor Mendel, the Father of genetics.* Boston, MA: A Mariner Book, Houghton Mifflin Co.

Jaki, S. L. (1978). "Johann Georg von Soldner and the gravitational bending of light, with an English translation of his essay on it published in 1801." *Foundations of Physics*, 8, Nos. 11/12.

Kaur, T., Blair, D., Moschilla, J., Stannard, W., & Zadnik, M. (2017). "Teaching Einsteinian Physics at schools: Part 1, models and analogies for relativity." *Physics Education*, 52(6), 065012.

Kolchinsky, E. I., Kutschera, U., Hossfeld, U., & Levit, G. S. (2017). "Russia's new Lysenkoism." *Current Biology*, 27, R1037–1059.

Kuhn, T. S. (1962). *The structure of scientific revolutions.* (2nd edn.). Chicago IL: University of Chicago Press.

Molenda, M. (2008). "The programmed instruction era: When effectiveness mattered." *TechTrends*, 52(2), 52–58.

Pollack, A. J. (1992, November 17). "Cold fusion, derided in US, is hot in Japan." *The New York Times*, Section C, page 1.

Qureshi, S., Vishnumolakala, V. R., Southam, D., & Treagust, D. (2017). "Inquiry-based chemistry education in a high-context culture: A Qatari case study." *International Journal of Science and Mathematics Education*, 15, 1017–1038.

Redish, E. F. (2014). "Oersted Lecture 2013: How should we think about how our students think?" *American Journal of Physics*, 82 (6), 537–551.

Renn, J. & Hoffmann, D. (2005). "1905 – A miraculous year." *Journal of Physics B: Atomic, Molecular and Optical Physics*, 38, 437–448.

Ritter, S. K. (2016). "Cold fusion died 25 years ago, but the research lives on." *Chemical & Engineering News*, 94(44), 34–39.

Shuell, T. J. (1988). "The role of the student in learning from instruction." *Contemporary Educational Psychology*, 13, 276–295.

Skinner, B. F. (1953). *Science and human behaviour*. New York. Simon and Schuster.

Smith, A. & Waller, A. (1997). "Afterword: New paradigms of college teaching." In Campbell, K. & A. Smith (Eds.), *New paradigms for college teaching*. Edina MN: Interaction Book Co.

Trilling, B. & Fadel, C. (2009). *21st century skills: Learning for life in our times*. San Francisco CA: Jossey-Bass.

Watson, L. & Reigeluth, C. M. (2018). "The learner-centred paradigm of education." In West, R. E. (Ed.), *Foundations of learning and instructional design technology: The past, present and future of learning and instructional design technology*. EdTech books. https://edtechbooks.org/lidtfoundations/learner-centered_paradigm

# 3 The difficult birth of quantum physics

*David Blair*

**WHAT?**

This chapter traces the sequence of discoveries which convinced physicists of the need to replace classical theories of light and matter. At a level appropriate for high-school students, the chapter describes the successive discoveries that gave us our modern understanding of quantum physics.

**FOR WHOM?**

This chapter is intended for readers who wish to learn how Einsteinian physics emerged through a sequence of discoveries that led us to the frontiers of our current knowledge. The key formulae are extracted into boxes so that the main text can be more easily accessible.

## INTRODUCTION

This chapter highlights the enormous struggle from which the revolutionary new understanding of the universe called Einsteinian physics emerged. Einsteinian physics is based on two major theories of physics: the theory of special and general relativity that describes space, time, and gravity at the cosmic scale, and quantum physics that describes the interactions of matter and radiation at scale down to the smallest subatomic particles. The first evidence that light comes in discrete units of energy was found 27 years before the term photon was coined. It took the better part of a century before the detection of individual photons of visible light was commonplace, and the mysterious nature of quantum reality could be clearly observed in simple experiments. Newton's law of gravitation was in trouble as soon as Einstein's special relativity identified the speed of light as the speed limit of the universe. Special relativity triggered a ten-year intellectual tour de force by Einstein as he tried to incorporate gravity into relativity, culminating in general relativity his full theory of gravity. While Einstein presented the general theory of relativity with beautiful completeness, its understanding and verification posed enormous tasks for generations of physicists and astronomers as described in Chapter 4.

### *Two parallel histories: Quantum physics and general relativity*

In both general relativity and quantum physics, subtle observations drove the conceptual revolutions. Yet the new theories radically changed our understanding of the reality of nature. Traditionally, quantum physics and general relativity have stood apart, separated by a fundamental mismatch between their language, their mathematics, and their structure. Yet, in reality, nature does not know this separation. Quantum physics and relativity work together, and in history, they were born concurrently. Einstein worked simultaneously in both fields, and he continued to publish new work in both areas throughout his life. Besides, experimental observations link both theories intimately. The big bang, neutron stars, black holes, atomic clocks, and gravitational wave detectors are all systems that we can only understand with both relativity and quantum physics.

Despite these connections, it makes educational sense to separate the narrative of Einsteinian physics into two parallel histories. For this reason, in this chapter, we restrict ourselves to the difficult birth of quantum physics. The next chapter focuses on the birth of the theory of space, time, and matter. The birth of quantum physics had three phases: the first in which many physicists met Einstein's theoretical discoveries with scepticism. In the second phase, the new quantum reality implied by Einstein's discoveries became widely recognised by physicists, but it was vehemently resisted by Einstein. In the third modern phase, quantum technology has allowed us to clearly observe quantum phenomena. This enables students (and everyone) to obtain a much deeper understanding of quantum reality based on direct observations. We will focus mainly on the first two phases, starting with the birth of the photon.

## THE HISTORY OF QUANTUM PHYSICS

### *The birth of the photon*

In 1900, Max Planck was trying to explain the spectrum of light emitted from a furnace: how does the range of colours depend on temperature? Eventually, "in an act of desperation", Planck realised that if he made the assumption that energy could only be absorbed or emitted in discrete lumps, or quanta, with energy proportional to frequency, then he could correctly predict the spectrum. He told his son, Erwin, "What I found out is either complete nonsense or else I have just made a discovery that is as important as Newton's law of gravitation" (Segrè, 2007).

$$E = hf$$

Planck's discovery required light energy quanta with energy $E = hf$, where $h$ is called Planck's constant = $6.63 \times 10^{-34}$ J.s. For light, the frequency $f$ is ~$10^{15}$ Hz, but the formula is valid for all forms of electromagnetic radiation from radio to gamma rays. See https://www.youtube.com/watch?v=Tz2D1aahd5A.

Planck's result was a big problem because Heinrich Hertz's experiments in 1886, followed by the development of radios, had conclusively shown the existence of Maxwell's predicted electromagnetic waves. While matter was quantised in the form of atoms, it seemed quite clear that light and all electromagnetic waves were continuous wave-like disturbances. Planck himself believed that his discovery could be reconciled with waves.

However, in 1905, Einstein published an obscurely titled paper in which he compared the *entropy* (the disorder) of a box of atoms with that of a box of electromagnetic radiation. He showed that both had the same mathematical form. From this, he concluded that "radiation... behaves... as though it consisted of a number of independent energy quanta". This meant that radiation must actually come in discrete quantised lumps, just like atoms. He used this to derive a formula for the *photoelectric effect* and discussed how ultraviolet light (with high frequency, and hence, high energy) could cause fluorescence and ionise gases by transferring the energy of each quantum to electrons. He won the Nobel Prize for this discovery 16 years later.

Still, given all the evidence for electromagnetic waves, Einstein's result was extremely difficult to accept. For many years, until further evidence was accumulated, it was opposed.

Max Planck discussed the possibility of rejecting Maxwell's concept of electromagnetic waves in 1909, stating that it "seems to be a step which in my opinion is not yet necessary".

By this time, Einstein had made another very important, but much overlooked discovery. In 1907, he showed that heat and sound vibrations in solids must also be quantised, and obey Planck's formula $E = hf$. Today, these quanta of sound and heat are called *phonons*. They underpin the theory of solids. All aspects of heat in solids, its flow, and its energy are now understood in terms of phonons.

Einstein continued to explore and promote the concept of light quanta. In 1909, he published another paper entitled "The Development of Our Views on the Composition and Essence of Radiation" in which he put forward the idea that particle-like energy quanta may have an oscillation component. This was the first suggestion of what later came to be called wave-particle duality.

In 1913, Planck, Nernst, and Rubens nominated Einstein to be a member of the Prussian Academy of Sciences. In their nomination, they stated that "the fact that he has now and then overshot in his speculations, as for example with his light quantum hypothesis, should not be held too strongly against him; without a certain amount of risk there cannot be any real innovation, even in the exactest of the natural sciences" (Correspondence 445 in Klein, Kox, & Schulmann, 1994).

Three years later, in 1916, Millikan, who was the first person to measure the charge of an electron, wrote "Einstein's bold, not to say reckless, light-quantum hypothesis, flies in the face of the thoroughly established facts of interference" (Millikan, 1916).

Millikan went on to experimentally validate Einstein's equation for the photoelectric effect, but he still vehemently rejected Einstein's conclusion. He wrote that Einstein's equation for light quanta "stands complete and apparently well tested, but without any visible means of support...Experiment has outrun theory, or, better, guided by erroneous theory, it has discovered relationships which seem to be of the greatest interest and importance, but the reasons for them are as yet not at all understood".

In 1917, Einstein asked this question, "Does the molecule suffer a push, when it absorbs or emits the energy [of a quanta]"? In other words, he asked whether what we now call photons carry momentum as well as energy. By considering gases inside a hot box, Einstein was able to show that if "a molecule ... absorbs or emits an amount of energy *hf* in the form of radiation by means of an elementary process, then momentum *hf/c* is always transferred to the molecule" (Einstein, 1917).

$$p = \frac{hf}{c}$$

Momentum $p$ of a photon is its energy $hf$ divided by the speed of light $c$.

For normal masses travelling at low-speed, momentum is mass times the velocity. In general, momentum determines the ability of objects to push and break things.

Einstein's (1917) conclusion completed the picture that photons have particle-like properties. Like atoms and all moving objects, photons carry both momentum and energy. We experience the energy of photons as heat when they hit our skin and create phonons that stimulate temperature sensing nerves. We experience the energy of molecules in the air as heat in exactly the same way. Similarly, we experience the momentum of molecules as pressure, while for light, the same pressure is called radiation pressure.

Einstein's conclusion made the conundrum of quantum physics quite clear. Photons behaved like particles with discrete energy. Yet their wave-like properties could not be disputed: they have wavelength, frequency, and speed just like sound waves.

$$c = f\lambda$$

For photons, the product of wave frequency $f$ and wavelength $\lambda$ is equal to the speed of light $c$. For all waves, wave speed = frequency × wavelength.

In 1923, Arthur Compton observed photons scattering off electrons in much the same way that billiard balls bounce off each other. This discovery left little doubt that photons really did act like projectiles (Figure 3.1).

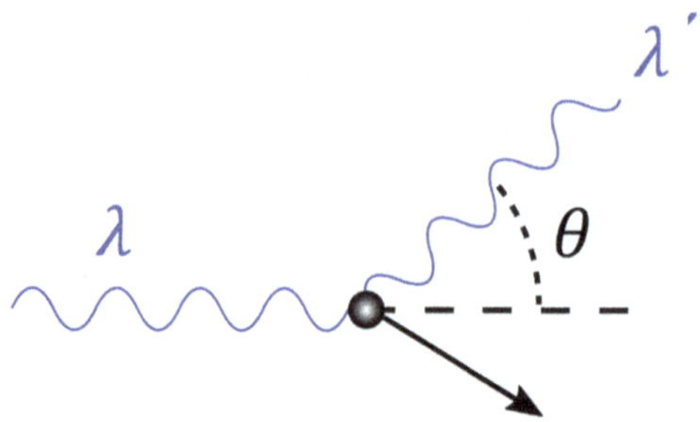


3.1
This standard image of Compton scattering shows a photon with wavelength λ coming from the left, hitting an electron, and a new photon with less energy emerges at an angle θ. Note that the snakey representation of photons in this diagram is an artistic model designed to show something with wave and particle-like properties. The electron should be equally fuzzy. (*Source: JabberWok / CC BY-SA (http:// creativecommons. org/licenses/ by-sa/3.0/)*)

$$\lambda = \frac{h}{p}$$

According to de Broglie's law, everything has a wavelength inversely proportional to its momentum. This is true for both photons and particles-such as atoms or bullets. If the momentum $p$ is large such as a speeding bullet, then the wavelength λ is infinitesimal, and the motion is dominated by *bulletiness*. If the momentum is low, collisions are not damaging, but *waviness* increases and quantum interference can be observed. Bulletiness invokes the ability of things to be destructive, while waviness invokes their ability to interfere.

The next year, in 1924, Louis de Broglie proposed a dramatic synthesis: not only do photons have particle-like properties but also matter like electrons or atoms should have wave-like properties. He proposed a universal equation that links wavelength and momentum: wavelength = Planck's constant/momentum. The greater the momentum, the smaller the wavelength. This meant wave and particle properties were common to everything.

De Broglie's synthesis should have been sufficient to quell the scepticism, but in 1925, two full decades since Einstein's first proposal, Niels Bohr was still able to joke "Even if Einstein sends me a cable that an irrevocable proof of the existence of light-quanta has now been found, the message cannot reach me, because it has to be transmitted by electromagnetic waves" (Mehra & Rechenberg, 1982, p. 554).

The term photon was proposed in a letter to Nature in late 1926. The descriptive name clearly aided the acceptance of the concept, and by 1927, the term photon came into common use.

The final step in uncovering the new reality was the validation of de Broglie's theory that matter has *waviness*. This also occurred in 1927 when Thompson and Davisson in separate experiments observed interference patterns made by electrons (Figure 3.2). Interference was the classic proof of waviness because this is what we observe when classical waves like ripples on a pond overlap, cancel, and add in destructive and constructive interference.

Having followed this series of discoveries, many physicists still remained sceptical. The fundamental reason was that the particle nature of photons implies that all quantum events are *statistical*. Statistical means that quantum events

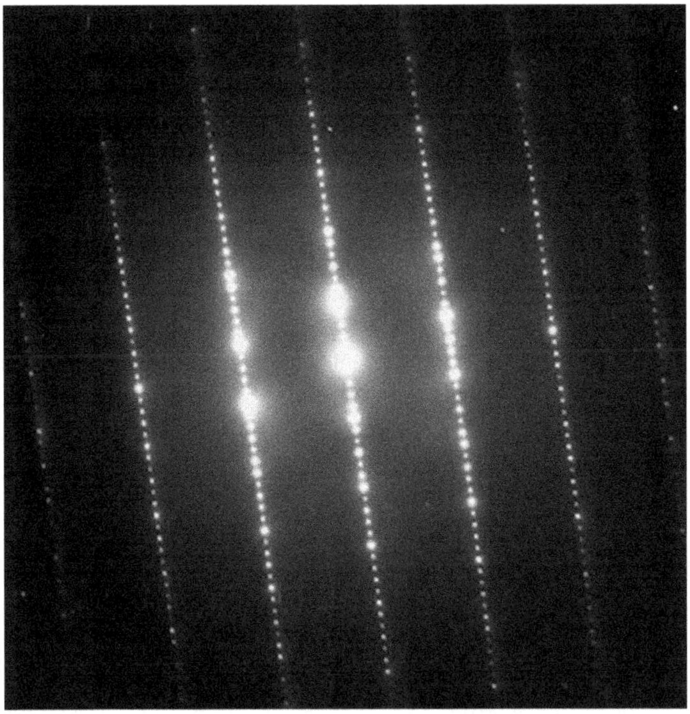

3.2
Electron diffraction image: electrons passing through a metal foil act like waves and create patterns of constructive (bright) and destructive (dark) interference. Selected area electron diffraction image is obtained in a JEOL 2000FX TEM. (*Source: Oysteinp at the English language Wikipedia / CC BY-SA (http://creativecommons.org/licenses/by-sa/3.0/)*)

are governed by the laws of chance. This means that they are not *deterministic*. One quantum event does not uniquely determine the next thing that will happen. Einstein was extremely troubled by this, even though it was his own discovery. He wrote to Max Born in 1924, "I cannot bear the thought that an electron exposed to a ray should by its own free decision choose the moment and the direction in which it wants to jump away. If so, I'd rather be a cobbler or even an employee in a gambling house than a physicist" (as cited in Born, 1968, p. 160).

### *Uncovering quantum reality*

Once we understand that light comes as photons, it is easy to see how the statistical conclusion must follow. Partial reflections from windows illustrate this conclusion. A window with a thin silver layer lets only half of the light pass, the other half gets reflected. How can this happen? The determinist who rejects a statistical universe might say "every photon must split in half". Each half photon would have half energy. But Planck's formula for photons says that the photon frequency is directly proportional to its energy. The half-photons with half energy would have half frequency. But frequency is related to colour. Hence, light that passed through the window would change its colour from say violet, (the highest frequency and highest energy we can see) to the deepest red (which is the lowest frequency we can see). Green objects would disappear into the infrared band. Experience shows that this does not happen. Reflections in windows don't change colour.

What we actually observe at a partial reflection is a reduction in intensity. The photons do not split. They have 50% probability of reflecting, with the other half transmitted. The modern photon detectors show that photons follow the rules of chance. Einstein thought that the statistics reflected our basic ignorance and believed that there must be a "hidden variable": something inside the photons that told the photons what to do. We will see below that his unsuccessful attempt to prove he was right was itself a major breakthrough.

Today, we know that there are no hidden variables. Nature is truly statistical. Light reflects off glass with a precisely determined probability that we normally call the *reflectivity*. Similarly, radioactive atoms disintegrate with a precisely determined probability that we measure as the *half-life*. Light appears bright at some locations (such as in an interference pattern) and dim at others because the probability of photons appearing at those locations is either high or low.

The observed statistical nature of the universe that caused Einstein such consternation drove physicists like Schrödinger, Heisenberg, and Bohr to develop sophisticated theories that codified nature's statistical rules for the quantum world. The theories were mathematical and abstract, but they worked extremely well, allowing precise predictions of experimental results. Many sophisticated and subtle experiments explored the nature of quantum interactions and confirmed the theories.

To provide some sense of a mechanism for understanding quantum behaviour, Schrödinger and Bohr developed *the Copenhagen interpretation* of quantum physics. According to this interpretation, quantum systems do not have a definite physical state prior to measurement; rather, the act of measuring forces the system into one of several possible states. Quantum systems are described by a mathematical entity called the *wavefunction* that encodes the waviness of everything in the system. Others prefer to use the term *probability wave* to stress its abstract nature and lack of physical reality. When a measurement takes place, the probability wave mysteriously *collapses*, coinciding, say, with the detection of a photon at a particular location on your retina. The amplitude of the wavefunction determines the *probability amplitude*. The actual probability of, say, a photon or an electron being measured in a detector is proportional to the square of the probability amplitude.

### *Feynman path integrals*

In 1948, Richard Feynman introduced his *path integral formulation* of quantum mechanics. In a nutshell, the method says that the probability of a photon appearing at a detector can be determined by taking the vector sum of *all possible paths* that the photon could follow. Feynman's formulation imposed a straightforward mathematical method for predicting and calculating the outcome of quantum processes. While mathematically equivalent to the other theories, Feynman's approach dispensed with the wavefunction, and instead, simplified quantum mechanics to rules for calculating the probability of particles appearing at a location. The calculations are done by using vector arrows called phasors as discussed in Chapter 16. The beauty of Feynman's method is that we can place it in the context of simple observations such as the partial reflections off windows. Therefore, we can turn Feynman's ideas into classroom exercises that teachers can introduce to students at the middle school level.

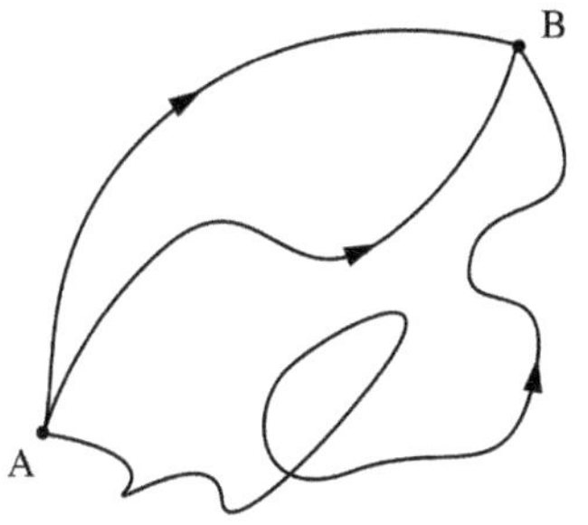


3.3
All possible paths: an example of three of the infinitely many paths from A to B. (*Source: Drawn by Matt McIrvin / CC BY-SA (http://creativecommons.org/licenses/by-sa/3.0/)*)

Following Feynman's approach, the rules of quantum reality can be summarised as follows:

1. The quantum nature of everything leads to all measurements and observations being statistical.
2. Quantum objects such as photons or molecules do not have local reality. Reality is determined when an object is registered by a measuring device. While the trajectories of balls can be observed, the trajectories of quantum objects can only be imagined until they are registered at the point of measurement.
3. The quantum rule for determining the probability of a quantum object appearing at a location consists of taking a sum of all possible paths to the location, following the rules of vector addition. Figure 3.3 shows an example of all possible paths.

In Feynman's Auckland lectures, he said, "Relax and enjoy it and realise that nobody understands it...My students don't understand it. That is because their professors don't understand it" (Feynman, 1985). It is interesting to reflect on this point. Feynman points out that many things that are normally accepted are equally "screwy" when you think deeply about them such as action at a distance by electricity, magnetism or gravity. As physicists, we get used to new concepts, until they no longer seem screwy, and we use their predictive power to explain other things.

When a single concept unifies many phenomena, we are satisfied. For example, Newton's law of gravity provided a unifying explanation for gravity on the Earth and the motions of asteroids, moons, planets, and stars. However, behind Newton's unification, there was still the mystery of action at a distance that he described as "an absurdity" (quoted in Ohanian, 1976, p. 75). Similarly, the electromagnetic force and the quantum of electric charge unifies our understanding of electrostatic and magnetic forces, and the attraction and repulsion of atoms that determine the properties of all materials. However, many school curricula do not recognise this unification and separate "contact forces" between solids as if they were different.

Today, most people readily accept Newton's "absurdity" without concern, but find curved space and quantum phenomena to be absurd. The reason for our difficulties in accepting a new reality was identified by Einstein in 1948. He said, "common sense is the collection of prejudices acquired by age 18" (as cited in Bell, 1951). We learn a language of reality as children, and at the same time, we learn spoken language. The language of reality is a conceptual framework defined by names, words, ideas, and beliefs. The spoken language stays with us for the lifetime, and so does the conceptual language of reality. This is our "common sense". It is difficult to replace one language with another. Thus, common sense is merely our first learnt language. This is the reason that Einsteinian physics, our best understanding of reality, must be taught at an early age.

### *Crucial quantum concepts*

Four other aspects of quantum mechanics are crucial to the modern world of quantum science. They are *stimulated emission*, which powers lasers; *Bose-Einstein condensation* in which quantum mechanics becomes macroscopic; the *Heisenberg uncertainty principle*, and *quantum entanglement*. We shall discuss these topics below.

#### *Stimulated emission*

In his 1917 paper "On the quantum theory of radiation" in which Einstein demonstrated that photons must have momentum, he pointed out that for radiation (that is, photons) and atoms to be in thermal equilibrium (as happens in a hot gas), there must be two separate processes happening. In one case, atoms must spontaneously and randomly emit photons. It is the process that happens in fluorescent lights when electrons spontaneously change energy levels to emit photons.

In the second case, Einstein showed that there must be a process where a photon can *stimulate* atoms to emit photons. This he called *stimulated emission*. Imagine a photon passing by an atom in which an electron is in a higher energy level, ready to decay. The photon's electric field stimulates the atom to decay. In this case, the new photon is coherently connected to the incoming photon: for every incoming photon, two photons depart, and they travel in the same direction, in phase with each other. This doubling means that the light is *amplified*. Years later, Light Amplification by Stimulated Emission of Radiation, the LASER, was invented. We use Einstein's stimulated emission every time a barcode is scanned at a supermarket check-out! This short video gives further insight into stimulated emission: https://www.youtube.com/watch?v=YHmGNDMV1cY.

#### *Bose-Einstein condensates*

If gases are cooled down to a low enough temperature, the speed of the atoms becomes very low, so that every atom has very low momentum. De Broglie's law says that the wavelength of everything increases as the momentum reduces. There must be some temperature where the wavelength becomes as large as the spacing between the atoms. In 1926, Einstein, in collaboration with Satyendra Bose, showed that in liquid helium at very low temperature, there would be a sudden transition to a state where a large fraction of all atoms suddenly condense into the state of lowest possible energy called the ground state, where the quantum wavelength would be maximum.

This was the beginning of the study of *Bose-Einstein condensates*: systems where the probability waves overlap and somehow merge their individual existences into a single coherent *macroscopic wavefunction*. Suddenly, like flocks of birds and schools of fish, the system starts to behave like a single coherent entity. Astonishing phenomena arise such as superfluidity, the frictionless flow of liquids, and superconductivity, the frictionless flow of electric currents. Magnetic resonance imaging devices in hospitals use superconductivity to make extremely powerful magnets for imaging human bodies. See https://www.youtube.com/watch?v=shdLjlkRaS8.

*Uncertainty principle*

In 1927, Heisenberg realised uncertainty was an intrinsic part of quantum physics. Later, he summarised, "It is impossible to determine accurately both the position and the direction and speed of a particle at the same instant".

Mathematically, Heisenberg showed that the product of the uncertainty of position and uncertainty of momentum has to be larger than Planck's constant. If you determine the position of a particle, precisely you can have no knowledge of its momentum. If you know its momentum, precisely its position is indeterminate.

$$\Delta x \Delta p \geq h/4\pi$$

In the Heisenberg uncertainty principle, $\Delta x$ is the uncertainty in position and $\Delta p$ is the uncertainty in momentum.

Quantum uncertainty can be introduced via the observer effect: that measurements always disturb the system being measured. For example, photons used to make a measurement have momentum. The momentum kicks the system being measured and makes it uncertain. But the uncertainty principle is much more than the observer effect. Quantum uncertainty is intrinsic as a result of the waviness of everything. It does not require a measurement. Today, physicists are learning to manipulate quantum uncertainty, so as to increase measurement sensitivity in one variable while sacrificing sensitivity in another. In gravitational wave detectors, this technique, called quantum squeezing, has become very important for increasing sensitivity to gravitational waves. See https://www.youtube.com/watch?v=a8FTr2qMutA and https://www.youtube.com/watch?v=TQKELOE9eY4.

*Quantum entanglement*

The last essential topic in the birth of quantum physics came about when Einstein with Podolski and Rosen attempted to falsify quantum physics. Their 1935 paper famously pointed out the *EPR Paradox* (Einstein, Podolsky, & Rosen, 1935). It predicted what Einstein called "spooky action at a distance", that violated what he considered obvious: the local reality of everything. Local reality describes a principle in physics that assumes, first, that objects have physical properties (such as velocity or position) independent of whether or not a measurement is performed, and second, that objects can only be directly influenced by their immediate surroundings.

Einstein considered two interacting particles that could be described by a single wavefunction. Most commonly today, such pairs are created by causing a photon to split into two lower energy photons. The new photons are called entangled pairs. As long as there is no measurement, they are described by the same wavefunction, that is, they have some sort of shared waviness. If there is a measurement of one of the photons, the waviness of that photon disappears. The click of the light detector registers its existence. Astonishingly, this measurement immediately determines the measurement outcome of the other photon, even if the two photons have moved far apart in space.

Today, the phenomenon of quantum entanglement has been tested over huge distances. The "spooky" outcome is always observed. Einstein's attempt to prove that quantum mechanics was incomplete, in fact, came to allow a later generation of physicists to prove that quantum reality is non-local. The measurement of the reality of one photon at one location makes the reality of its entangled pair instantly known, even if it is at another location very far away. Yet in general, photons have no position and no momentum until they are measured. Things can be simultaneously in all possible states such as taking all possible paths until the moment of measurement.

Quantum entanglement is central to the design of quantum computers that may revolutionise computing in the future, and also, has allowed gravitational wave detectors to improve their sensitivity by changing the randomness of the photons used in the measurements. For quantum superposition, see https://www.youtube.com/watch?v=7B1IICxVdkE; for images of quantum entanglement, see https://www.youtube.com/watch?v=wGkx1MUw2TU; and for two explanations of entanglement, see https://www.youtube.com/watch?v=DbbWx2COU0E and https://www.youtube.com/watch?v=ZuvK-od647c.

## CONCLUSION

From a few small clues, Einstein and others realised that sound, heat, and light come in energy packets called phonons and photons. Einstein realised that photons must have momentum like bullets, and soon de Broglie proposed that everything in nature combines waviness and bulletiness. These discoveries have changed our view of reality, and in turn have allowed humanity to invent new technologies like phones, cameras, and solar panels. Today, we can detect single quanta and observe the quantum world in exquisite detail. Astonishing new technologies like quantum computers use quantum entanglement, while gravitational wave detectors use entanglement to change the statistical uncertainty of measurements to discover black holes far away in the universe.

The quantum reality of the universe is relevant and accessible, while the story of its discovery is rich and captivating, and can be told at all levels of school education. Chapter 4 tells an equally rich story. This is the parallel story of the difficult birth of gravitational wave astronomy.

## ACKNOWLEDGEMENTS

I wish to thank Marcus Aspelmeyer for a public lecture entitled "What is a Photon" that inspired some parts of this chapter, and Thibault Damour for long conversations that helped me unravel much of the history.

## REFERENCES

Bell, E. T. (1951). *Mathematics, queen and servant of the sciences*. New York. McGraw-Hill.

Born, M. (1968). *Physics in my generation* (2nd ed.). New York: Springer.

Einstein, A. (1917). Zur Quantentheorie der Strahlung [on the quantum theory of radiation]. *Physikalische Zeitschrift*, *18*, 121–128. https://doi.org/10.1038/138483a0.

Einstein, A., Podolsky, B., & Rosen, N. (1935). Can quantum-mechanical description of physical reality be considered complete? *Physical Review*, *47*(10), 777–780. https://doi.org/10.1103/PhysRev.48.696.

Feynman, R. (1985). *QED: The strange theory of light and matter*. Princeton, New Jersey: Princeton University Press.

Klein, M. J., Kox, A. J., & Schulmann, R. (Eds.). (1994). *The collected papers of Albert Einstein, volume 5: The Swiss years: Correspondence, 1902–1914*. Princeton, New Jersey: Princeton University Press.

Mehra, J., & Rechenberg, H. (1982). *The historical development of quantum theory*. New York: Springer.

Millikan, R. A. (1916). A direct photoelectric determination of Planck's "h". *Physical Review*, *7*(3), 355–388.

Ohanian, H. (1976). *Gravitation and spacetime*. New York: W.W. Norton.

Segrè, E. (2007). *From X-rays to quarks: Modern physicists and their discoveries*. Dover: Dover Publications.

# 4 The difficult birth of gravitational wave astronomy

*David Blair and Magdalena Kersting*

**WHAT?**

This chapter traces the sequence of discoveries which led to our current best theory of space, time, and matter. The presentation starts with special relativity and the foundations of general relativity, and discusses some of the key tests of the theory before focussing on the century-long quest to detect one of the most spectacular predictions of Einstein – gravitational waves.

**FOR WHOM?**

This chapter is intended for readers who want to understand the historical development of relativity, and the key ideas that led to gravitational wave astronomy, which allows us to listen to the gravitational sounds of the universe.

## INTRODUCTION

This chapter tells the story of Albert Einstein's theory of relativity, the "greatest feat of human thinking about nature, the most amazing combination of philosophical penetration, physical intuition, and mathematical skill" as physicist Max Born (1968) described Einstein's achievement. The story begins in 1905 – the miraculous year in which Einstein established the foundations of a new scientific worldview – and ends more than a century later. This story encompasses the struggle with ideas that preceded Einstein's theory, the difficulties in understanding the theory, the experiments that validated its predictions, and the heroic struggle that culminated in the detection of gravitational waves in 2015. This quest has given us a new reality: a reality in which space is not only curved but rippling unpredictably; a reality in which we can listen to the gravitational sounds of the cosmos, hear black holes colliding, and catch neutron stars coalescing. Gravitational waves have opened a new frontier of our knowledge and marked the beginning of a new astronomy in which we explore the universe with gravitational waves.

## SPECIAL RELATIVITY AND EINSTEIN'S HAPPIEST THOUGHT

In 1905, the same year in which he predicted photons, Einstein published his special theory of relativity. Special relativity is a theory of the structure of space and time, and its development took place in the context of intense investigations by some of the most brilliant mathematicians and physicists of the late 19th and the early 20th century, including Henri Poincaré, Hendrik Lorentz, and Hermann Minkowski. Shortly before his death, Einstein said, "There is no doubt that the special theory of relativity, if we regard its development in retrospect, was ripe for discovery in 1905" (Einstein as cited in Born, 1968). Indeed, the formulation of special relativity is a direct consequence of Maxwell's theory of electromagnetism. Maxwell's equations describe the propagation of electromagnetic waves. These equations predict that light propagates at a constant speed *c* in vacuum.

Einstein took the constant speed of light as the first postulate of special relativity and combined it with a second postulate, the principle of relativity. According to this principle, the laws of physics are the same for all inertial observers. Albeit radical, Einstein embraced the implications of his two postulates: absolute motion, absolute space, and absolute time do not have any physical significance; it depends on the observer and the choice of coordinates whether or not two events occur at the same time.

Einstein's theory became quickly accepted, especially after a series of lectures on the mathematical formalism of special relativity that Minkowski gave in 1908. Minkowski (1909) famously proclaimed, "The views of space and time which I want to present to you arose from the domain of experimental physics, and therein lies their strength. Their tendency is radical. From now onwards space by itself and time by itself will recede completely to become mere shadows and only a type of union of the two will still stand independently on its own".

Despite its success, Einstein's special theory of relativity did not include the physics of gravity. It was clear that Newton's law of gravity was in direct conflict with special relativity. According to Newton, the force of gravity acts instantaneously; according to Einstein, there exists a universal speed limit in our universe. Nothing can move faster than light, not even gravity! Besides, there was experimental evidence that Newton's force model might be incomplete. In the 19th century, astronomers observed tiny deviations in the orbit of planet Mercury around the Sun that they could not explain. Despite many attempts, astronomers were not able to match observational data to Newton's theory.

In 1907, Einstein had his "happiest thought": "I was sitting in a chair at the patent office in Bern, when all of a sudden, a thought occurred to me: If a person falls freely, he will not feel his own weight. I was startled. This simple thought made a deep impression on me. It impelled me toward a theory of gravitation" (Einstein as cited in Pais, 1982). Although his happy thought put him on the right track, it took eight more years and help from his friends and colleagues for him to develop his new theory of gravity.

Einstein recognised that in free fall, weight disappears, just like an astronaut floating freely and weightlessly in outer space. If an astronaut's rocket accelerates, the astronaut feels a force towards the floor, just like gravity. The *equivalence principle* captures the observation that locally, it is not possible to distinguish between gravity and acceleration. A freely falling observer feels weightless, measuring no local effect of gravity at all.

During his struggle to combine special relativity with a theory of gravity, Einstein had to grapple with conceptual obstacles as well as mathematical difficulties. Einstein received enormous assistance from his friend and former classmate Marcel Grossmann who became a professor of mathematics (Einstein & Grossmann, 1913). General relativity, the theory that eventually emerged, is considered by many to be one of the most beautiful and elegant human creations. In a stroke of genius, Einstein identified gravity as arising from changes in the shape of spacetime. Seeing that the shape of spacetime changes geometry, we can say that gravity is geometry: the curvature of spacetime explains gravitational phenomena.

The idea that space could be curved was not new to mathematicians at the turn of the 20th century. A century before Einstein and Grossmann linked gravity to the geometry of curved spacetime, mathematician and geometer Carl Friedrich Gauss realised that the shape of three-dimensional space need not follow the two-dimensional geometry of a flat paper, which is the geometry set out in Euclid's book, *Elements*. Indeed, Euclidean geometry, in which parallel lines never meet, could be seen as a postulate, rather than an established fact. Gauss suggested that the geometry of our world should be experimentally tested. This suggestion led to a paradigm shift in geometry, and inspired by Gauss' work, mathematician Bernhard Riemann developed the mathematics of curved spaces during the first half of the 19th century. It was this geometry that Grossmann had studied extensively, and that provided the mathematical tools to develop a geometric theory of gravity.

General relativity displays a conceptual simplicity, and yet, it stretches our minds and completely changes our view of the universe. The conceptual simplicity is married to sophisticated mathematical notation and abstraction. Einstein's field equations are at the heart of the theory and describe the dynamic interplay between matter and spacetime. The mathematics is necessary for theoretical investigations, but is entirely unnecessary for teaching the theory in schools.

The interplay between matter and spacetime is summarised by the aphorism "matter tells spacetime how to curve, spacetime tells matter how to move". For spacetime to be able to curve, it must be elastic. For this reason, we often see deformed elastic membranes being used to represent curved spacetime. These images are deceptive because they do not emphasise the enormous stiffness of spacetime. The stiffness is represented by a constant of nature made up of the speed of light to the fourth power (a huge number), divided by the constant of gravitation $G$ (a tiny number): the stiffness of spacetime is $\frac{c^4}{8\pi G}$, a number equal to about $10^{43}$. Einstein's theory then resolves into a set of equations of the form: mass-energy = stiffness of spacetime × curvature of spacetime.

**_Table 4.1_ An overview of the fundamental features of Einstein's theory of relativity**

| *Special relativity* | *General relativity* |
|---|---|
| 1. **Spacetime:** Spacetime has four dimensions consisting of three dimensions of space, and one dimension of time. Neither space nor time is an independent entity; they are inseparably linked.<br>2. **Light postulate:** In empty space, light always propagates at a definite speed called *c* (which to very good approximation is 3 x $10^8$ metres per second). This speed is independent of the state of motion of the emitting source or observer. The speed of light is the speed limit of the universe: nothing can exceed this speed.<br>3. **Principle of relativity:** The laws of physics are the same in all inertia frames of reference.[1]<br>4. **Relativity of simultaneity:** The simultaneity of events is not absolute, but depends on the observer.<br>5. **Time dilation:** For observers in an inertial frame, a clock that is moving relative to them is measured to tick more slowly than a clock that is at rest in their inertial frame. | 1. **The principle of equivalence ("Einstein's happiest thought"):** Locally, it is not possible to distinguish between a gravitational field and an accelerated system.<br>2. **Principle of relativity:** The laws of physics are the same for all frames of reference.<br>3. **Reference frames:** An inertial reference frame is a system in free fall.<br>4. **Gravity:** Gravity is not a force, but a manifestation of the curvature of spacetime. Einstein's field equations describe how matter influences spacetime to create curvature. What we normally think of as gravity is the force that has to be applied to prevent free fall.<br>5. **Einstein's first law of motion (by analogy with Newton's first law):** In the absence of external forces, matter follows geodesics in spacetime. The geodesic is the curved space analogue of a straight line. |

The equations tell us that though being flexible and able to curve, spacetime is the stiffest thing in the universe. That is why many effects of general relativity are tiny and difficult to measure. Even your muscles, weighing just a few kilograms, can easily overcome the gravitational pull of our planet with a whopping mass of $6 \times 10^{24}$ kilograms! Table 4.1 summarises the fundamental ideas that make up special and general relativity.

## PREDICTIONS OF GENERAL RELATIVITY

Einstein's field equations describe how massive objects curve space and how the gravitational field influences the flow of time. In our solar system, the deviations from flat-space geometry are extremely small, but sufficient to

1. The principle of relativity was often taken to mean that there was no way of defining or measuring your absolute motion. Today, we know that the universe does give us a universal reference frame in the form of the cosmic microwave background radiation left over from the Big Bang. Many modern treatments of special relativity do not use the two postulates, but introduce special relativity via Lorentz transformations, or by the energy-mass relation $E = mc^2$.

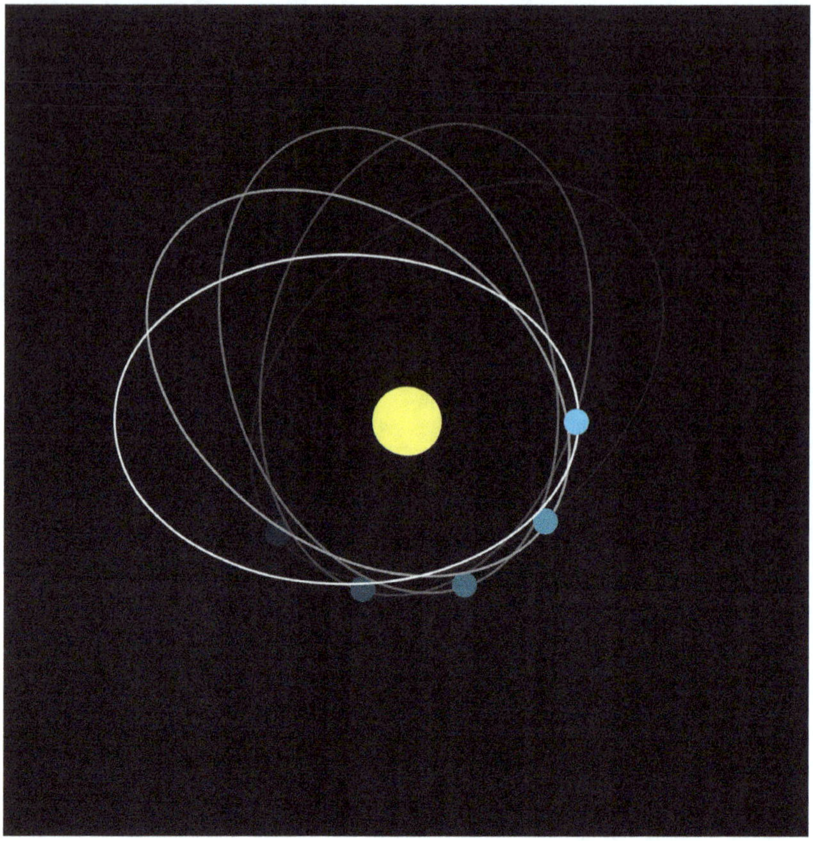

4.1
The correct description of the orbit of Mercury was an early triumph of general relativity. Within his curved spacetime framework, Einstein was able to explain deviations from Mercury's orbit as predicted by Newton. (*Source: Copyright by Rainer Zenz / Public domain*)

demote Euclidean geometry from being an exact description to a mere approximation of reality. Equipped with his field equations, Einstein was able to solve the riddle of Mercury's orbital pattern. Orbits in Newtonian gravity are defined by the area an orbiting body sweeps out, and by the perimeter of the orbit. In flat space, for a circular orbit, the area to perimeter ratio is exactly equal to half the radius, but in curved space, you can fit extra area within the same circle, so the perfect ratio is violated. Without the perfect ratio, orbits cannot stay fixed in the sky: they *precess* and create a daisy petal pattern (Figure 4.1). The correct description of the orbit of Mercury was an early triumph of general relativity, but it was the next test of general relativity that made Einstein truly famous.

### *The bending of light*

Since the early 1800s, astronomers had suggested that starlight might be deflected by the Sun, thinking that gravity should deflect light as if it was a stream of bullets. A bullet travelling at the speed of light and passing the Sun in flat Euclidean space would be deflected by 0.87 arc seconds. Since the 1890s, astronomers had searched unsuccessfully for this effect by trying to photograph stars near the Sun during a solar eclipse. If they had succeeded, they would have been surprised because if space is curved, there are two separate effects: the

Newtonian effect that many had expected, and an equal-sized effect due to the straight-line motion of light through curved space. This combined effect doubles the prediction to 1.75 arc seconds.

An eclipse in 1919 provided an ideal opportunity to test Einstein's prediction. Overcoming many difficulties, Sir Arthur Eddington led an expedition to the west coast of Africa and succeeded in photographing a group of stars close to the Sun. Eddington and his team measured the changed position of the stars, not with great accuracy, but sufficient to proclaim that Einstein was correct: space is curved by the Sun. (Chapter 11 discusses this topic in detail.)

Eddington's expedition was the turning point in Einstein's career. The curvature of spacetime was no longer just a mathematical curiosity; it explained fundamental phenomena in nature. The proclamation on the front page of the New York Times stated, "Lights askew in the heavens....men of science agog", and finally "A book for 12 wise men...no more in the whole world could comprehend it, said Einstein...". Almost overnight, Einstein had become an international celebrity. Although this observational confirmation of general relativity was a scientific triumph for Einstein, the "twelve wise men" claim marked the beginning of the widespread perception that Einstein's theory was almost incomprehensible to laypeople.

***Singularities and waves: The first solutions of Einstein's equations***

In early 1916, within months of the publication of Einstein's theory, physicist and astronomer, Karl Schwarzschild found a solution to the field equations of general relativity in the simple case of spherical symmetry. Schwarzschild's solution describes the curvature of space and the warping of time around a spherical mass. However, if the density of the central mass is high enough, Schwarzschild's solution features a singularity, a point of infinite density (Box 4.1), surrounded by an event horizon that marks the region of space in which the gravitational field is so strong that no light can escape. Physicists found it impossible to believe that such a solution could represent reality. It was considered to be a weird mathematical artefact and physically meaningless.

In mid-1916, months after Schwarzschild's solution, Einstein found another solution to his field equations. This solution predicted gravitational waves and fluctuations in spacetime that would travel at the speed of light. However,

**BOX 4.1: THE SCHWARZSCHILD RADIUS OF A BLACK HOLE**

The Schwarzschild radius is the physical parameter $r_S$ that appears in the Schwarzschild solution of Einstein's field equations:

$$r_s = \frac{2GM}{c^2}$$

This parameter describes the event horizon of a spherically symmetric, non-rotating, and electrically neutral black hole with mass M; here, G is the gravitational constant and c is the speed of light.

**BOX 4.2: GRAVITATIONAL WAVE POWER**

Gravitational wave power from two masses $M$ (in kilogrammes) orbiting each other at a distance $L$ (in metres), with orbital frequency $\omega$ measured in radians per second:

$$\text{Power} \approx \frac{G}{c^5} \cdot M^2 L^4 \omega^6$$

Imagine trying to create gravitational waves by placing huge masses (say $10^5$ kilogrammes) on a large turntable (say 100 metres diameter) and spinning it to the point of destruction (say 10 radians per second): the output power would be less than $10^{-30}$ watts, an utterly negligible amount of power.

Einstein was not the first to propose the existence of gravitational waves. In 1905, the year that Einstein published his special theory of relativity, Henri Poincaré proposed the existence of gravitational waves, while Oliver Heaviside had suggested the same thing 12 years earlier. Much like Einstein in the development of his spacetime formalism, Heaviside and Poincaré had been inspired by Maxwell's theory of electromagnetism. In analogy to accelerating electrical charges that emit electromagnetic waves, they hypothesised that accelerated masses could produce gravitational waves. For example, a pair of stars orbiting each other could create these ripples in space. However, any source of gravitational waves that Einstein could imagine would emit utterly negligible gravitational wave power (Box 4.2) because the immense stiffness of spacetime means that ordinary masses like stars and planets make only tiny amounts of curvature. Einstein concluded that the waves were of academic interest only.

Not only did the effects of gravitational waves seemed too tiny to be ever measurable, but Einstein also started to doubt the validity of the mathematical derivations. The mathematical complexity of the theory led people to ask whether gravitational waves were not merely ripples in the curved coordinates, like ripples in lines of longitude, rather than ripples in space itself. In 1922, Eddington was able to show that Einstein, indeed, had misinterpreted parts of his solution. It seemed that some forms of gravitational waves could propagate at any speed, which led to Eddington's famous joke that gravitational waves "propagate at the speed of thought" (Eddington, 1922). Einstein himself was confused, and in 1935, he submitted a paper proposing the non-existence of gravitational waves. To Einstein's consternation, the paper was rejected by the editor. The confusion around gravitational waves remained for a further two decades.

It was not until 1957, two years after Einstein's death, that Richard Feynman settled the question of the reality of gravitational waves with a simple thought experiment (DeWitt & Rickles, 2011). Feynman described two masses A and B spaced some distance apart, with gravitational waves passing. "Let one mass A carry a stick which runs past touching the other [mass] B. I think I can show that the second, in accelerating up and down, will rub the stick, and therefore by friction make heat".

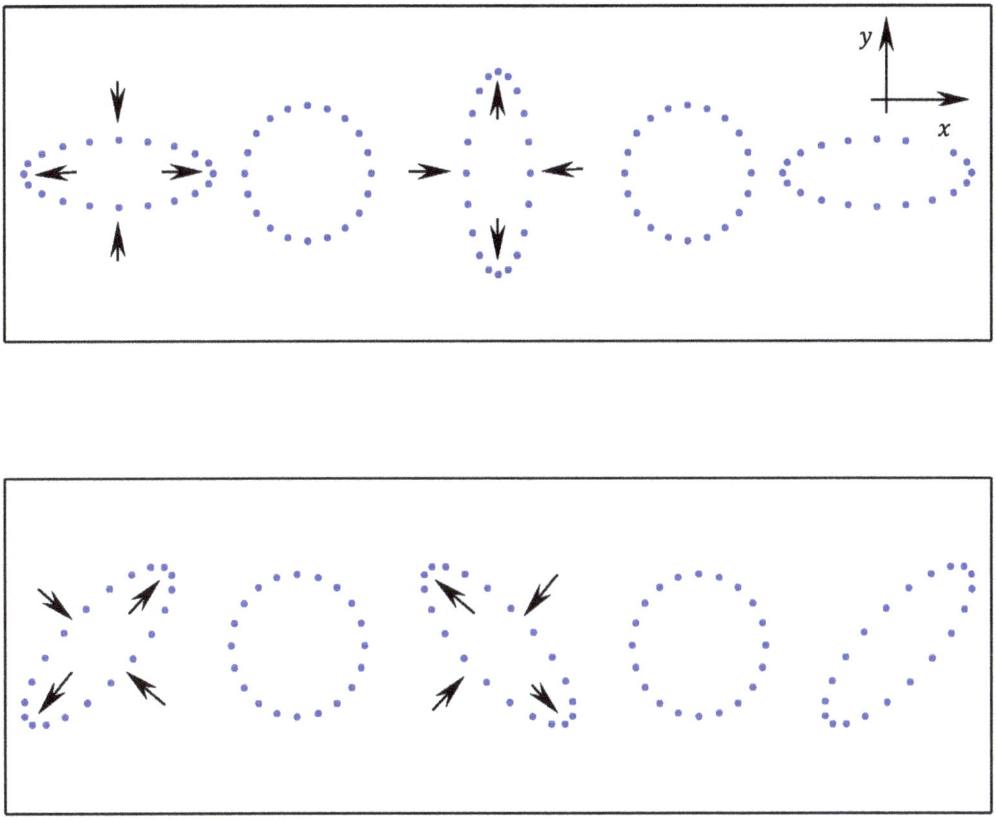


4.2
The figure illustrates the effect of a passing gravitational wave on a ring of freely falling test particles. Gravitational waves come in two polarisations (often called "plus" and "cross" polarisations) that are oriented 45 degrees from each other as shown in the top and bottom panels. (*Source: Copyright by Giles Hammond, Stefan Hild & Matthew Pitkin (2014) Advanced technologies for future ground-based, laser-interferometric gravitational wave detectors, Journal of Modern Optics, 61 / CC BY 3.0 http://creativecommons.org/licenses/by/3.0/*)

Feynman then went on to demonstrate this effect using an equation derived by theoretical physicist Felix Pirani. Finally, Feynman concluded, "In view therefore of the detailed analysis showing that gravity waves can generate heat… I conclude also that these waves can be generated and are in every respect real".

Feynman had shown that gravitational waves make real measurable changes in the distance between objects (Figure 4.2). We can think of this change in distance in terms of inertia. Inertia is the coupling of matter to space. Usually, we think of space as unchanging. Then, inertia is experienced as resistance to acceleration through space – heavy masses like cars are difficult to accelerate when you push them. But what if space expands or contracts? Then, the coupling of matter to space acts the other way – the matter experiences an acceleration provided by the expanding space. This is a real acceleration creating real changes in distance between objects. This phenomenon can be measured by light; just like light is used to measure the speed of cars on a road or even the expansion of the universe, which is also an expansion of space.

## THE QUEST TO VALIDATE GENERAL RELATIVITY

When the conceptual difficulties of the reality of gravitational waves were overcome, physicists began to take a much greater interest in general relativity after a long hiatus, partly caused by the Second World War. Full validation of general relativity would require evidence of objects with event horizons in line with Schwarzschild's solution (black holes) and observation of gravitational waves. Physicists and astronomers were inspired by Fritz Zwicky's prediction in the 1930s that ultra-compact *neutron stars* might be formed in supernova explosions. Neutron stars consist mostly of neutrons, and their density is comparable to that of an atomic nucleus. Robert Oppenheimer, who went on to become the father of the atom bomb, combined quantum mechanics with relativity to predict a *critical mass* for such balls of neutrons. Oppenheimer and his colleague, George Volkoff concluded that "actual stellar matter after exhaustion of thermonuclear sources of energy will, if massive enough, contract indefinitely" (Oppenheimer & Volkoff, 1939, p. 374). It seemed that nature allowed the existence of neutron stars, but the greater their mass, the smaller they became, and at a certain point, they would collapse. A few people imagined that such things might actually exist, but there was no evidence. It was not until 1964 that the evocative name *black hole* first appeared in print in Science News, a year after theoretical physicist Roy Kerr had found an exact mathematical solution to spinning black holes. Kerr's solution was immensely important because all black holes are likely to spin.

Besides, Kerr's solution predicts a new type of gravitational phenomena, called *gravitomagnetism*. Gravitomagnetism is analogous to magnetism, the force that is created when electric charges move. In Einstein's theory, we can think of mass as the gravitational charge of gravity in analogy to the electric charge of the electromagnetic force. Following this reasoning, we expect spinning objects to create a gravitomagnetic force. Massive spinning objects seem to drag space around them similar to the way a floating spinning ball drags water into rotation, and electrons flowing in loops of wire create magnetism. Kerr's solution also provided a mechanism for energy to be extracted from black holes: the rotational energy of the hole can be extracted because it brings surrounding matter into rotation. All future observations of black holes would be measured against Kerr's solution.

Even before neutron stars were discovered, Freeman Dyson showed that enormous amounts of gravitational wave power would be emitted if two neutron stars collided. If only Einstein had believed in the reality of Schwarzschild's solution, he might have considered two black holes orbiting each other, instead of two average stars. Then, a few lines of algebra would have shown him that the wave power produced when two black holes coalesce is roughly $c^5/G$, a vast number representing a power greater than that of all the stars in the universe combined. The algebra just needs substitution of the Schwarzschild radius and orbital speeds into Einstein's formula for the power of gravitational waves (Box 4.2). School students can do this calculation. The result tells us how nature can convert the mass-energy of black holes into pure gravitational energy in the form of ripples of space. In the 1960s, these insights provided optimism about the detectability of Einstein's waves. They motivated a growing band of gravitational wave physicists for the next 50 years.

During the 1960s to 80s, general relativity was investigated separately in three different domains:

1. Astronomers discovered strange new objects – X-ray stars, pulsars and quasars that needed either neutron stars or black holes to explain them.
2. Space scientists conducted experiments to test general relativity in the solar system.
3. Experimental physicists in the USA and Russia began to consider how to build sensitive gravitational wave detectors.

### *Astronomers discover pulsars*

Pulsars are rapidly spinning magnetised neutron stars. In 1967, astrophysicist Jocelyn Bell was the first to detect a pulsar as a pulsing radio signal. Pulsars emit narrow beams of radio waves, and while rotating quickly, they sweep the sky like lighthouses to create a pattern of pulses. Each pulsar can act as an extraordinarily precise clock. By timing when the pulses arrive, radio astronomers can probe changes in the pulsar's motion. In 1974, Russell Hulse and Joseph Taylor discovered the first *binary pulsar*. This binary system of a pulsar and another neutron star orbiting around a common centre of mass was a veritable laboratory for probing general relativity.

The pulses came faster when the pulsar was approaching, and slower as it receded. Using this periodic pattern in the arrival time of the pulses, Hulse and Taylor could map the orbit of the binary system. The two stars orbited each other once every eight hours, inside a volume a bit larger than the size of the Sun. The orbital precession due to the curvature of space around the stars was enormous compared with the tiny precession of Mercury that first helped prove general relativity.

The precession caused by curved space has quite far-reaching effects in astronomy. The geometric perfection of orbits in Newtonian physics limits what can be learnt from observation of pairs of orbiting stars because it is complicated to distinguish between circular orbits seen obliquely and elliptical orbits seen face on. When the orientation is unknown, the masses and distances are ambiguous. This ambiguity makes it extremely difficult to make precision measurements in astronomy.

The binary pulsar quickly showed how observations of systems where spacetime is strongly curved resolve this ambiguity. Suddenly, just one observation and knowledge of Newton's constant of gravitation measured in a laboratory on the Earth determine the critical properties of distant star systems. The

**BOX 4.3: THE FIRST BINARY PULSAR**

The precession caused by curved spacetime allows the masses of the two neutron stars to be determined to four significant figures, very close to the precision that we know the gravitational force constant. This level of precision had been unprecedented in astronomy up to the discovery of the first binary pulsar:

$$M_1 = 1.441 M_\odot,\ M_2 = 1.387 M_\odot$$

Here, $M_1$ is the mass of the pulsar, $M_2$ the mass of its companion, and $M_\odot$ is the mass of our Sun.

physics of strongly curved spacetime also implies that gravitational wave signals carry with them distance information. Bernard Schutz, the author of the first chapter of this book, proved this point which we discuss below.

The binary pulsar allowed the first precise measurements of the masses of stars (Box 4.3). Besides, Hulse and Taylor were able to measure the pulse delays due to the radio pulses passing through the curved spacetime of the other unseen neutron star, and the gravitational time dilation as the pulsar moves closer and farther from its companion. Most importantly, the measurements of Hulse and Taylor showed that the orbital radius was shrinking by 3.1 millimetres every orbit exactly as predicted by Einstein's formula for gravitational wave emission. This observation was a triumph for general relativity for which Hulse and Taylor won the Nobel prize in 1993. The shrinking orbit was indirect evidence that gravitational waves exist in accordance with general relativity.

Moreover, simple extrapolation of the orbital decay implied that in 300 million years, these two stars would be brought together by gravitational waves. The stars would spin about each other faster and faster, until in their last minute, they would emit a rising tone of gravitational waves from 50 hertz to 1 kilohertz in a waveform called a *chirp*. It would be an extremely powerful source of gravitational waves. Freeman Dyson had hypothesised such a source in 1962 before neutron stars were even discovered. The proof of energy loss from the binary pulsar was the first proof that powerful gravitational wave sources *must* exist in the universe. Yet, it would still be a long way to the direct detection of gravitational waves, when such chirps became a regular occurrence.

### *Space scientists test general relativity in the solar system*

While general relativity could be tested in distant astronomical systems, space scientists wanted to probe general relativity in the solar system directly. In 1971, Irwin Shapiro reported an experiment in which he bounced radar pulses off planets when their line of sight passed close to the Sun. The warping of time in the Sun's gravitational potential should cause the pulses to be increasingly delayed as the line of sight moved closer to the Sun. This, he duly observed (Shapiro et al., 1971). A few years later, as discussed above, Hulse and Taylor found the same phenomenon in the binary pulsar.

NASA planned two projects designed to test general relativity around the Earth that would test the warping of time, the curvature of space, and gravitomagnetism. The latter would confirm the second necessary component of general relativity for gravitational waves to exist.

All types of waves involve an exchange of energy between two components: an elastic component (like the stretching of a spring), and a dynamic component (like the movement of mass). In electricity, the dynamic component is magnetism, and because of its existence, we have electromagnetic waves. If gravitational waves exist, both the elastic and dynamic components of gravity must exist, too. Thus, we should be able to measure gravitomagnetism, the dynamic component of gravity that occurs in the Kerr solution.

The two NASA space experiments would use the mass and rotation of the Earth to probe both of these components. Schwarzschild's solution predicts the static effect of the Earth's mass: mass distorts both space and time. Kerr's solution predicts the dynamic effect from a rotating mass. NASA's first experiment, called Gravity Probe A, was led by Robert Vessot. Gravity Probe A consisted of an atomic clock launched into a 10,000 kilometres trajectory above the Earth. The warping of time, or *gravitational time dilation*, was clearly observed. This observation confirmed that time speeds up as you leave the Earth's gravitational potential as per Schwarzschild's prediction (Box 4.4).

Gravity Probe B was to be an exquisite quartz gyroscope that would maintain its spin axis so precisely that it could discern the static curvature of space around the Earth (Figure 4.3) which causes the 40,000 kilometres length of the orbit to differ from its Euclidean value by 28 millimetres. More importantly, this gyroscope would measure the dynamic gravitomagnetic component from the Earth's rotation. This experiment, led by Francis Everitt, was a marathon. It took 40 years and more than 100 PhD researchers to complete. Eventually, it confirmed both the curvature of space and that the spinning Earth creates the

### BOX 4.4: GRAVITATIONAL TIME DILATION NEAR THE SURFACE OF THE EARTH

The fractional change in time $\Delta t$ is given by

$$\Delta t / t = g / c^2 \text{ per metre of height}$$

where $g$ is the acceleration due to gravity.

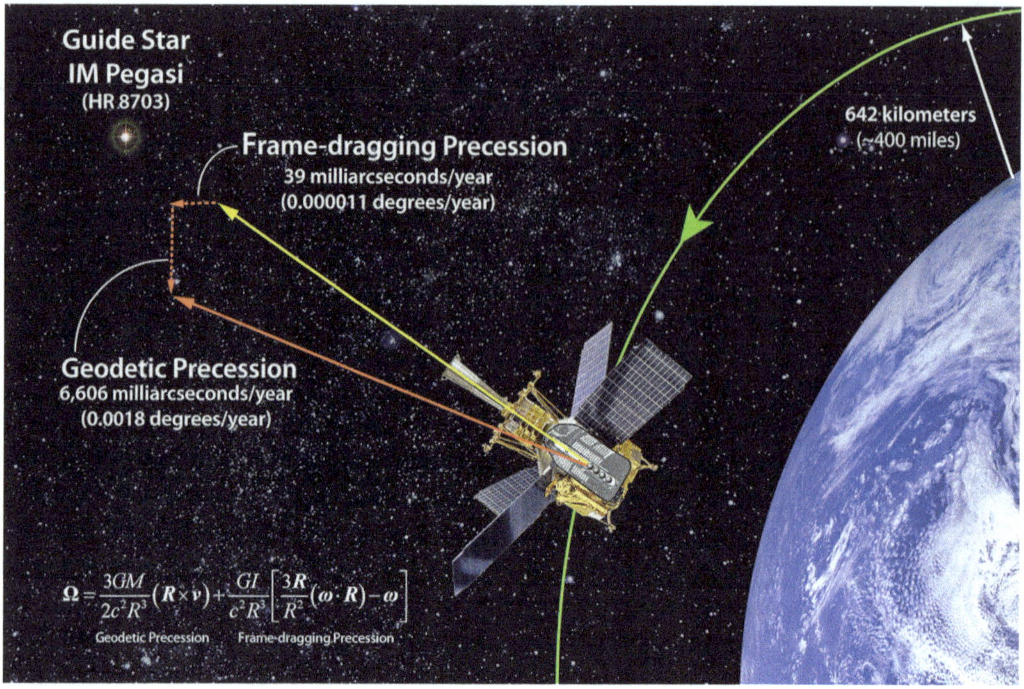


4.3
Gravity Probe B was a gyroscope in the Earth's orbit designed to measure the hitherto unverified prediction of general relativity – gravitomagnetism, as well as the spatial curvature caused by the Earth. Note that gravitomagnetism, also called frame dragging, acts at right angles to the curved space precession, causing a tiny change in the direction of the gyroscope to accumulate over thousands of orbits. (*Source: Copyright by Gravity Probe B Team, Stanford, NASA / Public domain*)

predicted "magnetic" component of gravity that is needed if gravitational waves are to exist. The results were published in 2011. By this time, binary pulsar observations meant that few physicists doubted the existence of gravitational waves.

Meanwhile, in the height of the cold war, the US military launched the Vela satellites, designed to pick up secret nuclear weapons tests in space. The satellites did not detect any tests, but instead, they discovered powerful bursts of gamma rays hitting the Earth about once a day. The source of these bursts was a mystery. Over many decades, it became clear that there were two types of bursts, called long and short bursts. While the long ones seemed to be associated with the collapse of massive stars into black holes, it took 40 years to discover the nature of the short bursts. Eventually, the observation of gravitational waves solved the mystery of the short bursts quite dramatically as we shall see below.

### *Experimental physicists design gravitational wave detectors*

The third domain of investigation was the quest to detect the gravitational waves that Einstein had thought to be of academic interest only. In the 1960s, Joseph Weber was the lone pioneer of gravitational wave detection. He investigated two types of detectors that closely followed Feynman's thought experiment of a pair of masses joined by a sliding stick. The key idea was to measure the energy imparted when the spacing between the two masses was changed by a passing gravitational wave (see Figure 4.2). Weber's plan was to treat the two masses as two halves of a solid bar. Then, the gravitational wave would squeeze and stretch the bar, leaving behind a tiny amount of acoustic energy. The second idea, first put forward by Russian scientists Gertsenshtein and Pustovoit (1963), was to use a laser-powered Michelson interferometer that could measure tiny changes in the spacing of widely spaced mirrors caused by gravitational waves. Such devices could be arbitrarily large. Weber's student Robert Forward tried out this idea and discovered the enormous technical challenges to make such a device having high sensitivity.

Weber used a pair of widely spaced detectors and searched for simultaneous excitations that could indicate gravitational wave signals arriving from space. In 1969, Weber announced that he had discovered gravitational waves. Within a few years, at least ten similar detectors had been built in research centres across the world. But nobody could repeat Weber's results. His results were statistical, and like many other researchers across many fields, Weber deceived himself with the statistics. The shame of Weber's false reports overshadowed the inspiration. The two key ideas Weber introduced are:

- Detectors must be near perfect oscillators with minimum *energy dissipation* because dissipation is always connected to noise and measurement uncertainty.
- Widely separated detectors are needed to suppress the effects of interference from other sources such as earthquakes, which travel slowly compared to the light speed of gravitational waves passing through the Earth.

Weber's ideas underpinned the entire research program, which eventually created billion-fold improved detectors that detected gravitational waves. Detectors were always widely spaced on the planet. They all involved innovative new technologies that enabled every component to be an almost perfect oscillator with such low-energy dissipation that it would ring for millions or billions of cycles.

The idea of improved technologies for building better and better gravitational wave detectors emerged before Weber's work had been discredited. Detectors had to be able to measure minuscule stretching and shrinking of space due to gravitational waves because the enormous stiffness of spacetime means that vast amounts of energy correspond to tiny motions. Space is perhaps thirty orders of magnitude stiffer than steel. Even for the most optimistic sources – colliding black holes – the stretching and shrinking would be tiny, vastly smaller than the size of an atom.

Concepts for greatly improved detectors were put forward by William Fairbank at Stanford University and Rainer Weiss at the Massachusetts Institute of Technology. Fairbank proposed cooling huge metal bars to near absolute zero and use superconductivity to make extremely sensitive microphones to pick up vibrations induced by gravitational waves. Weiss proposed building laser interferometers kilometres in length. Theoretically, both approaches offered sensitivity billions of times better than Weber's detectors.

In the 1970s, Vladimir Braginsky pointed out something that shocked the small community of detector designers. Braginsky realised that the huge instruments they were designing, consisting of tonnes of cryogenic, superconducting metal bars or tens of kilogrammes of mirrors, were actually ruled by quantum mechanics. Simple calculations proved that the Heisenberg uncertainty principle that physicists imagined being relevant only to the realm of atoms would also apply to the planned measurements of mirrors and metal bars. This insight required a re-evaluation of all proposed detectors because the uncertainty principle would limit their sensitivity. This, in turn, led to a new field of research in quantum measurement and quantum engineering.

By the 1990s, physicists had created cryogenic metal bars with superconducting sensors that were able to measure motions approaching $10^{-20}$ metres. Five widely spaced detectors detected bursts of gravitational waves for many years, but the universe was silent. The signals they could have detected would have had to come from the Milky Way, and by this time, astronomers were fairly certain that the Milky Way could not have sufficiently frequent events involving black holes or neutron stars.

In the meantime, another community of scientists had created 10–40 meters length prototype laser interferometers. Like all the detectors, these devices had taken years to understand and bring into sensitive operation. For laser detectors, the teams had to discover how to create, suspend, and control near-perfect mirrors and noise-free lasers, and detect tiny motions. The laboratory-scale instruments were not in themselves nearly sensitive enough, but proved the possibility of high sensitivity if expanded from metres to kilometres in length.

Kip Thorne championed the development of the kilometre-scale detectors, showing how these detectors could specifically detect the only type of gravitational wave source known to actually exist: coalescing pairs of neutron stars like the binary pulsar. The logic of the design follows simple mathematics: knowing that the binary pulsar in our galaxy will coalesce in 300 million years, how many galaxies like our own would we need to listen to, to hear on average one binary pulsar coalesce every year? The answer is easy: 300 million galaxies. To reach so many galaxies would require enough sensitivity to detect neutron stars coalescing hundreds of millions of light years away. In fact, the numbers are not quite so pessimistic because subsequently, many more binary pulsars were discovered in the Milky Way, including one pair that will coalesce in about 80 million years. The careful analysis allowed the necessary sensitivity to be specified and gave the laser interferometers a clear sensitivity target.

Two groups, the Virgo collaboration and the LIGO collaboration, obtained funds for enormous observatories in Europe and the USA. The detectors in these observatories were remarkably complex instruments, needing huge teams to design and operate them. It took many years, intense effort by hundreds of physicists, and a billion dollars first to construct the detectors, and then to learn how to operate them and make them sensitive. The first large-scale LIGO detectors began to operate in 2002, but it took eight years of excruciating efforts to bring them to design sensitivity. Yet, even at full sensitivity, the universe was again silent. No signals were detected. Although disappointing, this silence did not come as a surprise. By this time, there were estimates of the rates of coalescence of neutron stars, and the chance of one occurring close enough to be detectable was rather small.

### *A new astronomy*

The Virgo and LIGO teams had been careful to emphasise that the first interferometer detectors had only a small chance of success. They had always foreshadowed the need for another step in sensitivity, called Advanced LIGO and Advanced Virgo. Only these detectors would have a high chance of success. Nevertheless, the joke spread that "gravitational wave detection is always ten years in the future." The Advanced LIGO detectors came into operation in September 2015. They were better isolated from seismic vibrations and had mirrors that had much lower acoustic losses, as well as more powerful lasers, making them about three times more sensitive than the previous detectors, and able to measure signals at lower frequencies.

On the fourteenth of September 2015, just before the official start of observations, but when all systems were operating, the rising tone of a chirp suddenly appeared in the two LIGO detectors. It was the unmistakable sound of two objects spiralling together faster and faster. The strength of the signal, the rate of frequency increase, and the wave pattern are enough to define the masses and the distance. The teams had been expecting pairs of neutron stars, but the frequency rose much too rapidly for neutron stars. It had to be a pair of huge black holes, each more than thirty times the mass of the Sun. The signal seemed just too good to be true! No black holes of this mass had ever been identified.

It took the collaboration months from the first detection to announce the results because of fear that it could have been a fake signal injected into the data, or some form of interference. The memory of Weber's false announcements still haunted the community. When it was announced in February 2016, it brought worldwide headlines, and in 2018, Weiss, Thorne, and Barish received the Nobel Prize for the discovery that had taken a team of 1000 physicists and several decades to accomplish.

The gravitational wave detectors were the most sensitive quantum instruments ever created. The clear gravitational wave signal from the two coalescing black holes represented a power output in gravitational waves 50 times larger than the power output of all the stars in the universe. The waves had travelled for a billion years, and yet when they reached the Earth, their power per square metre represented *the most powerful burst of energy ever detected from outside the solar system* – pure gravitational energy. Yet the signal captured by the detectors was *the smallest amount of energy ever recorded*, corresponding to a motion ten thousand times smaller than the diameter of a proton.

The signal precisely described the moment that two black holes merged into one, followed by the faint ringing of the newborn black hole. At the point of coalescence, the two holes travelled at a good fraction of the speed of light. They emitted three solar masses of energy in less than one second. The chirp had travelled for a billion years while life was steadily evolving on the Earth. Finally, it arrived just at the time when the life forms on our Earth had invented detectors able to hear it.

Since that momentous day, similar signals have been detected frequently.

In 2017, 55 years after Freeman Dyson's prediction of gravitational waves from pairs of neutron stars, two such stars in their final minute of inspiral were detected. The signal arrival time at each of three detectors located the burst in the sky, and optical telescopes were soon able to pinpoint the flash of a *kilonova* in galaxy NGC4993. This time the gravitational waves had travelled for 130 million years to reach the Earth. Remarkably, the gravitational wave distance estimate of the event, which ultimately depends only on the laboratory measurement of the gravitational force constant *G*, matched the distance estimate from the recession speed of the galaxy due to the expansion of the universe. The remarkable ability of gravitational waves to map and measure the universe, first predicted by Bernard Schutz, was demonstrated.

Equally exciting was the detection of a burst of gamma-ray photons by the Fermi gamma-ray observatory satellite, which arrived 1.7 seconds after the gravitational wave chirp. Given that the journey time of both signals was 130 million years, the difference in speed between light and gravity had to be less than four parts in $10^{16}$. Assuming that the "magnetic engine" that created the burst took a second or so to form, we can treat this as clear confirmation that gravitation travels at the speed of light to quite extraordinary precision. Einstein was right again!

The discovery of radio waves marked the beginning of a century in which humanity explored the enormous breadth of the electromagnetic spectrum. The recent discoveries mark the start of a similar exploration of the gravitational wave spectrum. Gravitational waves allow us to listen to the birth and death of black holes, and to know them from the way they create gravitational waves. In 2019, we were able to identify a black hole by how it interacts with background light. A set of microwave telescopes called the Event Horizon Telescope imaged the shadow created by the supermassive black hole in galaxy M87 (Figure 4.4).

The next significant steps in gravitational wave astronomy will be detectors in space (https://lisa.nasa.gov/) designed to measure much lower frequency gravitational waves, and new detectors on the Earth. Both are designed to listen to gravitational waves from the entire universe. We expect to hear black holes coalescing every five minutes, and neutron stars coalescing even more often, as well as stellar-mass black holes spiralling into supermassive black holes. The evolving and slowly dying universe will be audible in real-time, thanks to general relativity and gravitational waves.

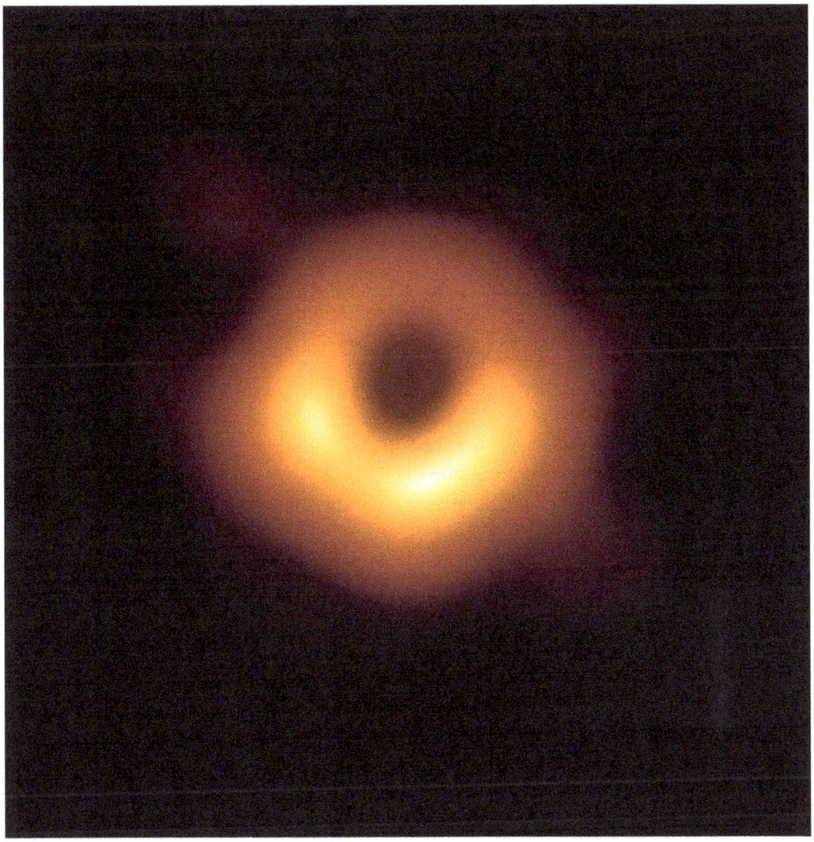

4.4
The Event Horizon Telescope captured the first-ever picture of a supermassive black hole at the core of galaxy M87. This black hole has a mass ~7 billion times the mass of the Sun. (*Source: Copyright by Event Horizon Telescope / CC BY 4.0 (https://creativecommons.org/licenses/by/4.0)*)

## CONCLUSION

A century of studying Einsteinian physics has given us a deep understanding of general relativity and quantum physics that has allowed us to create instruments to hear the universe in gravitational waves. Before the first observation of two merging black holes in 2015, gravitational waves had been the last prediction of general relativity that had not been tested directly. We now know that spacetime is dynamic, continually rippling and vibrating, and carrying vast amounts of energy like invisible and harmless tsunamis travelling at the speed of light.

Gravitational wave astronomy is still a young field of research that gives scientists insights into cosmic regions not accessible by conventional telescopes. The combined measurement of gravitational waves and electromagnetic waves opens up exciting new possibilities in multi-messenger astronomy, in which objects are observed simultaneously by different forms of radiation or other messengers from the universe. Besides, gravitational waves promise new insights into the nature of spacetime because they allow us observing black holes very close to the singularities where the theory of relativity might break down. We have just begun this brand-new exploration of a brand-new spectrum. We are at a frontier, facing the unknown.

Einsteinian physics is a triumph for humanity that everyone deserves to share. Besides its enormous scientific potential and ongoing discoveries, Einsteinian physics and the emerging discipline of gravitational wave astronomy have the opportunity to motivate students and encourage interest in STEM topics. While scientists listen ever deeper into space, we hope that Einstein's ideas will entice students and instil an appreciation for the astonishing story of our existence, a story that can only be told in the language of Einsteinian physics.

## REFERENCES

Born, M. (1968). *Physics in my generation* (2nd ed.). New York: Springer.

DeWitt, C. M., & Rickles, D. (Eds.). (2011). *The role of gravitation in physics: Report from the 1957 Chapel Hill Conference (open access).* Communicated by Jürgen Renn, Alexander Blum and Peter Damerow, Berlin: Max Planck Research Library for the History and Development of Knowledge.

Eddington, A. S. (1922). The propagation of gravitational waves. *Proceedings of the Royal Society A, 102*(716), 268–282.

Einstein, A., & Grossmann, M. (1913). Entwurf einer verallgemeinerten Relativitätstheorie und einer Theorie der Gravitation [Outline of a Generalized Theory of Relativity and of a Theory of Gravitation]. *Zeitschrift Für Mathematik Und Physik, 62*(225).

Gertsenshtein, M. E., & Pustovoit, V. I. (1963). On the detection of low frequency gravitational waves. *Soviet Physics JETP, 16*(2), 433–435.

Minkowski, H. (1909). Raum und Zeit [Space and Time]. *Jahresbericht Deutscher Mathematischer Verein, 18*, 75–88.

Oppenheimer, J. R., & Volkoff, G. M. (1939). On massive neutron cores. *Physical Review, 55*(4), 374–381.

Pais, A. (1982). *Subtle is the lord : The science and the life of Albert Einstein.* Oxford: Oxford University Press.

Shapiro, I. I., Ash, M. E., Ingalls, R. P., & Smith, W. B. (1971). Fourth test of general relativity: New radar result. *Physical Review Letters*, 26(18).

## Section 2

# Instructional approaches to teach Einsteinian physics

### *Special relativity*

# 5 Dynamics first

## *A novel approach to relativity*

*Friedrich Herrmann and Michael Pohlig*

**WHAT?**

This chapter describes the main features of an introductory course about relativity at high-school level that focusses on relativistic dynamics. A part of the course is suitable for middle school students while the complete programme is intended for secondary school students.

**FOR WHOM?**

This chapter is intended for readers who are interested in a novel approach to relativity that avoids the topic of changing reference frames, which students often find difficult.

### INTRODUCTION

We describe the basics of a course on special and general relativity. The concept has been tested and is used at numerous secondary schools: parts of it already in the lower secondary school, the complete programme in the upper secondary school. The course is part of the *Karlsruhe Physics Course*. It can be downloaded from the Internet in various languages (Herrmann and Pohlig, 2016). A bilingual English-Chinese version was published recently (Herrmann and Pohlig, 2018).

We would like to emphasise that we do not intend to describe the details of our course in this chapter, we are merely presenting some ideas that underlie its development. Some of the topics we address here are also discussed in some articles of the column *Historical burdens on physics* (Herrmann and Job, 2019).

### PARADIGM CHANGE

#### *Additional postulate of special theory of relativity*

The laws of special theory of relativity, or special relativity for short, are derived from those of classical mechanics along with one extra postulate. Traditionally, and for historical reasons, the choice was made for the invariance of the speed of light upon a change of the reference frame.

As one becomes aware that this choice, as a starting point, is only one of the several possibilities, one discovers that completely new perspectives arise for the development of a teaching concept. We have decided to introduce the mass–energy equivalence as an additional postulate instead of the invariance of the speed of light:

Postulate of special relativity: mass–energy equivalence

With this choice, one arrives more quickly at that part of special relativity that we consider being the most important one, namely relativistic dynamics.

The traditional choice of the invariance of the speed of light has a rather incidental cause: when special relativity came into being, light was the only known system that behaved relativistically. Einstein's work – both his famous publication of 1905 (Einstein, 1905) and his textbook *Grundzüge der Relativitätstheorie* (Einstein, 1970) – begins with a detailed, and one can say somewhat tiring, part on relativistic kinematics.

One might imagine what the course of history would have been if the first relativistic observation had been that a cup of hot coffee is heavier than a cup of cold coffee (or that the corresponding observation had been made with particles in an accelerator). The presentation of the theory of relativity in our textbooks would certainly be very different from what it is today.

Of course, the mass–energy equivalence is not supported by our everyday experience (neither is the invariance of the speed of light). But one can make it easily plausible and discuss its consequences even in beginners' classes. It leads to surprising and, at first, unbelievable statements; but it does not lead to the cognitive conflicts one has to deal with in the traditional approach to relativistic kinematics which question our basic convictions about space and time.

***Reference frames and reference frame changes***

In the traditional approach to special relativity, the following topics are dealt with before relativistic dynamics is addressed:

The invariance of the speed of light
Clock synchronisation
The relativity of simultaneity
Time dilatation and length contraction
Velocity addition
Lorentz transformation
Minkowski diagram

The problem with such an approach is that one begins with the most difficult and confusing part of the theory: the relationship between space and time.

Anyone concerned with special relativity knows the endless discussions about the twin paradox. In our experience, the difficulties result from the recurring changes of the reference frame.

Certainly, students can learn a lot of physics by analysing the same process in different reference frames, but we should not forget that we are dealing with beginners, and it is better to stick to the old rule that says:

> Choose a suitable reference frame at the beginning, that is, a reference frame in which the description of your problem becomes as simple as possible, and don't change it anymore.

And above all, don't change the reference frame in the middle of dealing with your problem (as is usually done when discussing the twin paradox). By the way, in classical mechanics and electrodynamics, too, one can create the greatest confusion if one chooses the reference frame improperly and changes it in the middle of the discussion.

This is why we decided not to make reference frame changes the main topic of the lessons about special relativity and to avoid them as far as possible. Above all, the impression should not be created that special relativity is essentially a theory of reference frame changes – an impression that some presentations certainly arouse. Even the name relativity gives that impression.

***The observer***

Closely related to the question of the choice of the reference frame is the problem of the so-called observer. The observer seems to be particularly important in two areas of physics: in quantum physics (where the observer always appears as the one taking a "measurement") and in the theory of relativity.

An observation is always made from a certain perspective. It thereby emphasises something that does not play a particular role in the phenomenon to be described.

We believe the observation should not be in the foreground as long as the understanding of a process is the objective. This is especially true when teaching at school, that is, for beginners.

It is true that we get all the information about the world by observing and measuring. But the idea we form of the world is quite different from what we observe. So, if we wanted to explain the shape of the Earth to someone, we would certainly not start with the shadow of the obelisk in Alexandria, but simply say: the Earth is a sphere.

> Physics lessons should primarily give a picture of what nature is like – not how it is perceived by an observer.

## RELATIVISTIC DYNAMICS

### *The use of the term mass*

Mass is a physical quantity that, until not so long ago, did not cause any problems. It was known which properties could be described with it; it could be measured.

With Einstein's special theory of relativity, this only changed insofar as the mass of a body became dependent on its velocity, temperature, and other variables. It was no longer a quantity that had a characteristic value for a body or a particle.

A body, a particle, a field, or any other structure, thus has a mass that depends, among other things, on its velocity. The value that mass assumes when the centre of mass of the particle or body is at rest is called its rest mass (symbol $m_0$). Even more appropriate would actually be the less common term proper mass, because when the centre of mass is at rest, this does not mean that the parts or particles of the system are at rest.

It is that simple, or, unfortunately, one must say: it could be that simple.

For there is an area of physics in which another use of the term mass has established itself: particle physics. A particle has a well-defined rest mass. The rest mass is characteristic of the particle species. Among various other parameters, such as electric charge, spin, lepton number, etc., it is considered the main characteristic. It seems to constitute the identity of the particle. For this property, a compact, plausible name was needed, and particle physicists simply called it mass. Thus, in particle physics, the term mass refers to only part of the quantity that describes the inertia of a particle.

However, this custom also spread beyond particle physics and resulted in several misunderstandings and ambiguities. What is to be understood by the mass of a resting macroscopic body? Is it the mass that would be measured with a (very accurate) scale, or is it the sum of the masses of the elementary particles that constitute the body? This is a question that particle physicists probably do not ask, but we teachers do.

We have, therefore, decided to use the term mass (symbol $m$) exclusively for the quantity that measures gravity and inertia, no matter what kind of object is considered and in what state it is. Thus, a hot cup of coffee has a larger mass than the same cup of coffee when it is cold. A photon has a mass, and a magnetic field has a mass (a litre of the magnetic field near a neutron star has a mass of some hundred grams).

> We use the term mass as a measure of inertia and gravity – not as a parameter that characterises a specific type of particle.

By the way, if one follows this use of the term mass, it makes no sense to say that mass is a "form of energy" or that mass can be converted into energy.

### *The identity of mass and energy*

First, let us look at the term mass–energy equivalence. It is a pity that a simple fact is expressed so unclearly. Energy and mass are the same physical quantity. When measured in kilogrammes, it is called mass; when measured in joules, it is called energy. The word equivalence is certainly not wrong, but why not say directly:

> Mass and energy are the same physical quantity.

If one were to ask someone who has never seen the equation $E = mc^2$ to express this fact in a formula, they would probably write something like this:

$$E = k \cdot m. \tag{5.1}$$

The factor $k$ tells us how the units joule and kilogramme are converted into each other. As the definitions of the units kilogramme and joule are independent of the choice of the reference frame, $k$ is a universal constant.

Its value is obtained by a measurement and one finds that

$$k = 9 \cdot 10^{16}\ \mathrm{J/kg}.$$

But what's wrong with writing

$$E = mc^2? \tag{5.2}$$

Every student learns in mathematics that a linear relationship between the variables $x$ and $y$ is written as:

$$y = m \cdot x.$$

On the right side, first is the factor of proportionality and second is the independent variable. The unbiased student might interpret the famous equation (5.2) this way: the energy is proportional to the square of the speed of light – and not that energy and mass are the same physical quantity. One might object that this can easily be explained to the students. Of course, it can. But doesn't the statement become clearer if one writes $E = k \cdot m$? Would the iconic character of equation (5.2) survive if it were formulated in this way?

### *The laws of dynamics*

In the *Karlsruhe Physics Course*, the extensive quantities, such as energy, momentum, electric charge, and entropy are introduced as basic quantities. Especially, momentum and entropy have a very direct and vivid interpretation. Momentum is a measure of the "amount of motion", that is, what one would colloquially call "impetus" or "drive". Entropy measures almost perfectly what would colloquially be called the amount of heat (not to be confused with the rather difficult concept of heat that has established itself in physics).

Therefore, in the context of relativity, it is natural to ask in the first place for the dependence of different quantities on momentum. Momentum is our independent variable. We give momentum to a body or particle and ask how it reacts to it: how does its mass (= energy) behave? What happens to its velocity? In other words: we ask for the functions *E*(*p*) and *v*(*p*).

To derive *E*(*p*) (Box 5.1), we take over as much as possible from non-relativistic physics. In addition, we only require the identity of mass and energy, that is, we assume the validity of equation (5.1). We thus obtain:

$$E(p) = \sqrt{E_0^2 + kp^2}. \qquad (5.3)$$

Here, $E_0$ is the rest energy, corresponding to the rest mass $m_0$. The red line in Figure 5.1 shows the graphic representation of equation (5.3).

Two limiting cases are of particular interest.

For small momentum values, equation (5.3) changes to:

$$E(p) = E_0 + \frac{kp^2}{2E_0} = E_0 + \frac{p^2}{2m_0}$$

We obtain the classical kinetic energy, increased by the rest energy (grey line in Figure 5.1).

If the momentum is very large, such that $E_0^2$ can be neglected in comparison with $kp^2$, equation (5.3) turns into:

$$E(p) = \sqrt{k}p \qquad (5.4)$$

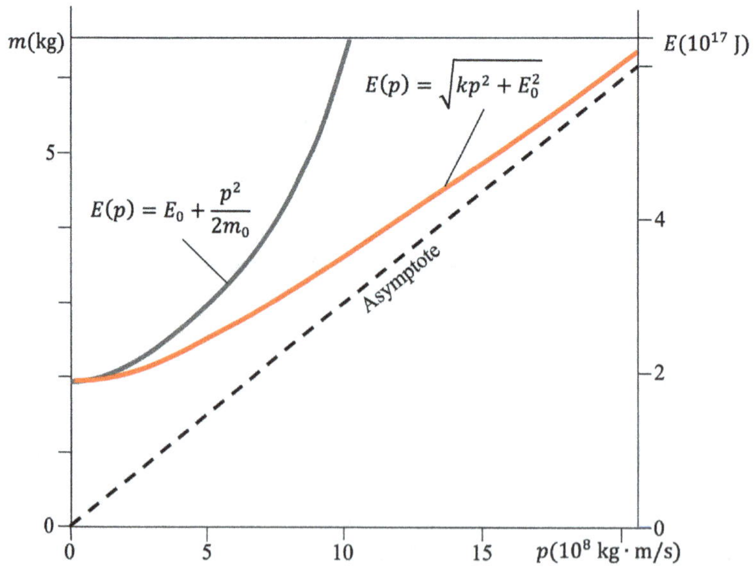


5.1 Relationship between momentum and mass/energy (*red line*). For large values of the momentum, the curve approaches the asymptote (*dashed line*), for small values the classical quadratic relation (*grey line*) is observed. (*Source: Copyright by Friedrich Herrmann, Michael Pohlig*)

**BOX 5.1 RELATIONSHIP BETWEEN ENERGY AND MOMENTUM**

We want to stay with the equation

$$\boldsymbol{p = mv,}$$

but we can no longer assume that $m$ is constant, that is, independent of the velocity.

We start with the change $dE$ of the energy that results from a change of the momentum $dp$ (Feynman et al., 2006: 15–10):

$$\boldsymbol{dE = vdp.}$$

With $\boldsymbol{p = mv}$, we obtain:

$$\boldsymbol{dE = \frac{p}{m}dp.}$$

Replacing $m$ with $E / k$, and reordering returns:

$$\boldsymbol{EdE = kpdp.}$$

We thus obtain:

$$\boldsymbol{d\left(E^2\right) = kd\left(p^2\right)}$$

and

$$\boldsymbol{E^2 = kp^2 + C,}$$

where, $C$ is the constant of integration.

The value of $C$ can easily be determined, because for $p = 0$, the energy $E$ assumes the value of the rest energy $E_0$. Thus, $C$ must be equal to $E_0^2$. We therefore get:

$$\boldsymbol{E^2 = E_0^2 + kp^2}$$

and for the sought-after relationship between energy and momentum, we get:

$$\boldsymbol{E(p) = \sqrt{E_0^2 + kp^2}.}$$

see the dashed line in Figure 5.1.

For bodies whose rest mass is 0 kg, equation (5.4) applies for all values of the momentum, not only for large values (Figure 5.2). Thus, in the highly relativistic limiting case, energy and momentum are proportional to each other. This shows that there is a similarity between these quantities which becomes even clearer when we solve equation (5.3) according to $E_0^2$.

$$E_0^2 = E^2 - kp^2 \tag{5.5}$$

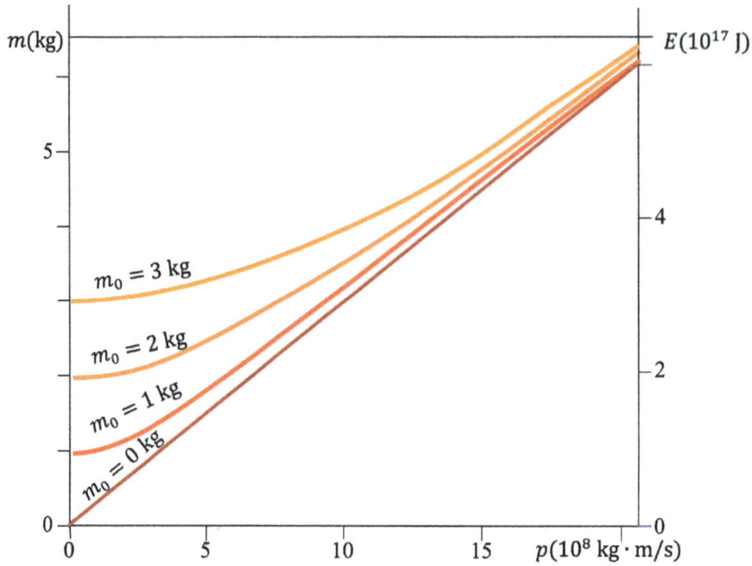


5.2
Energy momentum relationship for four different rest masses. For photons (rest mass zero), the relation is linear. (*Source: Copyright by Friedrich Herrmann, Michael Pohlig*)

We thus have the rule:

If the momentum of a body changes, its energy changes in such a way that the difference $E^2 - kp^2$ retains its value. This value is the square of the rest energy $E_0$.

Now our second question: how does the velocity of a body depend on its momentum? As presented in Box 5.2, we get:

$$v(p) = \frac{kp}{\sqrt{k^2 m_0^2 + kp^2}}. \tag{5.6}$$

**BOX 5.2 RELATIONSHIP BETWEEN VELOCITY AND MOMENTUM**

We solve $p = mv$ for $v$, then apply equations (5.1) and (5.3) and obtain:

$$\mathbf{v(p) = \frac{p}{m} = \frac{kp}{E} = \frac{kp}{\sqrt{E_0^2 + kp^2}}.}$$

If we replace the rest energy with the rest mass, we get:

$$\mathbf{v(p) = \frac{kp}{\sqrt{k^2 m_0^2 + kp^2}}.}$$

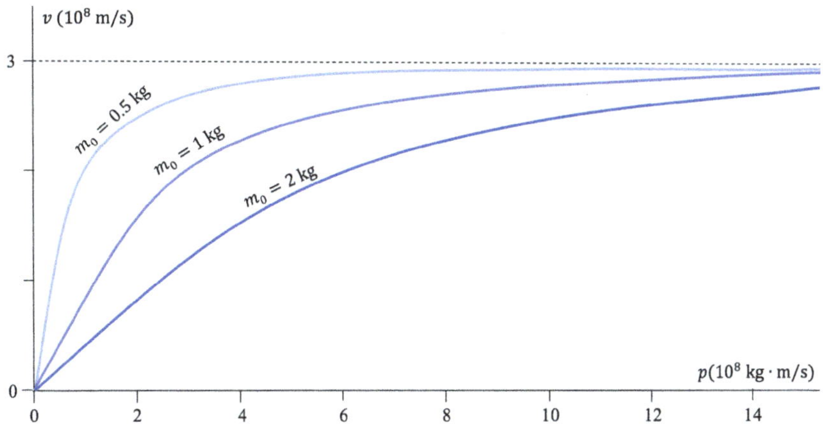


5.3 Dependence of the velocity on the momentum for different rest masses. (*Source: Copyright by Friedrich Herrmann, Michael Pohlig*)

Figure 5.3 shows the dependence of the velocity on the momentum for different rest masses. From equation (5.6) it follows that the velocity of a body approaches a terminal value as the momentum increases. It may be written as:

$$\lim_{p\to\infty} v(p) = \lim_{p\to\infty} \frac{kp}{\sqrt{k^2 m_0^2 + kp^2}} = \sqrt{k}.$$

Up till now, $k$ only played the role of a conversion factor, but now it gets a physical meaning. Its value is the square of the *terminal speed*. Since $k$ is a universal constant, its square root, that is, the terminal speed, is also a universal constant (and thus also an invariant upon a change of the reference frame).

The terminal speed is same for all bodies and is independent of the reference frame.

This can be seen in Figure 5.3. The diagram also shows that the smaller is the rest mass of a body, the "faster" it approaches the terminal speed.

Let us come back to equation (5.6). We see that if one supplies momentum to a body, its velocity initially increases linearly with the momentum, while its mass almost does not change. This is the Newtonian limiting case. But when its momentum has become very large, its velocity no longer changes, but its mass increases.

Let us ask how mass changes with velocity. As shown in Box 5.3, we get the well-known formula:

$$m(v) = \frac{m_0}{\sqrt{1 - \frac{v^2}{k}}}$$

But what is the value of $k$ and, thus, the value of the terminal speed? So far, nothing has been said about it. The answer to this question can only be obtained

**BOX 5.3 RELATIONSHIP BETWEEN MASS AND VELOCITY**

In equation (5.3), we replace $E$ by $km$, $E_0$ by $km_0$, and $p$ by $mv$

$$\boldsymbol{k^2 m_0^2 = k^2 m^2 - k m^2 v^2}.$$

We transform

$$\boldsymbol{m_0^2 = m^2 - m^2 \frac{v^2}{k} = m^2 \left(1 - \frac{v^2}{k}\right)}$$

and get:

$$\boldsymbol{m(v) = \frac{m_0}{\sqrt{1 - \frac{v^2}{k}}}}.$$

by a measurement. There are several ways to do that: either one increases the momentum of a particle until its velocity no longer changes (in a particle accelerator) and then measures its velocity, or one measures the velocity of photons, that is, particles of rest mass zero. Photons always move with the terminal speed.

Because of the great importance of the terminal speed, one gives it its own symbol

$$c := \sqrt{k}$$

The measurement results in:

$$c = 3 \cdot 10^8 \text{ m/s}$$

and therefore,

$$k = 9 \cdot 10^{16} \frac{\text{J}}{\text{kg}}.$$

***Table 5.1*** **Compilation of our most important equations**

| *Energy = mass* | $E = c^2 m$ |
|---|---|
| Dependence of energy on momentum, in general | $E(p) = \sqrt{E_0^2 + c^2 p^2}$ |
| Dependence of energy on momentum, relativistic limiting case | $E(p) = cp$ |
| Dependence of velocity on momentum | $v(p) = \frac{cp}{\sqrt{c^2 m_0^2 + p^2}}$ |
| Dependence of mass on velocity | $m(v) = \frac{m_0}{\sqrt{1 - \frac{v^2}{c^2}}}$ |

The constant $c$ is also called *speed of light*. But our derivation shows that light does not play a particular role in special relativity. That is why we prefer to call $c$ terminal speed.

In Table 5.1, we have recompiled the most important equations, but replaced the expression $\sqrt{k}$ with C.

## ABSOLUTE SPACE AND ETHER

### *Absolute space and ether in traditional teaching*

When we start with the topic of spacetime in our lessons, we have to reckon with the fact that the students bring along certain ideas from previous lessons about Newtonian physics that hamper our intentions.

In the following, therefore, we are concerned with the question of how to convey a viable idea of "empty space" as we are teaching classical mechanics, that is, a geometrically defined region of space in which there is nothing: no particles, no fields, no radiation. Physically we are dealing with the concepts of absolute space and ether.

In classical mechanics, students typically learn two facts: (1) there is no "absolute space" and (2) there is no ether. Moreover, it is said that the Michelson–Morley experiment has proven that ether does not exist.

The concepts of absolute space and ether have a long and intricate history. Although their definitions were, and still are, not clear enough to distinguish unambiguously between the two concepts, it can be said that Newton's absolute space merged into our modern concept of spacetime, and the old ether into the "vacuum" of quantum field theory.

We want to address some aspects of the historical development to show how the above-mentioned ideas could get into the classroom.

### *Newton's absolute space*

Newton had the problem that when speaking of a change in motion, he had to be able to say that a body is moving relative to something. There must be such a reference, since changes of momentum and velocity are measurable. The entity to which the momentum changes referred, however, was invisible and no property of it was known other than that it became effective as a reference for acceleration processes. Newton called this entity "absolute space". For a long time, absolute space was an undisputed concept of mechanics.

Criticism of this idea came up much later. Ernst Mach accused Newton with sharp words insisting that absolute time and space were only "things of thought". The only true movement was a movement relative to something else, and by that he meant other bodies. As a reference, he proposed the totality of the fixed stars, an idea that George Berkeley had already put forward in 1721, during Newton's lifetime.

For Newton and others, however, such an idea must have been quite intolerable: it represented an action-at-a-distance effect over gigantic distances. After all, Newton (1692) had already clearly classified such effects within the framework of the normal law of gravity as unacceptable.

However, Mach's criticism seemed plausible to many people, and Newton's absolute space was discredited and has remained so to a large extent to this day.

From the perspective of Einstein's general theory of relativity, however, it must be said that Newton's idea of an absolute space was a healthy and progressive idea; an idea that is quite close to our modern idea of spacetime (Giulini, 2002).

As early as 1909, Minkowski referred to an absolute spacetime. And this is how Einstein said it in 1922, in his book *Foundations of the Theory of Relativity*. In particular, the law of inertia seems to force the attribution of physical-objective properties to the spacetime continuum. While from Newton's point of view it was consistent with pronouncing the two terms *tempus absolutum* and *spatium absolutum*, from the point of view of the special theory of relativity one must speak of *continuum absolutum spatii et temporis est* (Einstein, 1970).

Of course, Newton had no idea yet of the intertwining of space and time, and the curvature of space by energy and momentum as well as by their flows or currents. But the basic idea was there.

When spacetime is introduced in this way, the idea that an absolute space does not exist does not even arise.

### *The Old Ether*

The ether has a long history. The discussion goes back to antiquity, and in modern times almost all important physicists commented on it and tried to explore its properties. But this discussion seemed to come to an end by one single almost inconspicuous sentence. In his famous publication, *On the Electrodynamics of Moving Bodies* Einstein formulated:

> The introduction of a 'luminiferous ether' will prove to be superfluous inasmuch as the view here to be developed will not require an 'absolutely stationary space' provided with special properties, nor assign a velocity-vector to a point of the empty space in which electromagnetic processes take place (Einstein, 1905).

After the publication of his great theories, Einstein was almost venerated as a saint. So, when he formulated that ether was superfluous, it became a creed: "An ether does not exist".

In addition, the Michelson–Morley experiment seemed to confirm this statement. This argument, however, is based on a misunderstanding. We want to show this with a thought experiment. We imagine that Michelson and Morley did not do their experiment with light, but with material objects. Here is a not quite realistic story:

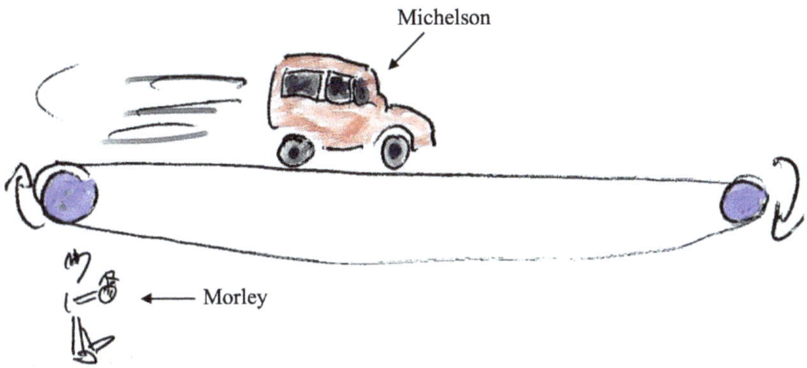


5.4
Would Michelson and Morley conclude that the conveyor belt does not exist? (*Source: Copyright by Friedrich Herrmann, Michael Pohlig*)

Michelson drives a car at high speed on a conveyor belt that is at first at rest (Figure 5.4). The velocity of the car relative to the belt is almost equal to the terminal speed *c*. Now, Morley, standing next to the belt, switches the conveyor belt on, in such a way that it moves in the same direction as the car. He notices that the car is still moving nearly with the speed *c*. If he lets the belt move in the opposite direction, the velocity of the car then also remains almost equal to *c*.

What would be the conclusion of Michelson and Morley? Would they conclude that the conveyor belt does not exist? Surely not. They would conclude that there is a terminal value for the velocity and that velocities cannot simply be added when the reference frame is changed. This statement would have led to special theory of relativity. Remember that in the case of the real Michelson–Morley experiment, the conclusion was: a carrier of the light wave does not exist. Thus, we emphasise: the Michelson–Morley experiment does not prove the non-existence of an ether.

Soon Einstein himself withdrew his original statement. He declared:

> According to the general theory of relativity, a space without ether is unthinkable (Einstein, 1920).

But the easy to repeat sentence "ether does not exist" was still there in the world. Neither the appearance of quantum electrodynamics nor that of dark energy or the Higgs field could change anything.

## FLAT SPACETIME

### *The twin paradox*

The twin paradox plays a major role in the teaching of special relativity and is part of every textbook or course on special relativity. In its common form, it goes somewhat like this: two twins match their clocks. Then one of them moves away at a constant speed, eventually the person reverses and returns at a constant speed. When comparing the clocks again, the twins find that more time has passed for the one who did not move than for the other.

The advantage of this version of the story in which one of the twins does nothing (or in other words: is at rest in a free-floating reference frame) and the other moves at a constant speed is that it is easy to calculate the time difference.

Here, in our opinion, something is proved by a complicated calculation but results in little understanding. For this reason, we prefer a different setting. The story is about Willy and Lilly, the protagonists of the *Karlsruhe Physics Course*. Willy and Lilly have met. They separate and drive their cars somewhere around in the area and then meet again. They now compare which distance they have covered (by looking at their mileage counters) and what time the journey has taken. Of course, in general, the distances are different. But they also notice that their clocks show different times. The time discrepancy is usually not noticed, because: (a) our clocks are not accurate enough, and (b) the journey was not long and fast enough.

We tell this simple story and we describe the observations as a fact. One has to live with them.

When two persons with their clocks separate (event A) and later meet again (event B), their clocks display different times.

For the person (clock) that does not move at all (in a free-floating reference frame), the most time passes. For a person (clock) that moves at (almost) the terminal speed, (almost) no time passes.

***Clocks in the gravitational field***

There is another well-known twin story. Usually, it is not called a paradox. Willy and Lilly are in a skyscraper, both on the middle floor. They synchronise their clocks and then separate: Lilly goes up, Willy goes down. After a while, they return to their starting point. They notice that Lilly's watch is showing more time than Willy's. More time has passed for Lilly since her first rendezvous than for Willy. It is now often stressed (especially in connection with the discussion of the GPS navigation system) that this is an effect of general relativity, unlike the normal twin paradox.

We think such a classification is inappropriate. Of course, one can argue about which effects have to be attributed to general relativity and which not. Is the mere fact that we take the equivalence of inertial and gravitational mass seriously a general relativity statement? Does the fact that one passes into an accelerated reference frame mean that one is dealing with general relativity? Probably not. It is more appropriate to define the distinction between general relativity and non-general relativity effects in this way: everything that can be described in the flat Minkowski space does not belong to the category of general relativity effects.

If one sticks to this criterion, the time difference between clocks at different altitudes is not an effect of general relativity. It also exists in a homogeneous gravitational field and for such a field, the Riemann tensor is zero everywhere. In other words: one can transform the gravitational field away by going into a freely falling reference frame. If this is done, it turns out that the skyscraper paradox becomes identical to the classical twin paradox.

In class, this fact is explained by means of a short story: apart from Willy and Lilly, Milly is also present. Lilly goes up in the skyscraper; Willy goes down – as before. Milly, on the other hand, jumps up in such a way that she just lands again for the second rendezvous. So she can describe the whole situation in a free-floating reference frame, that is, the simplest reference frame there is. It turns out that for her the different aging of Willy and Lilly is exactly the well-known twin paradox situation. By the way, Milly ages most of the three.

Gravitational time dilation is not a general relativity effect.

## CURVED SPACE

### *The curvature of space and the curvature of spacetime*

In our opinion, treating general relativity comprehensively is not an option for the school. For students, a large part of what they learned in the previous mechanics lessons and what they have understood, to some extent (if we are lucky), would be questioned again or simply would become invalid. There would no longer be gravitational forces. Newton's law of gravitation would no longer be usable. Even the concept of movement disappears. In spacetime, there are no more trajectories, only worldlines. A body moves on a trajectory, but not on a worldline. (Actually, it is often said or written that a body moves on a worldline, but this certainly will cause more confusion than understanding.)

We have, therefore, decided not to operate with curved spacetime. We choose the world time of cosmology as the time coordinate and introduce as general relativity effect only the curvature of the ordinary three-dimensional space. We have to pay for this with the restriction that we are no longer allowed to make certain reference frame changes because this would disturb our separation between space and time.

In our teaching, geodesics play an important role. Note that these are geodesics in the three-dimensional position space – not to be confused with the worldlines of free-floating bodies, that is, the geodesics in spacetime.

To explain and visualise the concept of a geodesic, we use the software *Geodesics* (Herrmann and Halupczok, 2017), Figure 5.5. It produces geodesics in the more perspicuous situation of a two-dimensional curved space. A car whose steering is set to "straight-ahead" travels over a hilly landscape.

We limit ourselves to the treatment of spatial geodesics.

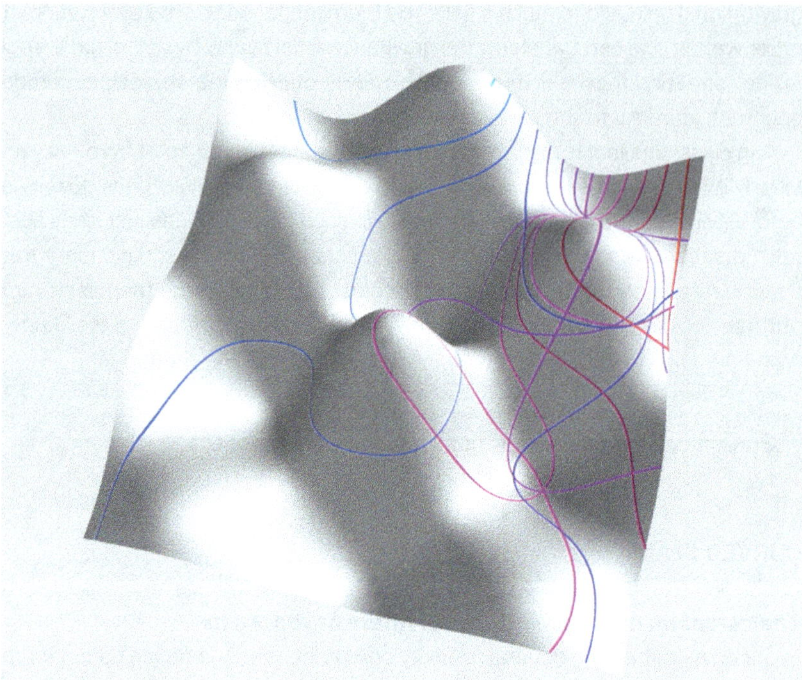

5.5
With the software Geodesics, a hilly "landscape" is defined. Starting from an edge, the programme then draws geodesics that start orthogonally to the edge. (*Source: Copyright by Friedrich Herrmann, Michael Pohlig*)

***The Schwarzschild metric***

An important teaching subject is the spacetime in the vicinity of spherically symmetric objects, that is, objects to which the Schwarzschild solution applies: the sun, a neutron star, a black hole.

The effect of the star on the behaviour of clocks at different distances can be easily explained and also quantitatively described. More difficult is the treatment of the curvature of space.

After some unsatisfactory attempts to quantitatively describe the curvature of space in the classroom using the Schwarzschild solution, we decided to treat the subject only qualitatively, but in our opinion more understandably.

We explain what happens to space by means of an analogous situation in a two-dimensional world (Figure 5.6). A farmer bought 4 acres of land to create a meadow for his sheep. He would like to have more land, but he can't find anyone to sell him. But he has an idea: he increases the area of the meadow by raising a hill on the 4 acres of land he bought.

It is obvious that by the action of the farmer:

- the pasture area increases, and
- the pasture becomes curved.

5.6
The meadow is not big enough (a). The farmer piles up a hill (b). This increases the pasture area and bends the pasture. (*Source: Copyright by Friedrich Herrmann, Michael Pohlig*)

Understanding this does not require coordinates, reference frames, or rulers that shorten or lengthen, or appear to be shortened or lengthened. Correspondingly we have, Figure 5.7:

The volume of a region of space is increased by mass/energy.

## CONCLUSION

The development of a teaching concept for the school, in our case for the secondary school, is a balancing act.

On the one hand, teaching at school differs fundamentally from popular science presentations. The latter can limit themselves to showcasing the spectacular, the impressive, and the surprises of the scientific results – one can almost say: to exhibit them like objects in a museum.

School teaching has to meet other requirements. The statements must be logically coherent. They have to fit into the previous teaching and form a foundation for future teaching, for example, at the university.

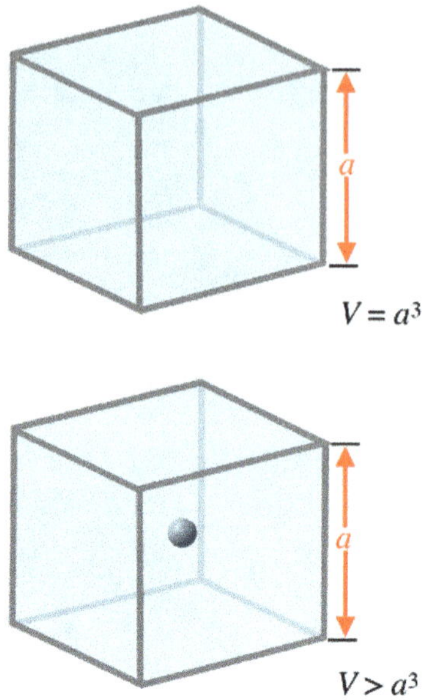


5.7
The volume of the cube, in which a heavy celestial body is located, is larger than $a^3$. (*Source: Copyright by Friedrich Herrmann, Michael Pohlig*)

On the other hand, we must make sure that we do not treat high-school students like university students and we do not overburden them. Let us not forget: one can calculate and prove without generating understanding.

It should also be borne in mind that most high-school students a priori have no particular interest in physics.

We have tried to develop a course under these constraints. We would like to emphasise once again that the above remarks do not represent the content of our course. We have only presented what we believe is different in our approach from that of other textbook approaches.

Here is a brief summary:

We avoid, as far as possible, changes of the reference frame.

We choose as an additional postulate of special relativity that energy and mass are the same physical quantity.

We consider the effects of general relativity only with regard to the three-dimensional position space, not to spacetime.

## FURTHER READING

Chapter 2: Time for changing paradigms in science and in education
Chapter 6: Event diagrams – supporting student reasoning in special relativity through thought experiments
Chapter 25: Toward Understanding Einsteinian Physics Education – Relativity Education as an Exemplar

## REFERENCES

Einstein, A. (1905). "Zur Elektrodynamik bewegter Körper." *Annalen der Physik*, 322(10), 891–921.

Einstein, A. (1920). *Äther und Relativitätstheorie.* Speech given on 5 May 1920 at the Reichs-Universität Leiden. Berlin: Springer.

Einstein, A. (1970). *Grundzüge der Relativitätstheorie.* 5th ed. Berlin: Akademie-Verlag.

Feynman, R. P., Leighton, R. B. & Sands, M. (2006). *The Feynman Lectures on Physics, Volume I.* (Definitive edition) Boston: Addison-Wesley.

Giulini, D. (2002). "Das Problem der Trägheit." *Philosophia Naturalis*, 39, 343–374.

Herrmann, F. & Pohlig, M. (2016). *The Karlsruhe Physics Course*. [pdf] Karlsruhe: KIT. Available at: http://www.physikdidaktik.uni-karlsruhe.de/ [Accessed May 1, 2019]

Herrmann, F. & Pohlig, M. (2018). *The Karlsruhe Physics Course.* 卡尔斯鲁厄物理课程. Guangzhou: 广东教育出版社微店.

Herrmann, F. & Halupczok, I. (2017). *Geodesics*. Karlsruhe: KIT.

Herrmann, F. & Job, G. (2019). *Historical Burdens on Physics*. [pdf] Karlsruhe: KIT. Available at: http://www.physikdidaktik.uni-karlsruhe.de/ [Accessed May 1, 2019].

Newton, I. (1692). *Original Letter from Isaac Newton to Richard Bentley.* Cambridge, UK: Trinity College Library.

# 6 Event diagrams – Supporting student reasoning in special relativity through thought experiments

*Floor Kamphorst, Elwin Savelsbergh, Marjolein Vollebregt, and Wouter van Joolingen*

**WHAT?**

This chapter presents event diagrams as a representational tool that allows students to visualise relativistic phenomena. The chapter puts particular emphasis on thought experiments that can help students obtain a deeper understanding of physical phenomena that are hard to imagine.

**FOR WHOM?**

This chapter is intended for readers who look for instructional models to teach concepts of special relativity at the secondary school level, and also, for those who wish to learn more about thought experiments as instructional tools.

## INTRODUCTION

"If I pursue a beam of light at the velocity *c* (velocity of light in a vacuum), what would I observe"? wondered Einstein at the age of sixteen. "I should observe such a beam of light as a spatially oscillatory electromagnetic field at rest. However, there seems to be no such thing, whether on the basis of experience or according to Maxwell's equations" (Einstein, 1949/1979, p. 53). This simple question and its paradoxical answer revolutionised how physicists look at the world. The mental exploration of this thought experiment eventually led Einstein to propose the light postulate, one of the basic principles of the special theory of relativity. Special relativity describes how space and time can no longer be understood as two separate concepts, but are one unified spacetime, even more, how this spacetime is observer-dependent.

Nowadays, special relativity is finding its way into secondary education, and it would be desirable that part of its revolutionary appeal would be retained in the secondary classroom. To this end, students need to gain an understanding of the relativity of simultaneity, time dilation, and length contraction. In addition, students need to understand how these phenomena relate to the basic principles of the theory: light propagation and observer dependence.

It is difficult to gain an understanding of these relativistic phenomena for three reasons. First, the phenomena only emerge if the same observation is described relative to different inertial frames[1]. Even then, the relativistic effects only become apparent if the relative velocities between the frames are high – much higher than velocities we encounter in daily life. Finally, special relativity is limited to idealised situations where acceleration and gravity are absent. Consequently, relativistic phenomena remain abstract and are hard to imagine.

Thought experiments are a powerful instrument for physicists to explore new phenomena and expand their understanding of the world. In a thought experiment, physicists can explore the consequences of relevant concepts in an idealised world (Gendler, 2004; Mach, 1976). For example, the consequences of an absolute speed of light and high relative velocities in a world without gravity. A thought experiment happens in three stages. In the first stage, the central question, theoretical basis, and what the idealised world looks like are identified. In the second stage, the thought experiment is performed by making deductive reasoning steps from the initial scenario to obtain an outcome. Finally, reflection on the entire process leads to the generalised implications of the thought experiment (Reiner & Burko, 2003).

Thought experiments can be a powerful instrument in education as well. When students perform a thought experiment, they reason with a "what if" scenario. This means that students must activate and communicate their ideas (Matthews, 1994). In this way, the thought experiment helps students overcome conceptual barriers and learn new, abstract concepts (Helm et al., 1985; Velentzas & Halkia, 2013). However, the thought experiment's educational power is often undermined by its story-like presentation. Let us take a closer look at two relativistic thought experiments. This will give us an idea of how they are typically presented in textbooks, and what reasoning is required to perform them.

**Einstein's train – The relativity of simultaneity**

Suppose a very long train travels along the rails at a constant velocity v. A lamp in the middle of the train is switched on. As soon as the light hits two automatic doors at the front and back of the train, these doors open. The question is whether the events of the opening doors are simultaneous in the reference frame of the train and of the embankment the train is passing by. Suppose a passenger finds themselves under the lamp and passes a traveller on the embankment at the instant the lamp is switched on. The passenger on the train will say that light covers equal distances to both doors and that they will open simultaneously. For the traveller on the embankment, light will have to travel a longer distance to the front than to the back door. Therefore, this observer will conclude that the doors do not open simultaneously. The overall conclusion of this thought experiment is that events which are simultaneous with reference to the train are not simultaneous with respect to the embankment, and vice versa.

(Landau & Rumer, 1960)

### Light clock on a train – Time dilation

Suppose the same train again, only this time the train is equipped with a light clock. This device measures time by counting the instants a light flash hits the bottom of the train. The light flash is emitted vertically, reflected by a mirror at the ceiling of the train and returning to the floor of the train again. The question is whether the time interval for one cycle of this process is the same for a passenger on the train and a traveller on the embankment. In the train frame, the passenger will observe the light flash travelling vertically up and down. Relative to the traveller on the embankment light will not only travel up and down but also in the direction of the moving train. The traveller will, therefore, conclude that the light flash covers a longer distance. The overall conclusion of this thought experiment is that the light clock in the train will run slower for a traveller on the embankment than for a passenger on the train, and vice versa.

(Landau & Rumer, 1960)

Both stories first describe an idealised world and pose a central question. After that, the outcome is immediately given. However, the reasoning that leads to the thought experiment's outcome remains implicit. Because of that, students are not stimulated to perform that reasoning themselves.

It can be difficult for students to perform all the required reasoning steps of a thought experiment. They need to reason consistently with an absolute speed of light from each of the two inertial frames. While they are doing that, they also need to assign time and place to events. Many students struggle with the concept of absolute light speed. Even after instruction, undergraduate students can only recite the light postulate, but they do not apply it correctly (Gousopoulos et al., 2016). Instead, students tend to reason with a spontaneous model of light propagation (Villani & Pacca, 1987). The pre-instructional models of secondary students describe light either with constant speed relative to the light source or relative to an absolute space (Kamphorst et al., 2019). Furthermore, performing a relativistic thought experiment is a difficult task in itself: students need to administrate all relative movement of the thought experiment scenario in their minds. As a result, they might not "see the forest for the trees" and fail to grasp the thought experiment's overall conclusion.

A representational tool may support students so they can perform thought experiments themselves. This tool should support their explicit reasoning with light propagation. The well-known Minkowski diagram seems a likely candidate. However, the abstract nature of this diagram makes it problematic for novice learners to perform thought experiments themselves. We propose that the event diagram is a more suitable tool to elicit student ideas, support consistent reasoning with light propagation, and derive relativistic concepts while performing thought experiments. Next, we take a closer look at event diagrams and present some tasks in which event diagrams support thought experiments.

## THE EVENT DIAGRAM

Event diagrams graphically represent spacetime and were first proposed by Scherr (Scherr, 2001). An event diagram shows the position of objects and events at several moments in time, from a specific reference frame. These event diagrams typically consist of one or two pictures. Scherr employed event diagrams in tutorials to familiarise undergraduate students with reference frames, spatial measurement, and the relativity of simultaneity. The diagram visually organises the information presented in the assignment, and is occasionally used as an answering format for students to draw the task outcome (Figure 6.1). However, Scherr's version of the event diagram does not support representing any intermediate reasoning steps.

We have added two features to the event diagram to make it a tool for student reasoning. First, our event diagrams consist of a series of snapshots taken at regular time intervals, much like a time-lapse movie. Second, the pictures are drawn on a grid. This makes it easy to measure position and distance in the ED (Figure 6.2). Together, these features make it possible to show the speed of moving objects and draw light propagation in the event diagram. The outcome of the thought experiment is intentionally not shown in the event diagram, so students are required to perform the thought experiment themselves. Students perform the thought experiment by drawing light propagation in the event diagram. For this, the speed of light needs to be given a value that makes sense in the context of the event diagram. In the tasks of this chapter, we set the speed of light to three squares per time unit. Through reasoning by drawing, students find out when a specific event occurs (that is, the doors opening, the light flash hitting the floor of the train) (Ainsworth et al., 2011). With teacher-supported reflection on the second stage of the thought experiment, students can draw the overall conclusion.

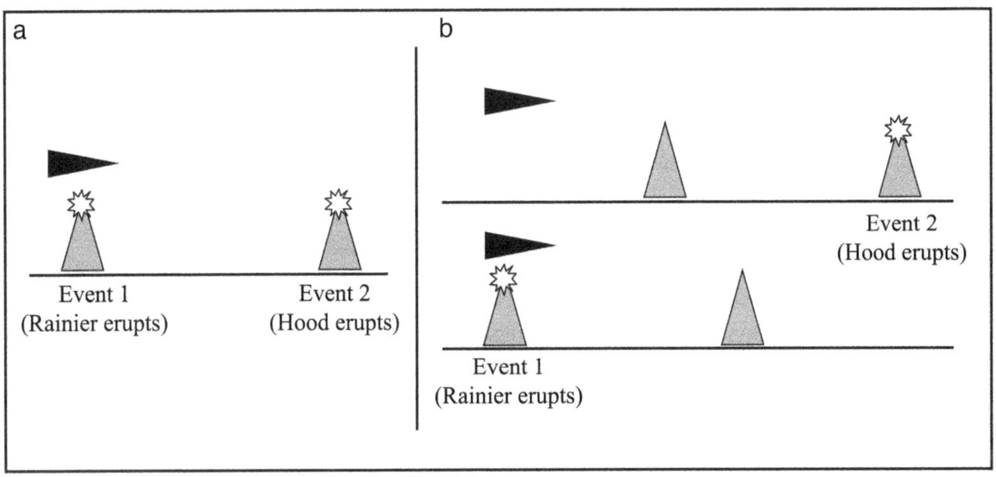


6.1
Drawn after the event diagram by Scherr (2001, p. 190). This diagram shows the position and time of objects (volcanoes and spacecraft) and events (erupting volcanoes). In the ground frame (a) these events are simultaneous, whereas in the spacecraft frame (b) event 2 happens before event 1. (*Source: Copyright by Floor Kamphorst*)

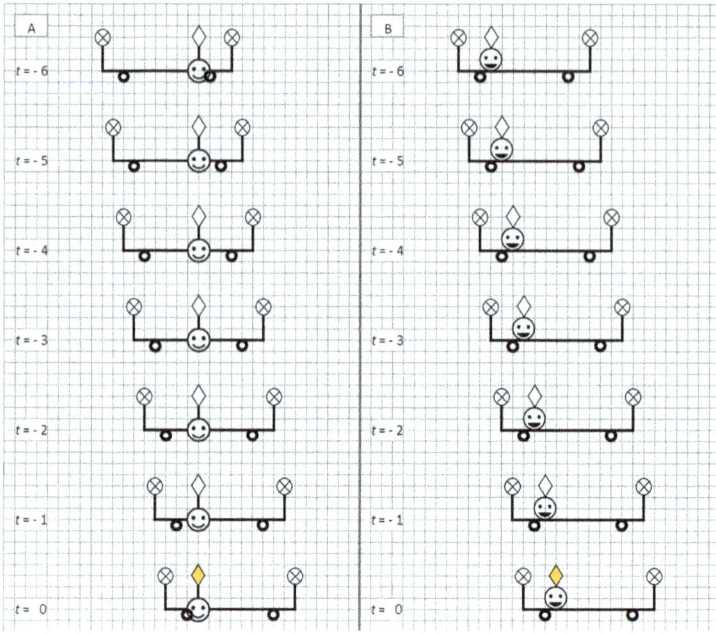


6.2
Event diagrams supporting a version of Einstein's train (described in task 1). Each snapshot shows the position of objects for subsequent moments. Each picture is given a unique time stamp. The observer is represented by the smiley face, the measuring device by the diamond. To show the key event, two light flashes arriving simultaneously at the measuring device, the diamond is painted yellow. The cart with two lamps (shown by the lamp symbols used in electric circuits) has a speed of one square per time unit, moving to the right. (*Source: Copyright by Floor Kamphorst*)

Compared to the traditional presentation of thought experiments, the event diagram stimulates students to reason with light propagation more explicitly. Time is represented stepwise in the event diagram. Therefore, students construct light propagation in a stepwise fashion as well (Kamphorst et al., 2019). To be more specific, students indicate the position of the light flashes in each snapshot. The snapshot in which the light flash and lamp share the same position indicates the time when the light was emitted, and students assign time and place to the event (see also the description of Task 1 and Figure 6.3). Students do not immediately reason about the time between two events; they find this as an outcome of the construction process instead.

Like all external representations, event diagrams are a simplified and idealised display of reality and are inherently limited. However, our tasks allow students to reflect on these limitations. This reflection enables them to see beyond the representation of the event diagram, and fathom the relativistic concepts separated from the representation.

## THOUGHT EXPERIMENTS SUPPORTED BY EVENT DIAGRAMS

Each of the following three tasks demonstrates a different feature of the educational power of the thought experiment. The task description follows the three phases of performing a thought experiment:

1. The situation of the thought experiment, central question and supporting event diagram.

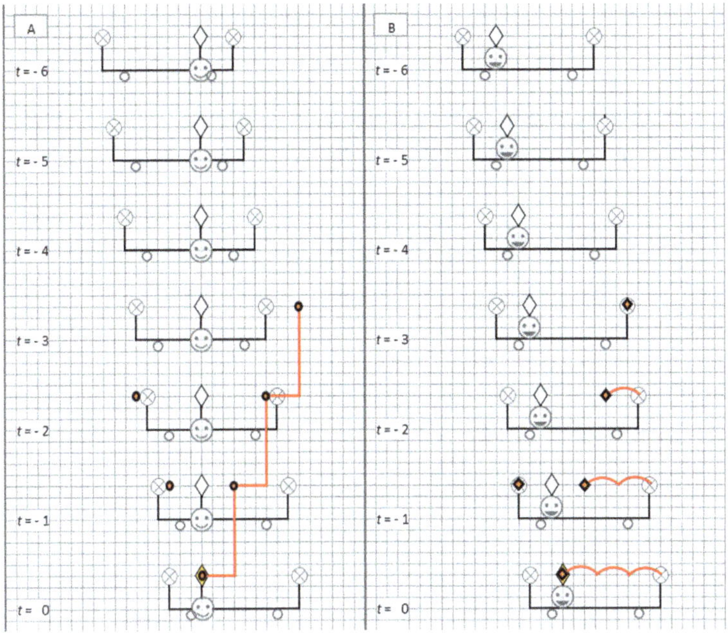


6.3
Event diagrams supporting task 1A (left) and task 1B (right). The two ways of drawing light are shown with dots and red lines in the event diagram, each leading to a different task outcome. The left event diagram shows light propagation drawn relative to the grid. Students construct their drawing by transporting the position of the light flash to the previous picture and count the displacement from that position. The event diagram on the right shows light propagation drawn relative to the lamp. These students count the distance the light flash has covered relative to the lamp. (*Source: Copyright by Floor Kamphorst*)

2. The expected student reasoning and task outcome.
3. How students can reach an overall conclusion and reflect on the limitations of the event diagram.

Each task is illustrated with a short excerpt of student discourse that shows what kind of student reasoning these tasks can achieve. These data were collected from a study in which the tasks were presented to groups of 2–4 students of 11th grade pre-university secondary education (16–18 years old).[2] The names of the students are fictitious.

### *Task 1. Thought experiment predicting the past: Eliciting reasoning with light propagation*

The aim of this task is to make students aware that light propagation is always measured relative to something. There are different possibilities for this reference frame. The task outcome is influenced by the choice of the reference frame in relativistic situations (that is, relative movement between light source and observer). This thought experiment is a variation of *Einstein's train*. An observer measures the simultaneous arrival of two light flashes at a measuring instrument. These light flashes are emitted by two distant lamps on a cart. The central question is: how long ago were the lamps switched on? To elicit the role of the reference frame for light propagation, we have developed two versions of this task. Students perform the thought experiment for a situation in which the lamps are moving relative to a stationary observer (Task 1A), and a situation in which the observer is positioned on the cart. In the latter situation, both lamps and observer move relative to the grid (Task 1B). These situations are represented in an event diagram (Figure 6.2). The final snapshot $t = 0$ at the bottom of the figure shows the instant the light flashes arrive simultaneously at the observer's measuring instrument. Snapshots above this picture show previous instants, where the light flashes were still travelling towards the researcher. It is up to students to draw the propagation of the light flashes, by indicating their position at subsequent moments in the event diagram. It is not specified in the task relative to *what* the speed of light should be applied. This allows students to explore their own ideas.

Students perform the task by drawing light propagation in the event diagram. Depending on their pre-instructional model of reasoning, students either draw light propagation with a constant speed relative to the grid or to the lamps (Kamphorst et al., 2019). These two ways of drawing light in the event diagram are shown in Figure 6.3. Each choice leads to a different task outcome (Table 6.1).

Comparing the outcomes of different students can be a starting point to address the choice of reference frame as a matter of concern. To this end, the teacher can ask questions that stimulate students to explicate *relative to what* light propagation is constant for *them*, for instance, by asking to explain how they drew light propagation in the event diagram, or how their drawing method differed from the method of a fellow student. Once students are aware of this role of reference frame, they can also become aware that an absolute speed

***Table 6.1*** **Outcome of thought experiment 1 for three light propagation models. The table shows the time step in which the lamps on the cart are switched on. Note that the pre-instructional models lead to the same outcome in both tasks, whereas the light postulate yields different outcomes in task 1A and 1B**

| *Light propagation model* | *Task 1A* | *Task 1B* |
|---|---|---|
| Relative to the grid | Left: between $t = -1$ and $t = -2$<br>Right: between $t = -2$ and $t = -3$ | Left: between $t = -1$ and $t = -2$<br>Right: between $t = -2$ and $t = -3$ |
| Relative to the lamp | Left: $t = -1$<br>Right: $t = -3$ | Left: $t = -1$<br>Right: $t = -3$ |
| Absolute speed of light (relative to the observer) | Left: between $t = -1$ and $t = -2$<br>Right: between $t = -2$ and $t = -3$ | Left: $t = -1$<br>Right: $t = -3$ |

of light is constant relative to *something else*: the observer. Also, students are introduced to an essential aspect of working with event diagrams: they should always specify relative to what reference frame they are reasoning with (light) propagation.

### *Students' reasoning with different pre-instructional models for light propagation*

Daniel and Lisa performed Task 1A, each with a different pre-instructional model. Lisa reasoned with a constant speed of light relative to the lamp. Students who reason with this model measure the distance between the lamp and the new position of the light flash.

*Lisa:* I thought that he [the light flash] covered nine squares here [$t = 0$, indicating the distance between the measuring device and the right lamp]. So, in the previous picture [$t = -1$], he covered three [squares] less, light propagates with three squares per time unit. So, I figured out that he would end up at six [squares, from the lamp].

*Teacher:* and before that?

*Lisa:* at the third [square from the lamp].

Daniel performed the task with a constant speed of light relative to the grid. Students with this way of reasoning focus on the position of light on the grid.

*Daniel:* The light is emitted from the light point [indicates the position on the grid where the lamps were when the light was emitted]. It does not matter whether the cart is moving. Light is already gone from that place, where it was emitted.

[...]

*Teacher:* Can you explain the difference between what you did and how Lisa performed the thought experiment?

*Daniel:* Yes, she has taken the movement of the cart into account, and I have not.

Daniel correctly identified that the moving cart caused the difference between the task outcomes he and Lisa obtained. This notion can form a starting point for the teacher to further explore the role of the reference frame for light propagation and introduce an absolute speed of light.

### ***Task 2. Thought experiment predicting the past: Deriving the relativity of simultaneity***

The aim of this task is to derive the relativity of simultaneity as a consequence of the light postulate. This task is based on the same version of Einstein's train as Task 1. Students perform the thought experiment in two different situations. The first situation describes stationary lamps with a moving observer (Task 2C). The

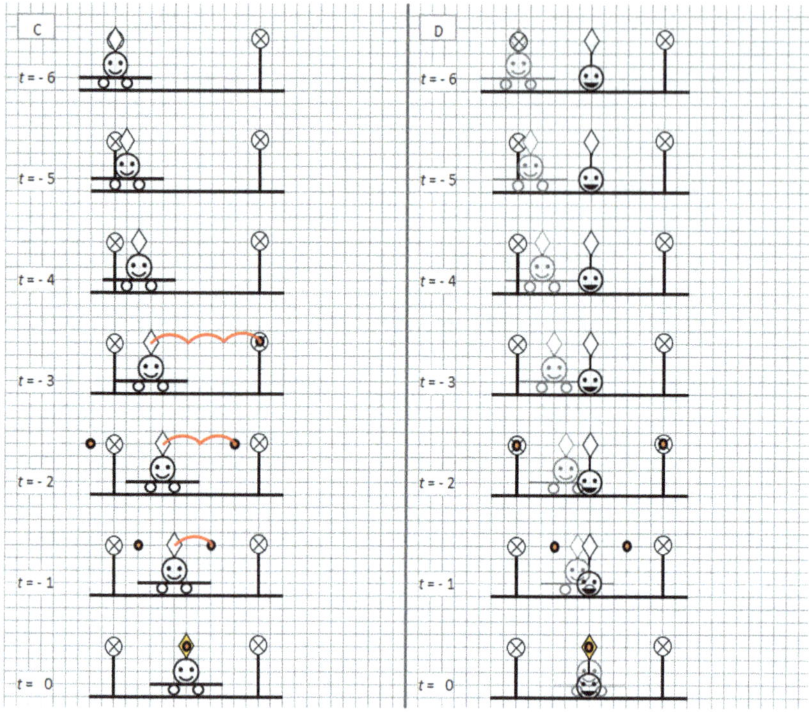


6.4
The two event diagrams supporting thought experiment 2. The light flashes drawn in the event diagram propagate with an absolute speed. Task 2C (left) is performed for the observer on the skateboard, and task 2D (right) for the stationary observer. In this figure, the observer on the skateboard is given a lighter colour. The red lines show how students construct their drawing, counting the distance the light flash has covered relative to the observer. The students who fall back to their pre-instructional reasoning will obtain the results of task 2D in both versions of the thought experiment. These students may explain their reasoning in the following way: "They [the light flashes] cover the same distance to arrive in the middle, so they [the lamps] should also have turned on at the same time" (Daniel). (*Source: Copyright by Floor Kamphorst*)

light flashes simultaneously arrive at the measuring device when the observer is midway between the two lamps. The second situation (Task 2D) adds a second observer to this scenario. The instant the light flashes strike the measuring device, both observers find themselves in the same position. However, the second observer is not moving relative to the lamps. These tasks are supported by EDs as well (Figure 6.4). The central question is: When were the lights switched on? A new element in this task is that students perform the thought experiment in two reference frames.

Students are invited to perform the thought experiment by drawing an absolute speed of light in the event diagram (Figure 6.4). Nevertheless, students might fall back to their pre-instructional model. The task outcomes for both ways of reasoning are given in Table 6.2. Students who correctly apply the absolute speed of light, find that for the moving observer (Task 2C), the lamps were

**_Table 6.2_ Outcome of thought experiment 2 for three light propagation models. The two pre-instructional models will give the same results.**

| *Light propagation model* | *Task 2C* | *Task 2D* |
|---|---|---|
| Absolute speed of light (relative to the observer) | Left: between $t = -1$ and $t = -2$<br>Right: $t = -3$ | Left: $t = -2$<br>Right: $t = -2$ |
| Pre-instructional model (relative to the lamp or to the grid) | Left: $t = -2$<br>Right: $t = -2$ | Left: $t = -2$<br>Right: $t = -2$ |

switched on at two different moments in time, while these events were simultaneous for the stationary observer.

The relativity of simultaneity is the generalised implication of observers assigning different times (and places) to the same event, which in its turn is a consequence of the light postulate. For students to grasp this far-reaching consequence, the teacher could ask how the two observers obtained different outcomes of the thought experiment, which outcomes would be obtained when the roles of the observers were reversed, or what these different outcomes would mean for two events in general. This task confronts students with a limitation of the event diagram: drawing light propagation in the event diagram will represent an absolute speed of light for only one of the two observers. This underlines the importance to specify for which observer they perform the thought experiment. Performing this task makes it clear why it is desirable to let the reference frame of the event diagram and the observer coincide.

### *Students reflecting on the outcome of the thought experiment*

Three students, Jeroen, Kelly, and Bart, talked about the task outcome of thought experiment 2. They each had a different interpretation. Jeroen explicated that the outcomes differ from his expectations. This difference made Kelly doubt the light postulate. In contrast, for Bart, the light postulate was the reason to accept the counterintuitive outcome of the thought experiment.

*Jeroen:* Strange that they [the lamps, seen from the two observers] turn on at different moments, while light [propagation] actually remains the same.

*Kelly:* I would wonder if it [light propagation] is really relative to the observer because here you have two observers and then you get a different outcome [of the TE]. [Points at the light flashes in the ED] ...that that [light propagation] is relative to the observer and not relative to the paper or the moment of turning on [of the lamps].

[...]

*Jeroen:* Because they [the observers] move with different speeds. [...] Light moves relative to the observer. So, if the one observer has a speed relative to the other, it [light propagation] will go in a different way. [...] It cannot be [a] different [outcome of the task]. Light moves relative to the observer, and that is what we did.

The following fragment illustrates how students reflect on multiple observers in the event diagram. Iris struggled with the two observers in one diagram. In this, she recognised that all observations depend on the reference frame. Only, this notion is still counterintuitive to her. Laura explained how this is inherent to the light postulate.

*Iris:* But see, they would not understand it [light propagation in the event diagram], because they both see something different.

[...]

*Laura:* No, because they both measure those three squares, if they would measure light propagation for themselves.

Both fragments show that students explore the consequences of an absolute speed of light and interpret their meaning.

***Task 3. Thought experiment light clock: Coming to a conceptual understanding of time dilation***

This task aims for students to gain a conceptual understanding of time dilation. Through performing this task, students come to understand how time dilation is a consequence of the light postulate. In addition, students understand how the extent of this phenomenon is influenced by the relative speed between the observers. The task follows the scenario of the light clock thought experiment. The central question is: How many time steps does light need to travel up and down between the mirrors of the light clock? First, students perform the reasoning steps in the train frame (Task 3E). Next, they describe the situation from the frame of the stationary observer on the embankment (Task 3F). Each event diagram shows the situation of the task in the reference frame of the observer that makes an appearance in the thought experiment (Figure 6.5).

Students draw light travelling up and down between the mirrors (Figure 6.5). The duration of the time interval can be determined in two ways. Either students count time in the event diagram, or they divide the travel distance of the light flash by its speed. Students collect the position of the light flash for all time steps and construct the light path relative to the observer in an ordinary *x*, *y*-graph (Figure 6.6). For the observer in the train, light travels up and down vertically. To the observer on the platform, light deflects in the direction of movement of the train. This is not clear to all students. These students draw light travelling vertically upward relative to the second observer as well. The event diagram is designed in such a way that light misses the mirror in that case. It might be helpful to remind these students that all events happen for all observers, although observers may assign different places and moments to them.

One of the difficulties in learning relativity is that it has no daily life reference. To understand why that is the case, students need to reflect on the role of relative velocity for the extent to which time dilation occurs. To this end, the teacher can ask what would happen to the time interval both observers measure

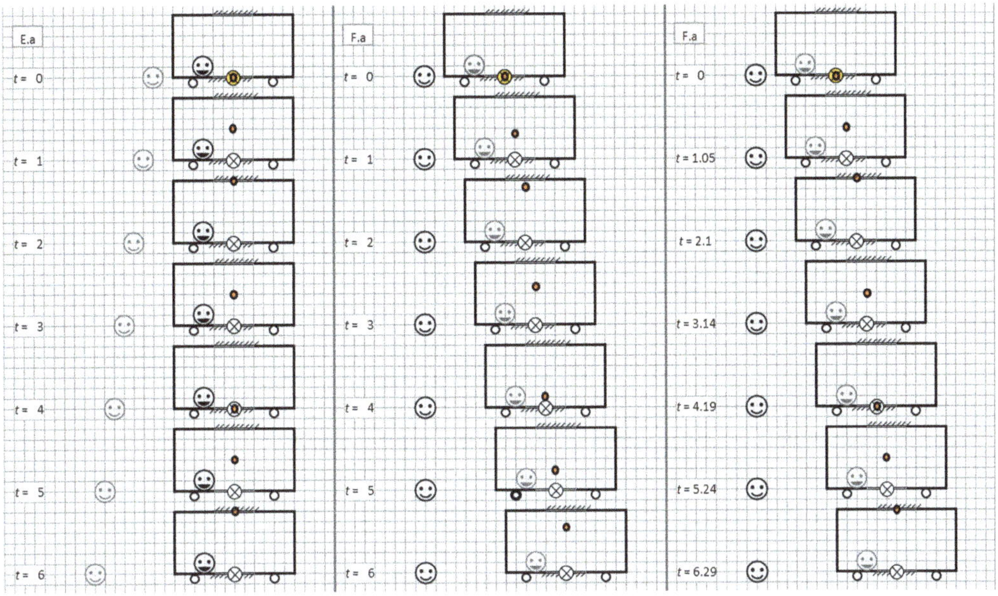


6.5
Three event diagrams to support the light clock thought experiment. Both observers are shown in the event diagram, the stationary observer is drawn in black. Diagram Task 3E (left) shows the situation for the observer with a stationary light clock; diagram task 3F (middle and right) shows the same situation from the reference frame of an observer studying a moving system. Light propagation is drawn relative to the stationary observer in the event diagram. In this event diagram, the mirrors are shown by the shaded lines, the instant the light flash is emitted by the yellow lamp at $t = 0$, at the top of the figure. The time indications in the student version of diagram task 3F are left blank; this allows students to deal with drawing light propagation in the thought experiment in two different ways: they either draw light that does not go up exactly three squares per time step (middle figure), or light that goes up three squares per snapshot and increases the time between snapshots (right figure). (*Source: Copyright by Floor Kamphorst*)

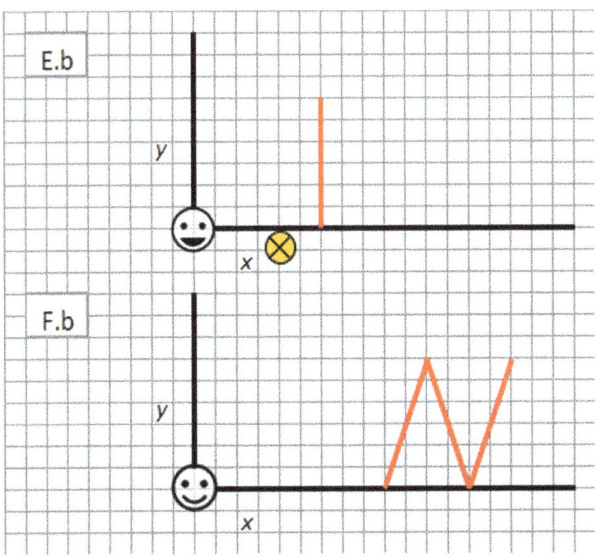


6.6
The *x*, *y*-graphs allow students to collect the data of the event diagram and find the path light travelled relative to the two observers. The lines in the graphs show the light path from $t = 0$ onward. (*Source: Copyright by Floor Kamphorst*)

***Table 6.3* Outcome of thought experiment 3: The time interval between light hitting the floor of the train twice from two different reference frames. The answers of how relative velocity influences the extent of time dilation are given as well.**

| *Sub-question* | *Task 3E* | *Task 3F* |
|---|---|---|
| How many time steps does it take for the light to travel up and down between the mirrors once? | $\Delta t = 4$ time steps | • $\Delta t > 4$ time steps<br>• By using Pythagorean theorem: $\Delta t = 4.19$ time steps.<br>• Alternatively, students might use a ruler to measure the distance light travelled. |
| What happens when the cart moves in the other direction? | $\Delta t = 4$ time steps | $\Delta t = 4.19$ time steps |
| What happens when the velocity increases? | $\Delta t = 4$ time steps | $\Delta t > 4.19$ time steps |
| And approaches the speed of light? | $\Delta t = 4$ time steps | $\Delta t \rightarrow \infty$ |
| What happens when the velocity would decrease? | $\Delta t = 4$ time steps | 4 time steps $< \Delta t <$ 4,19 time steps |
| At what velocity would $\Delta t$ be the same for both observers? | $v = 0$ | $v = 0$ |

if the relative velocity between the observers would be in the other direction, or if this velocity increases or decreases. If needed, these questions can be supported with their own event diagrams. The outcome of the thought experiment and these reflections is summarised in Table 6.3.

To help students come to a more comprehensive understanding of time dilation, teachers can extend this task. First, the situation can be reversed, performing the thought experiment when the observer on the platform has a light clock. After that, the two situations can be combined, showing that each observer will conclude the clock of the other one is running slow compared to his own. Besides that, the formula for time dilation can be derived from the task outcome. This formalises the concept of time dilation.

With the formal definition of time dilation, students can reflect on another limitation of the event diagram. According to this formal definition, the time interval in Task 3F takes on a different value than they have found before ($\Delta t = 4.24$ time steps). Apparently, light covers a greater distance than one would expect based on the drawing in the event diagram. This is a starting point to introduce length contraction.

### *Reflecting on the meaning of the difference in travelled distance*

Max and Niels performed the thought experiment of Task 3F. Max initially drew light travelling up in a vertical line. Niels corrected him. The two students then reflected on the task outcome.

*Max:* I am not sure if this is correct [...] because it goes against my intuition very strongly. That it [the light path] is a straight line, and then again that it is not a straight line.

[...]

*Niels:* For the other observer [Task 3F] it is [...] a zigzag pattern. In that case, [the observer in Task 3F] would think that light goes faster, or that it takes a longer time to go up and down.

*Teacher:* Which of the two is it?

*Niels:* It takes a longer time because light has a constant speed.

Max acknowledged the logic behind how the thought experiment should be performed. However, the task outcome remains counterintuitive to him. When they reflected on the meaning of the task outcome, Niels first reconsidered if he should use the light postulate to interpret the results. He came to a decision and showed the relation between the outcome of the thought experiment and the light postulate.

## CONCLUSION

To wrap up, we have shown how our tasks stimulate students to perform thought experiments by drawing light propagation in event diagrams. To achieve this, we have added extra snapshots to the diagram and drawn it on a grid. Because we only present the setup and central question of the thought experiment, students' reasoning is supported even further. The examples of student reasoning illustrate that this is a promising approach to bring the educational power of thought experiments to secondary students. These examples demonstrate that the tasks allow students to explore their ideas on light propagation, and become aware of the importance of reference frames for light propagation. Furthermore, students derive relativistic concepts and can explain how they relate to the light postulate. Not only can event diagrams productively support student reasoning but we have also shown how students can reflect on the limitations of the event diagram. In educational practice, these limitations can be used productively to introduce new aspects of relativistic reasoning and new relativistic concepts. Besides, we expect event diagrams can be used as a stepping stone to introduce the more formal Minkowski diagram.

## FURTHER READING

Chapter 5: Dynamics first – A novel approach to relativity
Chapter 7: Introducing relativity on rotated graph paper
Chapter 23: Towards a study module on general relativity in the Netherlands

## NOTES

1. An inertial frame is a coordinate system or reference frame that is not accelerating and in which the curvature of spacetime is negligible. This means that the frame is at rest relative to other frames, or in uniform motion and gravity is absent.
2. These interviews were conducted with a previous version of the event diagram in which the speed of light was set at a value of 2 squares per time unit. To prevent confusion, we adapted the numerical value of the speed of light in the quotes of Lisa (Task 1) and Laura (Task 2) to align with the newer version of the tasks presented in this chapter.

## REFERENCES


Ainsworth, S., Prain, V. & Tytler, R. (2011). "Drawing to learn in science." *Science*, **333**(6046), 1096–1097.

Einstein, A. (1949/1979). *Autobiographical notes.* Open Court Publishing Company, Chicago.

Gendler, T.S. (2004). "Thought experiments rethought—and reperceived." *Philosophy of Science*, **71**(5), 1152–1163.

Gousopoulos, D., Kapotis, E. & Kalkanis, G. (2016). "Students' difficulties in understanding the basic principles of relativity after standard instruction." In: J. Lavonen, K. Juuti, J. Lampiselkä, A. Uitto & K. Hahl (eds.), *Science education research: Engaging learners for a sustainable future*, Part 1 2016, ESERA 2015 Conference. University of Helsinki, 169–175.

Helm, H., Gilbert, J. & Watts, D.M. (1985). "Thought experiments and physics education." *Physics Education*, **20**(5), 211–217.

Kamphorst, F., Vollebregt, M. J., Savelsbergh, E. R. & van Joolingen, W. R. (2019). "Students' pre-instructional reasoning with the speed of light in relativistic situation." *Physical Review Physics Education Research*, **15**(2), 020123.

Landau, L. D. & Rumer, G. B. (1960). *What is relativity?* New York: Basic Books.

Mach, E. (1976). "On thought experiments." In: E. Mach (ed.), *Knowledge and error.* Springer, 134–147. Dordrecht.

Matthews, R. (1994). *Thought experiments.* New York: Routledge.

Reiner, M. & Burko, L. M. (2003). "On the limitations of thought experiments in physics and the consequences for physics education." *Science & Education*, **12**(4), 365–385.

Scherr, R.E. (2001). *An investigation of student understanding of basic concepts in special relativity.* University of Washington. Washington.

Velentzas, A. & Halkia, K. (2013). "The use of thought experiments in teaching physics to upper secondary-level students: Two examples from the theory of relativity." *International Journal of Science Education*, **35**(18), 3026–3049.

Villani, A. & Pacca, J. L. A. (1987). "Students' spontaneous ideas about the speed of light." *International Journal of Science Education*, **9**(1), 55–66.

# 7 Introducing relativity on rotated graph paper

*Roberto B. Salgado*

**WHAT?**

This chapter explores educational opportunities to present visual calculations in special relativity. The presentation relies on spacetime diagrams that are drawn on rotated graph paper and use a minimal amount of algebra.

**FOR WHOM?**

It is intended for readers who look for instructional tools to teach concepts of special relativity at the upper secondary school level.

## INTRODUCTION

Special relativity (Einstein, 1905) forces upon us a new view of time and space, with some consequences that are puzzling to our common sense intuition. For example, we measure an elapsed time between two flash events that an inertially moving astronaut says are simultaneous (the relativity of simultaneity). We measure her clock to tick more slowly (time dilation). We measure a meterstick in her ship to be shorter than a meter (length contraction). And she would report the same things about us (the principle of relativity). How can we understand these puzzling consequences in a precise and hopefully more intuitive way?

Unfortunately, the usual textbook treatment – with their formulas, carefully worded explanations, and diagrams of boxcars in relative motion – seems insufficient for building a sound intuition of relativistic phenomena. In my opinion, these approaches fail to make contact with the underlying geometry of spacetime. Like in many problems in kinematics, the drawing of a position-vs-time graph is a good strategy to visualise physical phenomena. However, in special relativity, the **position-vs-time graph –** called the **spacetime diagram** (Minkowski, 1908) – has some non-intuitive features because *the familiar position-vs-time graph in introductory physics is only an approximation to what is really happening.* We first introduce the problem of our flawed intuition with time, then suggest our solution: **visualising time**.

***The spacetime diagram and our nonrelativistic intuition of time***

Points in the spacetime diagram (a position-vs-time graph) are called **events** $(t,x)$, which are idealised short-lived occurrences in a small region of space. Let Alice be the "person at rest" in this diagram; so, she maintains the same value of $x$ (here $x=0$) for all times $t$, shown as the horizontal red line in Figure 7.1. On her position-vs-time graph, she draws the motions of other persons as sloped lines, called **worldlines**. For example, Terrence travels with velocity $v$ = 3 m/s with respect to Alice, as shown by the labelled sloped blue line.

Suppose Alice freezes her position-vs-time graph when her wristwatch displays $t$ = 1 s. *For each astronaut [each traveling with a different velocity], what is the time displayed on his or her wristwatch at the events on their worldlines when t = 1 s according to Alice?* (*On Alice's spacetime diagram, these events are the intersections of their worldlines with a vertical line corresponding to Alice's t = 1 s.*)

It turns out that special relativity (*which is well-supported by an abundance of experimental evidence*) says that Terrence's wristwatch displays *less than 1 second*. It is actually 0.999 999 999 999 999 949 seconds, which is about "$1-\left(5\times10^{-17}\right)$ seconds". The moving wristwatch is measured by Alice to tick more slowly than her wristwatch; it is called **time dilation**.

For practical purposes, this is simply "1 second". (The tiny discrepancy of $5\times10^{-17}$ seconds shows how long it takes for light to travel 15 nm, the length of about 140 hydrogen atoms.) For other larger velocities, the time dilation effect is more pronounced. Refer to Figure 7.2 and the Table 7.1 below. The formula used in the table will be explained in the examples section.

Since on Alice's frozen-at-1-second diagram, these various wristwatches display "less than 1 second", when do these different wristwatches actually display "1 second"? It turns out they lie on the red curve shown in Figure 7.3, called a **hyperbola**, whose asymptote corresponds to the velocity of light (drawn as a

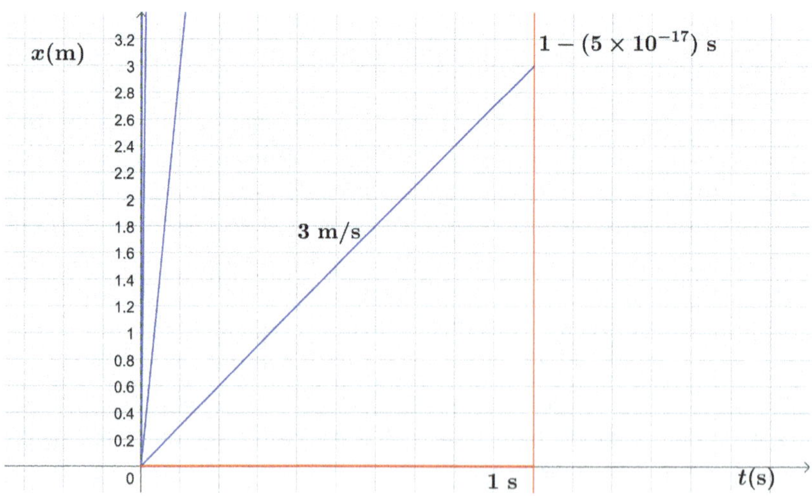


7.1
Alice's spacetime diagram (a position-vs-time graph) shows the worldline of Terrence, who travels at velocity $v$ = 3 m/s. When Alice's wristwatch displays "1 s", Terrence's wristwatch displays "$5\times10^{-17}$ s less than 1 s". The moving wristwatch is measured to tick more slowly; it is called time dilation. (*Source: Copyright by Roberto B. Salgado*)

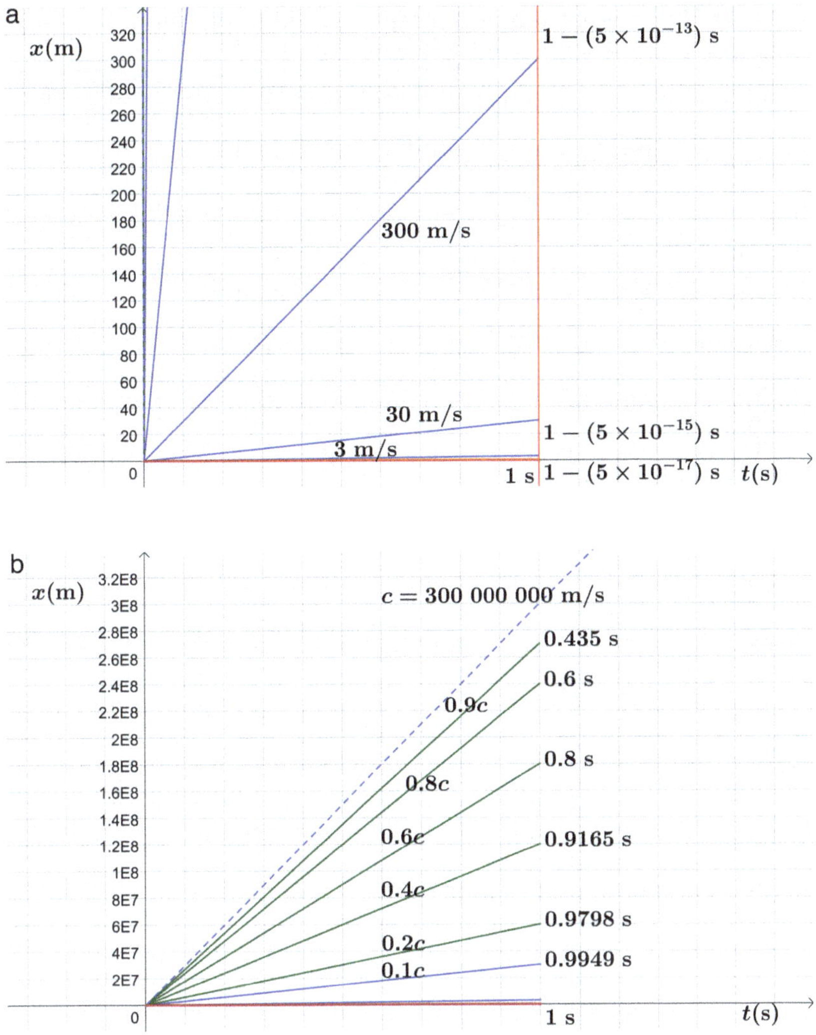


7.2
*(a) and (b)* Alice's spacetime diagram illustrates the worldlines of successively faster astronauts and their wristwatch displays (a). Note the distance scales needed to see them. The time dilation effect is more pronounced as the velocity approaches a significant fraction of the speed of light, $c$ = 299,792,458 m/s (b). (*Source: Copyright by Roberto B. Salgado*)

dashed line). *This hyperbola in spacetime geometry is like the circle in Euclidean geometry*. More precisely, the hyperbola locates events when and where all the wristwatches of **inertial** [constant velocity] astronauts from event O display the same elapsed time (say, "1 second"), just like the circle locates all points where the odometers along straight-line paths from the origin display the same distance travelled (say, "1 meter").

From a graph or the table, note that the hyperbola is practically indistinguishable from a straight line until we deal with velocities greater than about 0.1$c$. Since these velocities are beyond our everyday common sense experience, we have been unaware of these wristwatch discrepancies. But to understand relativity, we need to recognise that this is an essential feature of time, and we must shed misconceptions and refine our common sense.

***Table 7.1*** **Wristwatch displays for astronauts at different velocities when Alice's wristwatch reads 1 s**

| *Velocity of astronaut, with respect to Alice* | *Astronaut's wristwatch display, when Alice's wristwatch displays 1 s* |
|---|---|
| 3 m/s ( $\approx 10^{-8}c$) | $1-(5\times10^{-17})$ s |
| 30 m/s ( $\approx 10^{-7}c$) | $1-(5\times10^{-15})$ s |
| 3 000 000 m/s ( $\approx 0.01c$) | $1-(5\times10^{-5})$ s |
| 30 000 000 m/s ( $\approx 0.1c$) | 0.994 9 s $\approx 1-(5\times10^{-3})$ s |
| 60 000 000 m/s ( $\approx 0.2c$) | 0.979 8 s |
| 120 000 000 m/s ( $\approx 0.4c$) | 0.916 5 s |
| 150 000 000 m/s ( $\approx 0.5c$) | 0.866 s |
| 180 000 000 m/s ( $\approx 0.6c$) | 0.8 s |
| 240 000 000 m/s ( $\approx 0.8c$) | 0.6 s |
| 270 000 000 m/s ( $\approx 0.9c$) | 0.435 s |
| 296 794 533 m/s ( $\approx 0.99c$) | 0.141 s |
| 299 492 666 m/s ( $\approx 0.999c$) | 0.045 s |
| $\left(\frac{v}{c}\right)$, where $c$ = 299 792 458 m/s ($\approx 3\times10^{8}$ m / s) | $\sqrt{1-\left(\frac{v}{c}\right)^2}$ s |

*Note that in Figure 7.2, we have to rescale the position axis to get these larger-velocity motions to be drawn. For* $v = 0.1c$, *where the speed of light,* $c = 299{,}792{,}458$ m / s, *the discrepancy is 0.5%.*

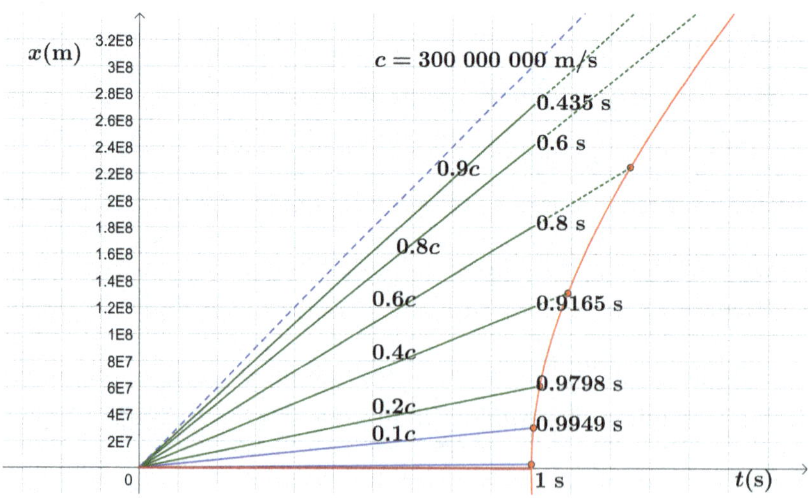


7.3
The red hyperbola marks where all wristwatches that met at event O would read " 1 second". (*Source: Copyright by Roberto B. Salgado*)

**Is there a way to visualise these elapsed times – without resorting to formulas?**

**Yes, we can visualise the ticks of the simplest clock: the light clock.**

### *Overview*

Our goal is to develop a visualisation of tick marks of time and space on a spacetime diagram with **light-clock diamonds**. In the next section, we will study a light clock, which is formed by a pair of mirrors at fixed separation that reflect light-signal flashes from a source located between the mirrors. A tick of the clock is marked when the reflected signals are received by the source. On the spacetime diagram in Figure 7.4, the worldlines of the light-signals (always along 45-degree lines) trace out a diamond shape – hence, the light-clock diamond. The light-clock diamonds for a moving astronaut Bob are also shown in Figure 7.4. The shape of the diamond encodes the velocity, with one diagonal of the diamond along the astronaut's worldline. **The key takeaway result is that the area of Bob's light-clock diamonds is equal to the area of Alice's light-clock diamonds.** The details of this construction are discussed in the section on Alice and her light clock.

With our visualisation, we can physically interpret units of time, units of distance, and velocity. We can calculate by drawing and counting, *without relying on the standard textbook formulas*. Once this is established, we are better equipped to tackle standard problems in special relativity. We will also be able to make contact with the standard textbook formulas, to handle more general cases.

Here is our plan. On a spacetime diagram, we are going to draw the worldlines – the spacetime paths – of an inertial astronaut Alice and the light-signals of her light clock. This establishes a standard of time and distance measurement for Alice that we can visualise. Using the principles of special relativity, we establish the equivalent standard for Bob, another inertial astronaut in relative motion with Alice. With the resulting visualisation of the tick marks for each frame of reference, we can then study various situations and compare the measurements that Alice and Bob make about them. Although we begin with graph paper in the usual orientation, we will soon realise that by rotating the graph paper by 45 degrees (to follow the worldlines of light-signals), we can more easily draw these diagrams quantitatively and do calculations by counting.

### *Visualising elapsed time with light-clock diamonds (a preview)*

In Figure 7.5, we present our light-clock diamond visualisation of the elapsed times of Figure 7.3. In this diagram, each diamond represents 0.1 s. So, *by counting* along Alice's worldline, we find $10\times(0.1\text{ s})=1\text{ s}$ elapses. Along the $0.6c$ worldline, $8\times(0.1\text{ s})=0.8\text{ s}$ elapses. Along the $0.8c$ worldline, $6\times(0.1\text{ s})=0.6\text{ s}$ elapses. Along the other worldlines shown here, it turns out we don't get a whole number of these diamonds, which makes our graphical

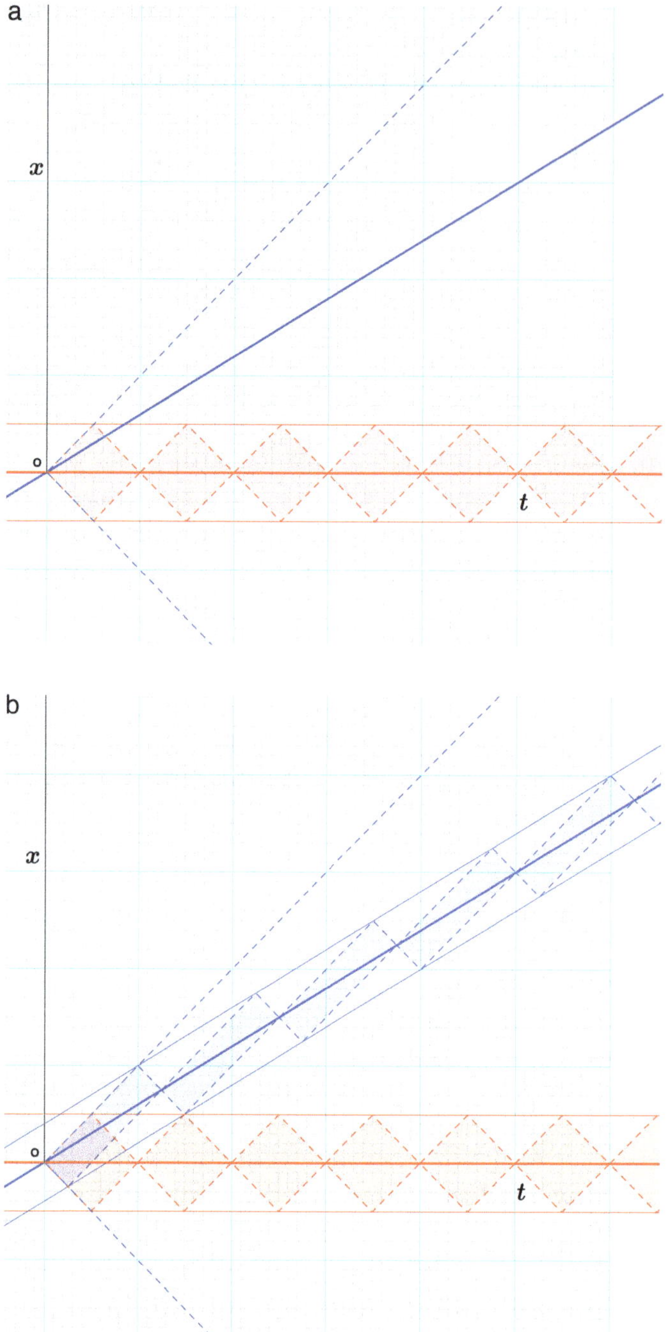


7.4
*(a) and (b)* Light-clock diamonds for Alice (a) and for another astronaut Bob (b). For each astronaut, the light-clock mirror worldlines are parallel to the astronaut's worldline. The light-signals being reflected by the mirrors trace out the diamonds. One diagonal of the diamond is along the astronaut's worldline. These diamonds will be used as tick marks for time and space. The construction of Bob's diamonds is shown in the text. The key takeaway result is that the area of Bob's light-clock diamond is equal to the area of Alice's light-clock diamond. (*Source: Copyright by Roberto B. Salgado*)

method a little harder to implement and appreciate. We would have to draw truncated diamonds or use finer resolution diamonds and have a computer to draw them for us.

Again, the remarkable feature of our light-clock diamonds is that **all of their areas are equal**. The rotated graph paper makes drawing these diamonds much

7.5
Visualisation of time using "light-clock diamonds", modelled on the light signals in a light clock. (*Source: Copyright by Roberto B. Salgado*)

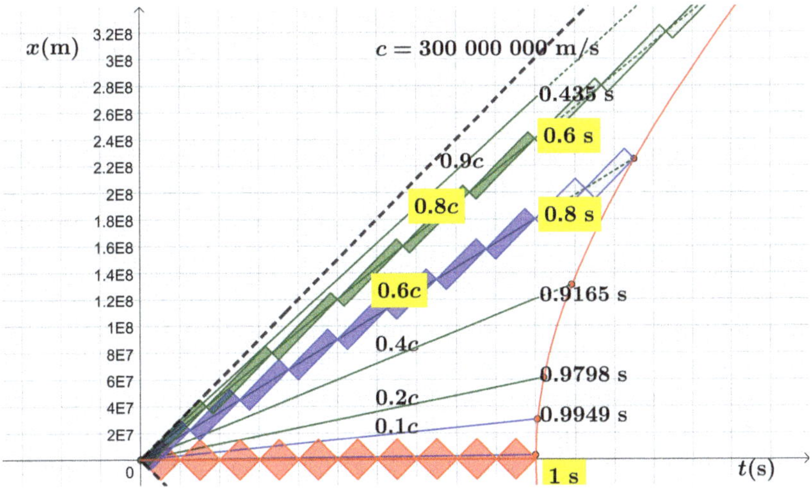


easier, as we will show. Once drawn, one can do calculations of relativity without explicitly using its standard formulas. We will do this in our example section later on.

### *Conventions*

In relativity, the time-axis usually runs upward. So, Figure 7.6 is a redrawing of Figure 7.5 in this convention.

7.6
In relativity, the convention is to have time run upwards on a spacetime diagram. Compare this with Figure 7.5. To visualise the motion according to Alice (like a movie), slide a slotted ruler as shown. Furthermore, define a unit of length called the light-second as the distance travelled by light in one second: 1 ls = $3 \times 10^8$ m. (*Source: Copyright by Roberto B. Salgado*)

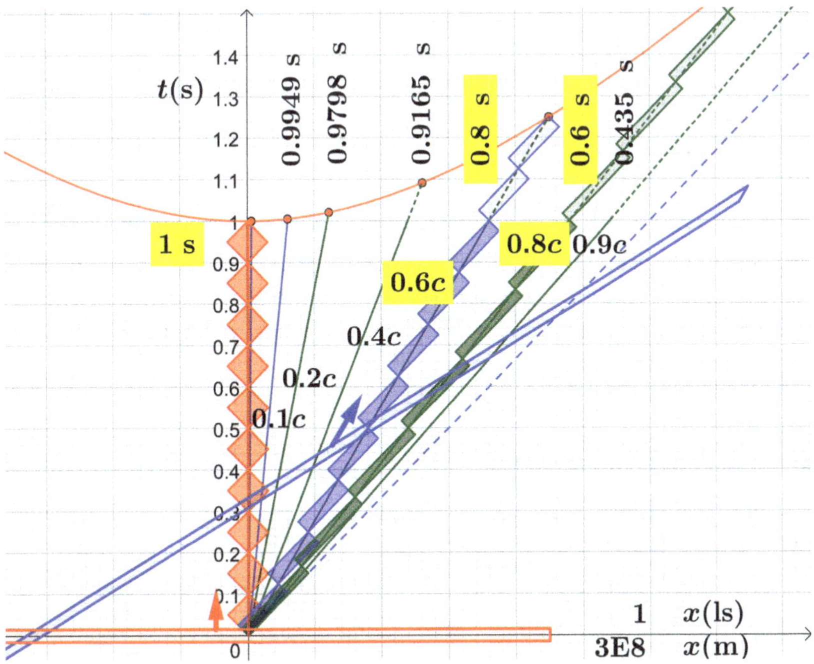

To visualise the motions in space as Alice interprets them, one can orient a slotted ruler to show an instant of time according to Alice [aligned with the horizontal diagonal of her diamond]. One can view the diagram like a movie for Alice by sliding the ruler upward, parallel to Alice's worldline.

To visualise the motions in space as Bob (the astronaut travelling with velocity, $v = 0.6c$) interprets them, one can orient a slotted ruler to show an instant of time according to Bob [aligned with the slanted diagonal of Bob's diamond, as shown]. One can view the diagram like a movie for Bob by sliding the ruler upward, parallel to Bob's worldline. (This is explained in the next section.) Furthermore, for convenience, let us define a unit of length called the **light-second**: $1 \text{ ls} = 3 \times 10^8$ m, which is the distance light travels in 1 s. With this choice of unit, light-signals are drawn at 45 degrees.

To understand the physical interpretation of these diamonds and the role of relativity, we need to spend a little time explaining how these diamonds are constructed and how various features are physically interpreted. This will prepare us to tackle the example problems later on.

### *The principles of special relativity*

We end this section with brief statements of the principles of special relativity. For more details, consult the works by Bondi (1965), Taylor and Wheeler (1966), and Moore (2016).

- **The relativity principle**: By studying the results of experiments done in a closed room, it is not possible to distinguish whether we are at rest in an inertial frame or we are moving with constant velocity in an inertial frame.
- **The speed of light principle:** All light-signals travel at the same speed, in all frames of reference. (This speed $c$ is 299,792,458 m/s, which can be approximated by $3 \times 10^8$ m/s and 1 ft/ns. [The nanosecond is one-billionth of a second: $1 \text{ ns} = 10^{-9}$ s.])

The relativity principle was first formulated in *Dialogue Concerning the Two Chief World Systems* (Galileo, 1632) with regard to mechanical experiments. With the two principles above, special relativity (Einstein, 1905) extended Galileo's relativity principle to include all of the laws of physics, in particular, optics and electromagnetism.

## ALICE AND HER LIGHT CLOCK

### *How to draw Alice's light-clock diamonds*

Take a sheet of graph paper in the standard orientation and draw a standard set of axes, with time $t$ running vertically upward and $x$ running to the right (see

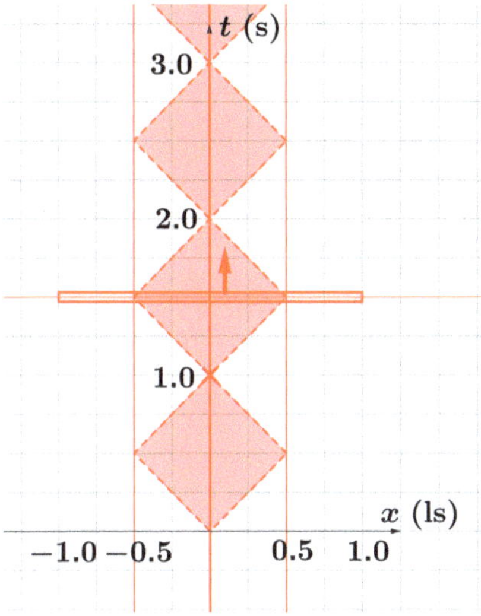


7.7
Alice's worldline and the worldlines of the mirrors and light-signals in her light clock. Her light clock ticks once every second. To visualise the motions of the light-signals and mirrors, use a slotted ruler aligned horizontally. (*Source: Copyright by Roberto B. Salgado*)

Figure 7.7). To describe Alice at rest in this frame, draw a vertical line through $x = 0$. This is Alice's worldline.

We will model Alice's wristwatch by a **light clock**, which is composed of a pair of mirrors at a constant distance away from her, (say) at $x = 1/2$ ls and $x = -1/2$ ls.

Draw the worldline of each mirror parallel to Alice's worldline, as a vertical line through these $x$ -values. At $t = 0$, Alice sends out momentary light-signals, which travel at the speed of light (1 ls/s) drawn at 45 degrees. These light-signals are said to be on the **future light-cone –** the spacetime paths of the emitted light-signals. When a light-signal meets a mirror, the light-signal is reflected, reversing only its spatial direction. Observe that all reflected light-signals from mirrors at the same distance away from Alice are received by Alice at the same reception event 1 s later, marking "1 tick" of her light clock. Allowing the light-signals to continue onward (or to send out new light-signals upon reception of the old ones), we generate a sequence of ticks of her clock. The light-signals have traced out "**light-clock diamonds**".

### *Alice's diamonds as her units of measurement*

The light-clock diamond has an upward diagonal, called the "**timelike diagonal**", which measures off "1 **tick** of time for Alice". It also has a sideways diagonal, called the "**spacelike diagonal**", which measures off "1 tick of space for Alice" or 1 "space-tick", or even better, "1 **stick**"! (See Figure 7.8.) Thus, the light-clock diamond represents a **standard unit of measurement** for Alice. (For the construction above, "1 tick" = 1 s and "1 stick" = 1 ls. Let us use "tick" and "stick" from now on.)

Furthermore, the diagonals of the diamond help convey a very important feature of spacetime: Since Alice's sense of space is *"perpendicular"* to her sense

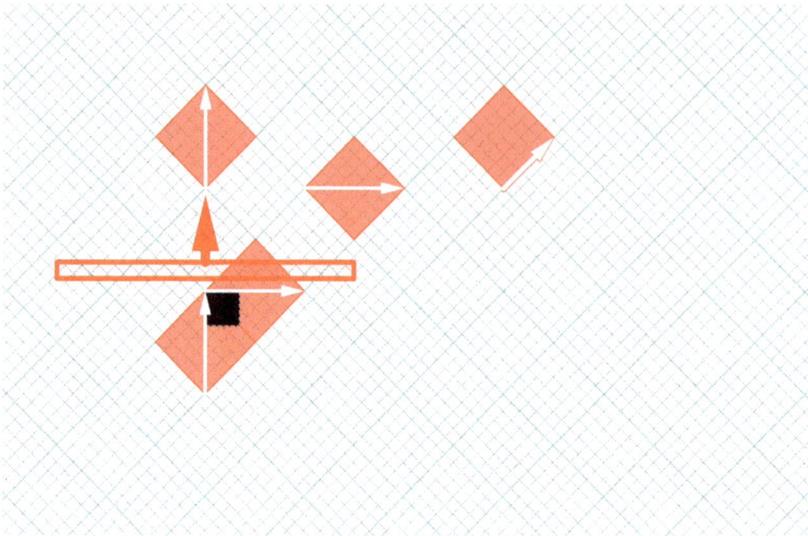

7.8
Alice's light-clock diamonds as units of measurement. One tick of time is along the timelike diagonal. One stick of space is along the spacelike diagonal. These diagonals are spacetime-perpendicular, which can be seen when two of her clock-diamonds share a common edge. To visualise the motion according to Alice, use a slotted ruler aligned with the spacelike diagonal. (*Source: Copyright by Roberto B. Salgado*)

of time, the diagonals of these diamonds encode what it means to be *"perpendicular" in the geometry of spacetime* (called "**spacetime-perpendicularity**").

With this notion, we refine how to visualise the motion according to Alice: *Use a slotted ruler aligned with the spacelike diagonal, and advance in time along the timelike diagonal*, as shown in Figure 7.8. These features will be important when we apply them to another astronaut Bob, in relative motion with Alice.

At this point, Alice can assign $(t,x)$-coordinates to all of the events on her worldline (where $x=0$) with the reading $t$ of her clock when it occurred, that is, $(t,0)$.

*But what about distant events, those events not on her worldline?*

It will turn out that by tiling all of spacetime with Alice's light-clock diamonds, we will set up a grid for Alice's coordinate system. Hence, from now on, we will use the graph paper rotated by 45 degrees.

## BOB AND HIS LIGHT CLOCK

The previous sections have prepared us for the important step to draw Bob's frame of reference on Alice's spacetime diagram, in particular, *how Bob's clock ticks are spaced along Bob's worldline.*

### *How to draw Bob's worldline*

Suppose Bob is an inertial astronaut who is moving with velocity, $v=(3/5)c$ with respect to Alice. Suppose further that Bob and Alice met briefly at the origin event O $(t_0,x_0)=(0,0)$.

**How should Bob's worldline be drawn?** Refer to Figure 7.9. Since $v=\dfrac{\Delta x}{\Delta t}=\dfrac{3}{5}$, we have $\Delta x=\dfrac{3}{5}\Delta t$. If $\Delta t=5$ ticks, then $\Delta x=3$ sticks. So, from

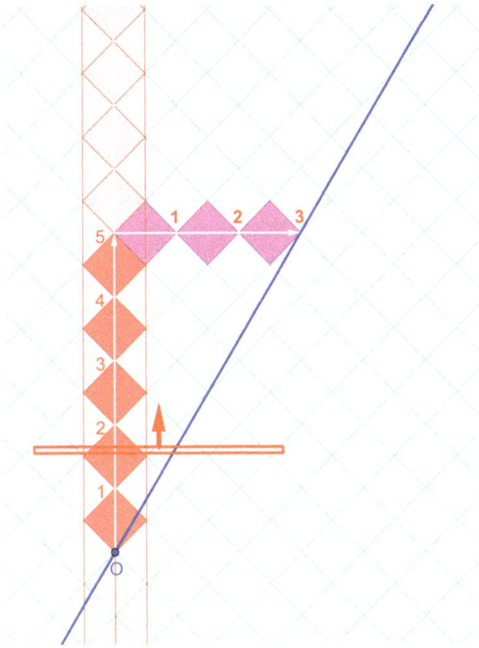

7.9
Drawing Bob's worldline for $v = (3/5)c$.
(*Source: Copyright by Roberto B. Salgado*)

event O, Bob's worldline will pass through $(\Delta t, \Delta x) = (5, 3)$. So, draw a line through $(0, 0)$ and $(5, 3)$. This is Bob's worldline.

(It turns out that the choice of $v = \left(\frac{3}{5}\right)c$ leads to calculations involving simple arithmetic with fractions. Other choices include $v = \left(\frac{4}{5}\right)c$ and $v = \left(\frac{5}{13}\right)c$, which lead to triangles that are associated with Pythagorean triples. The choices of $v = \left(\frac{1}{2}\right)c$ and $v = \left(\frac{9}{10}\right)c$ are not among these, and thus introduce distracting arithmetic.)

***How to calibrate Bob's light clock***

**How should Bob's light-clock worldlines and light-clock diamonds be drawn?** Since Bob's light clock travels with him at the same velocity, the mirror worldlines would be parallel to Bob's worldline. But here are two related questions: *in terms of Alice's diamonds, what is the separation in space between the mirror worldlines? And what is the spacing in time between the ticks of Bob's clock?* Refer to Figure 7.10 to see how these questions are related. We note that the ***shape*** of Bob's light-clock diamonds on Alice's spacetime diagram is determined by his velocity with respect to Alice. What determines the ***size*** of Bob's light-clock diamond?

We want our construction to be in accordance with special relativity.

- From the **speed of light principle**, all light-signals [regardless of the motion of the source] should be drawn at 45 degrees, parallel to the lines of the

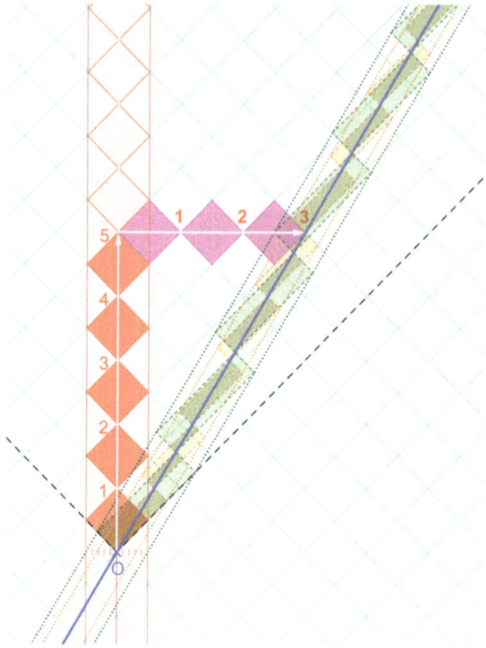


7.10
Bob's mirror worldlines are parallel to his worldline – but what is their separation? A related question is: how are Bob's ticks spaced in time along his worldline? (*Source: Copyright by Roberto B. Salgado*)

rotated graph paper. This means that **the edges of Bob's light-clock diamond should also be drawn parallel to the rotated grid lines.**

- Second, from the **relativity principle,** it is impossible to distinguish from experiments whether we are at rest in an inertial frame, or if we are moving with constant velocity in an inertial frame. **For our construction, we will have Alice and Bob who send light-signals to each other and require that their results are identical**, otherwise they could distinguish their state of motion between them.

Consider three of the possibilities to draw the mirror worldlines. (See Figures 7.11–7.13.) To check if the case is consistent with the relativity principle, let each inertial astronaut do a "signal-sending experiment": wait for 2 ticks after meeting at event O, then send a light-signal to the other. Do the results of their experiments agree?

In Figure 7.11, the spacing corresponds to "absolute time," where Bob's clock reads 5 ticks on Alice's diagram frozen-at-5-ticks – the situation for the non-relativistic position-vs-time graph. However, this setup violates the relativity principle since the results of the corresponding signal-sending experiment for two inertial astronauts differ: Alice receives Bob's signal at 3.2 ticks, whereas Bob receives Alice's signal at 5 ticks. For agreement, since Alice's 3.2 must be increased and Bob's 5 must be decreased, Bob's ticks must be longer in time. This hints at the "time dilation" effect, treated in the examples.

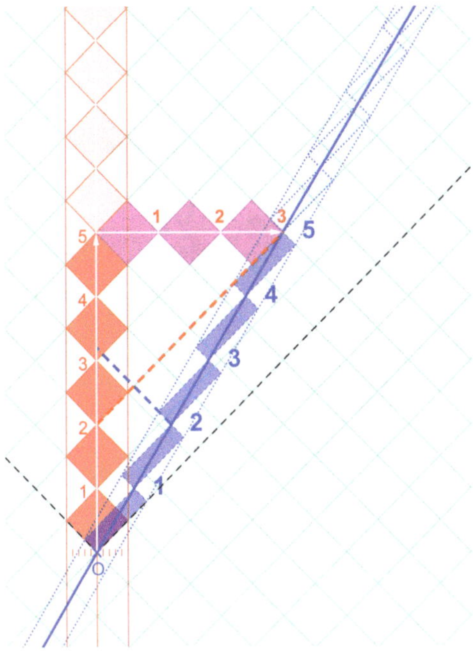


7.11
This spacing corresponds to "absolute time". However, the signal-sending experiments disagree since Alice receives Bob's signal at 3.2 ticks, whereas Bob receives Alice's signal at 5 ticks. Bob's ticks need to be longer in time than Alice's ticks (time dilation). (*Source: Copyright by Roberto B. Salgado*)

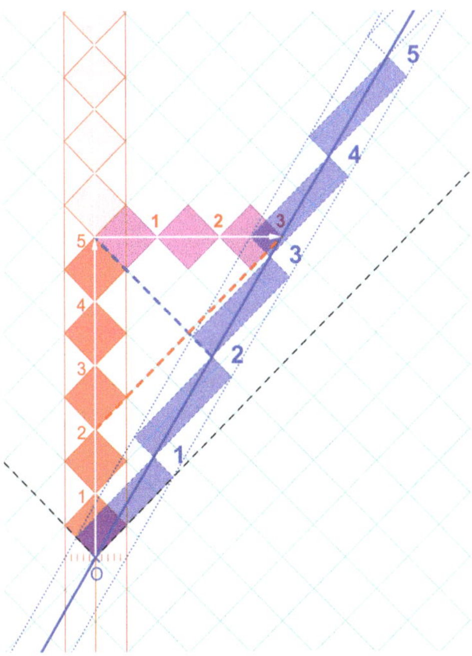


7.12
This spacing corresponds to "absolute length". However, the signal-sending experiments also disagree since Alice receives Bob signal at 5 ticks, whereas Alice receives Bob's signal at 3.2 ticks. Bob's mirror worldlines need to be less separated than Alice's mirror worldlines (length-contraction). (*Source: Copyright by Roberto B. Salgado*)

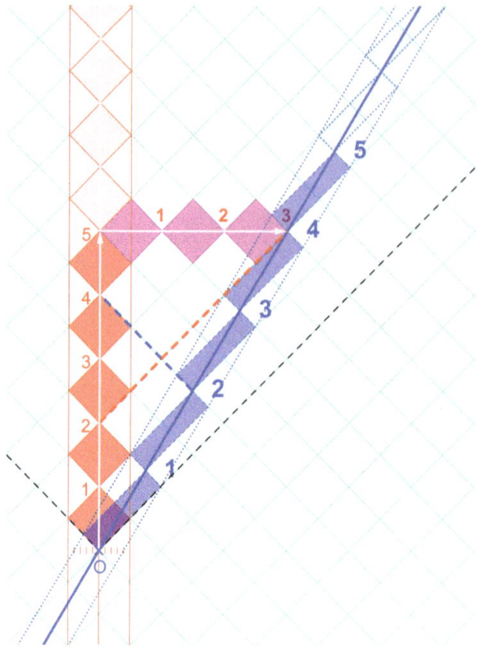


7.13
This spacing corresponds to special relativity – since their signal-sending experiments agree. Alice receives Bob's signal at 4 ticks and Bob receives Alice's signal at 4 ticks, in accordance with the relativity principle. See the text for how we arrived at this result. It turns out that this result can be constructed by noting that the area of Bob's light-clock diamond is equal to the area of Alice's light-clock diamond. (*Source: Copyright by Roberto B. Salgado*)

In Figure 7.12, the spacing corresponds to "absolute length", where Bob's light clock (which is moving according to Alice) has the same length as Alice's own light clock – which is also the situation for the non-relativistic position-vs-time graph. However, this setup also violates the relativity principle since the results of the corresponding signal-sending experiment differ: Alice receives Bob's signal at 5 ticks, whereas Bob receives Alice's signal at 3.2 ticks. For agreement, since Alice's 5 must be decreased and Bob's 3.2 must be decreased, Bob's ticks must be shorter in time, which means that the length of Bob's light clock (in sticks) must be shorter than Alice's. This hints at the "length contraction" effect, treated in the examples.

To help us determine the correct spacing of Bob's ticks as in Figure 7.13, Alice can use a radar method by Bondi (1965) [which inspired our signal-sending experiment]. (See Figure 7.14 ahead.) Suppose Alice broadcasts a one-hour video [in 15-minute segments], which will be watched by Bob and simultaneously rebroadcasted by Bob back to Alice. If Bob is at a distance, but at rest with respect to Alice, then (after a delay, since it takes some time for the signals to reach Bob), it *seems plausible* that Bob will watch the show [at normal speed] in 1 hour of his time. *But how would Alice know that if she couldn't just ask Bob?* Alice could study Bob's rebroadcast of her original one-hour video, as follows:

$$\text{Let } k = \frac{[\text{reception}] \text{ viewing interval}}{[\text{transmission}] \text{ broadcast interval}}.$$

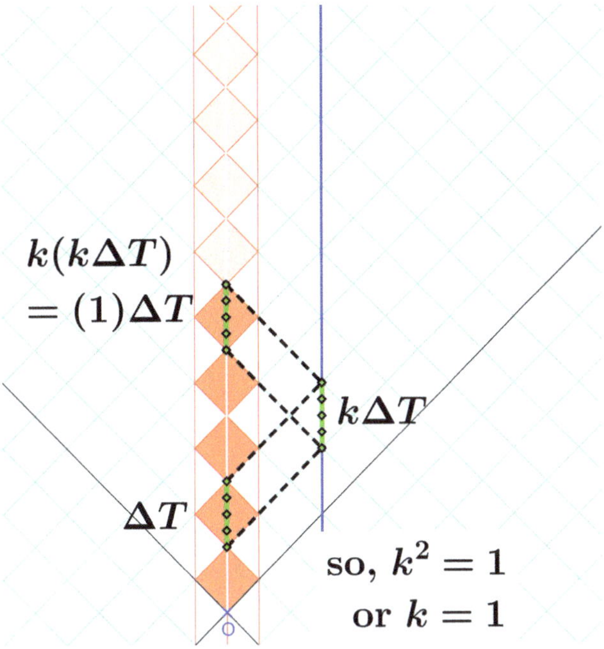


7.14
Bondi's radar method for Bob at rest. Each diamond represents 1 hour. We indicate the 15-minute intervals. From Alice's measurement of Bob's rebroadcast, Alice deduces that Bob watches her one-hour broadcast at normal speed in 1 hour. (*Source: Copyright by Roberto B. Salgado*)

For the case when Bob is at rest with respect to Alice (as in Figure 7.14), at some distance away, $k = 1$.

For the case when Bob is travelling away from Alice (as in Figure 7.15), $k$ must be greater than 1 since each successive part of the transmission takes longer to reach Bob. Thus, Bob would take $k$ of his hours to watch the one-hour video broadcasted by Alice in slow motion. For Alice to determine $k$, we arrange to have Bob immediately reflect or rebroadcast the received signals from Alice.

According to the relativity principle, Alice must receive the delayed broadcast from Bob at the same slowed rate. Thus, Alice would take $k$ of her hours to watch in slow motion a one-hour video broadcasted by Bob. So, Alice would take $k^2$ of her hours to watch in *super-slow* motion her originally broadcasted one-hour video. Since for v = (3/5)c, the rebroadcast is found to be – by counting – 4 of Alice's light-clock diamonds, we deduce that $k = \sqrt{4} = 2$. So, we can conclude that there must be two of Bob's light-clock diamonds between the start and end of his viewing Alice's video.

### *How to draw Bob's light-clock diamonds*

To construct Bob's ticks, divide Bob's viewing interval into $k = 2$ sections, marked by three events. Sketch the light-cone at each event (since we expect all light-signals travel at the speed of light). These light-cones mark off the corners of Bob's light-clock diamonds. Thus, we can draw all of the light-clock diamonds on Bob's worldline. (Refer to Figure 7.16a–c.)

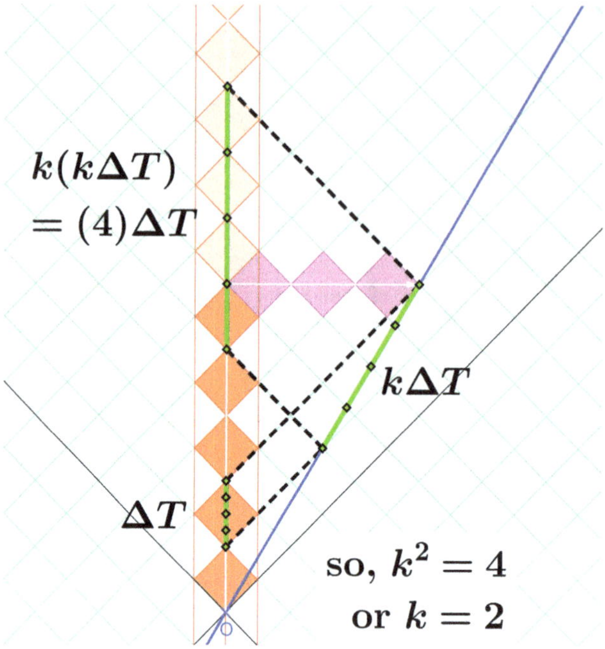


7.15
Bondi's radar method for Bob moving at $v = (3/5)c$. Each diamond represents 1 hour. From Alice's measurement of Bob's rebroadcast, we have (by counting) $k^2 = 4$. Thus, Alice deduces that Bob watches her one-hour broadcast in slow motion in two hours. (*Source: Copyright by Roberto B. Salgado*)

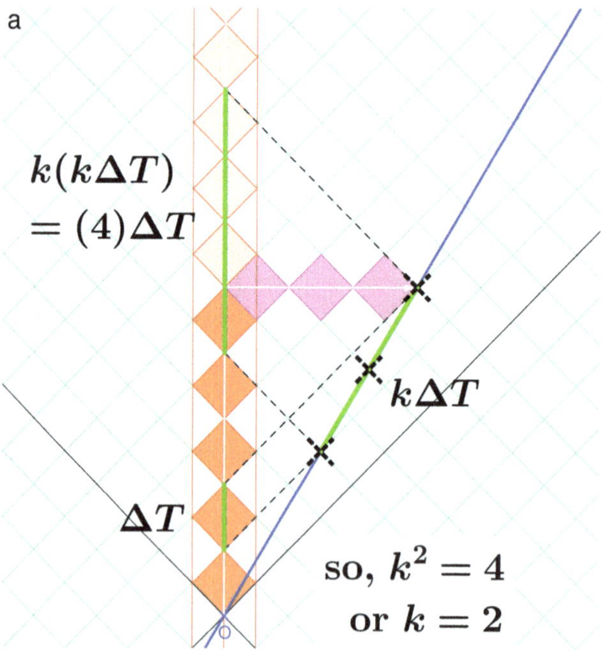


7.16
*(a)–(c)* Drawing Bob's light-clock diamonds, once $k$ is determined by Bondi's method. (*Source: Copyright by Roberto B. Salgado*)

7.16
*(continued)*

b

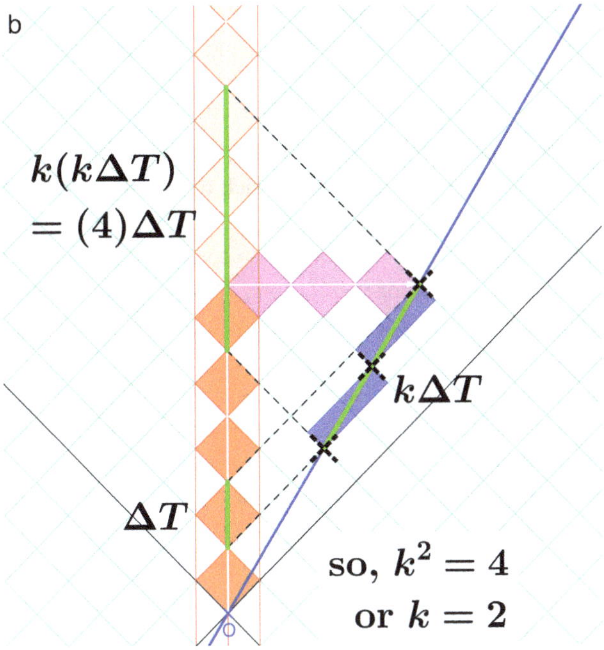


c

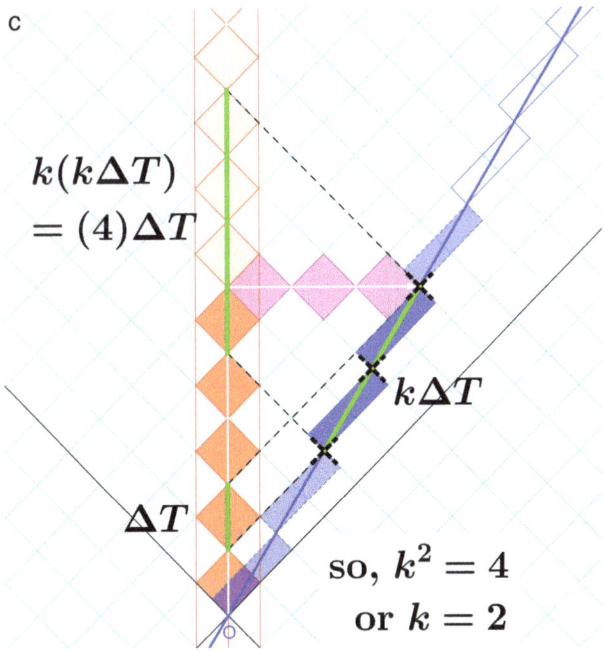

### *Bob's diamonds as his units of measurement*

Finally, with Bob's light-clock diamond, we can [as Alice did in the previous section] tile all of spacetime with his light-clock diamond to set up a grid for Bob's coordinate system.

(Refer to Figure 7.17.) Analogous with that of Alice's light-clock diamonds, the timelike diagonal of Bob's diamond measures off "1 **tick** of time for Bob", and the spacelike diagonal measures off "1 **stick** of space for Bob". In addition, these diagonals are spacetime-perpendicular to each other – in spite of how it appears to someone interpreting it with the Euclidean geometry. (*Note that the ordinary Euclidean trigonometry and its angles and the usual Pythagorean theorem do not apply in spacetime geometry.*)

To visualise the motions according to Bob, we use the same prescription as we do for Alice: *Use a slotted ruler aligned with the spacelike diagonal, and advance in time along the timelike diagonal*, as shown in Figure 7.17.

Note that the **areas of these light-clock diamonds are equal**, as we mentioned in the preview earlier. This is true for arbitrary velocities. A proof appears in Salgado (2016). (*Recall the shape of the light-clock diamonds was determined by the velocity. The further deduction of this equal-area property is determined by the relativity principle.*)

To suggest that this equality of area is true in general, please redo the construction for Carol, another inertial astronaut with velocity, $v=(4/5)c$. In that case, it turns out that $k=3$. Carol's diamond is shown in Figure 7.18, alongside the previous $v=(3/5)c$ case.

### *How to draw Bob's light-clock diamonds on rotated graph paper*

Let us take advantage of this equality of area property on rotated graph paper. (Refer to Figure 7.18.) Observe that Bob's light-clock diamond, with $k=2$ for

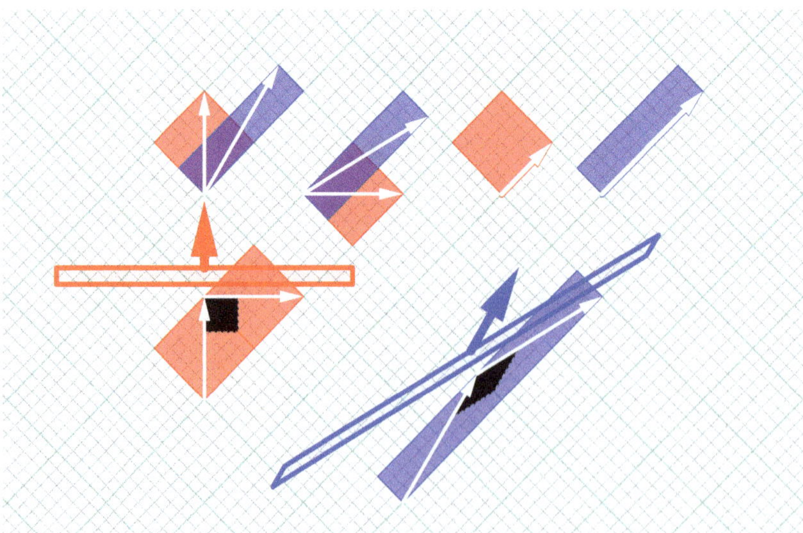

7.17
Bob's light-clock diamonds as units of measurement. Alice's diamonds are shown for comparison. One tick of Bob's time is along Bob's timelike diagonal. One stick of Bob's space is along Bob's spacelike diagonal. These diagonals are spacetime-perpendicular, which can be seen when two of his clock-diamonds share an edge. Note that the area of Bob's light-clock diamond is equal that of Alice's light-clock diamond. (*Source: Copyright by Roberto B. Salgado*)

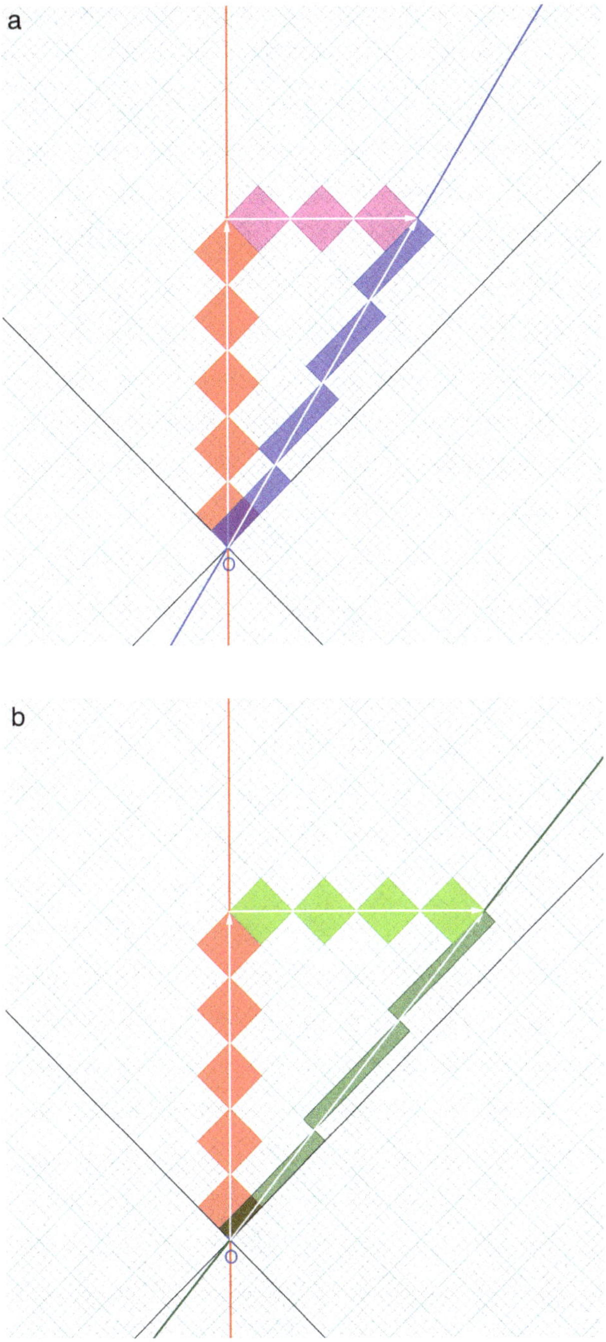


7.18
*(a) and (b)* Light-clock diamonds for Bob (with $k=2$ for $v=(3/5)c$) (a) and Carol (with $k=3$ for $v=(4/5)c$) (b). Using the equality of light-clock diamond areas, Bob and Carol's diamonds are obtained by deforming Alice's diamond by stretching by a factor $k$ in the forward direction and shrinking by the same factor $k$ in the backward direction. (*Source: Copyright by Roberto B. Salgado*)

$v=(3/5)c$, can be drawn by *stretching* Alice's forward edges by a factor of 2 and *shrinking* the backward edges by the same factor of 2, which keeps the area unchanged. Carol's light-clock diamond, with $k=3$ for $v=(4/5)c$, is obtained similarly. On rotated graph paper, this is easier with 6 × 6 subdivisions.

At this point, we can visualise some of the famous effects of special relativity. [Refer to Figure 7.19a and b.] **Time dilation** is visualised because the moving clocks take longer to tick according to Alice. She determined that Bob's clock elapsed 4 ticks, whereas her clock elapsed 5 ticks. But

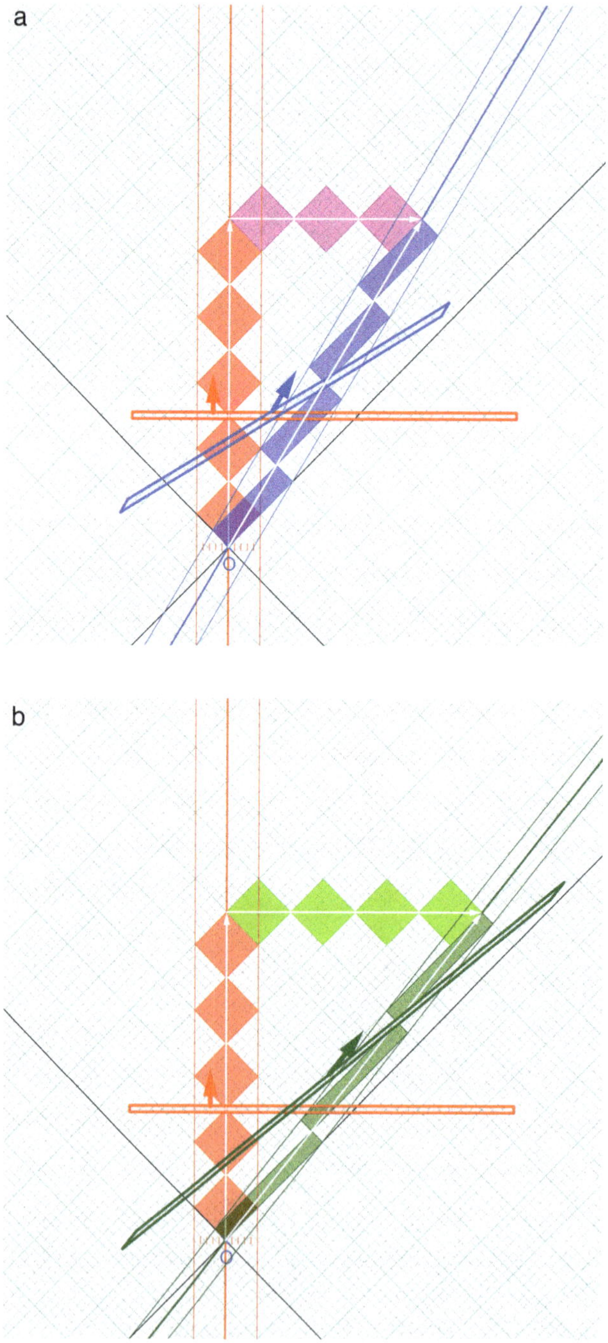


7.19
*(a) and (b)*Time dilation, relativity of simultaneity, and length contraction are visualised. (See the text.) (*Source: Copyright by Roberto B. Salgado*)

note that according to Bob [using his slotted ruler], Alice's clock takes longer to tick! Bob determined that Alice's clock elapsed 4 ticks, whereas his clock elapsed 5 ticks. (To see this, add a fifth tick for Bob. See Figure 7.20 ahead.) **Relativity of simultaneity** is visualised because Alice's two reflection events (along her diamond's spacelike diagonal) are simultaneous according to Alice, but not according to Bob. Similarly, Bob's two reflection events are not simultaneous according to Alice. Finally, we can visualise **length contraction**. (Refer to Figure 7.19.) Given the light-clock diamonds, draw the mirror worldlines by drawing parallels through the diamond corners, as shown. According to Alice [using her slotted ruler through event O], Bob's identical moving light clock is shorter than her own. But note that according to Bob [using his slotted ruler through event O], Alice's identical moving clock is shorter than his own!

## EXAMPLES

Now, we have established our methods; we proceed to do some standard textbook examples. These and other problems are treated in more detail in my paper. (*See the references and e-resources.*) While we have used units of seconds (s) and light-seconds (ls) up to this point, we can use more convenient units like years (y) and light-years (ly), as appropriate.

### *Example 1: Time dilation*

**Question:** After leaving inertial astronaut Alice at event O, another inertial astronaut Bob travels at velocity $v = \left(\frac{3}{5}\right)c$ according to Alice. (Refer to Figure 7.20.) Thus, according to Alice, after $5\ \mathrm{y}$ has elapsed on her clock, Bob is located $3\ \mathrm{ly}$ away, at event Q. How much time has elapsed on Bob's clock between events O and Q, both of which are on his worldline?

**Solution:** Construct Bob's worldline by counting Alice's light clock diamonds with $\Delta t = 5\ \mathrm{y}$ and $\Delta x = 3\ \mathrm{ly}$ to locate event Q. Construct Bob's light-clock diamonds using equality-of-areas. Starting from O and ending at Q, we find that (by counting) $4\ \mathrm{y}$ have elapsed on Bob's light clock.

**Interpretation:** Note that event Q is a distant event for Alice – she does not experience event Q. Bob, however, experiences both events O and Q. Although the elapsed time from O to Q according to Alice is $\Delta t_{OQ,\ \mathrm{Alice}} = \Delta t_{OP,\ \mathrm{Alice}} = 5\ \mathrm{y}$, the elapsed time from O to Q according to Bob is only $\Delta t_{OQ,\ \mathrm{Bob}} = 4\ \mathrm{y}$. Alice says that, between O and Q, Bob's moving clock ticks more slowly than her own clock – this is **time dilation**. The ratio $\gamma = \frac{\Delta t_{OQ,\ \mathrm{Alice}}}{\Delta t_{OQ,\ \mathrm{Bob}}} = \frac{\Delta t_{OP,\ \mathrm{Alice}}}{\Delta t_{OQ,\ \mathrm{Bob}}} = \frac{5}{4}$ is called the "time-dilation factor", which can be shown to be equal to $\frac{1}{\sqrt{1-\left(\frac{v}{c}\right)^2}}$. Thus, $\Delta t_{OQ,\mathrm{Bob}} = \frac{\Delta t_{OP,\mathrm{Alice}}}{\gamma} = \Delta t_{OP,\ \mathrm{Alice}} = \Delta t_{OQ,\mathrm{Bob}}\sqrt{1-\left(\frac{v}{c}\right)^2}$, which is the formula used in the Table 7.1. Since Bob is also an inertial astronaut, by the relativity principle, Bob should be able to say the same statement about Alice: Bob says that, between

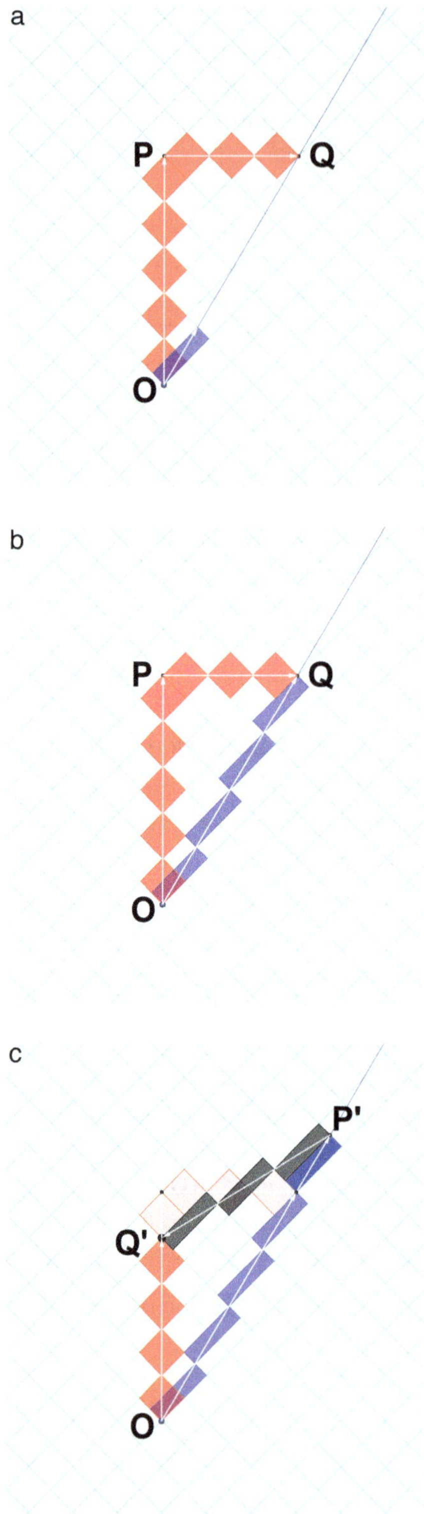


7.20
*(a)–(c)* Time dilation and its symmetry. (*Source: Copyright by Roberto B. Salgado*)

O and Q′, Alice's moving clock ticks more slowly than his own clock. To see this, carry out the analogous construction using Bob's diamonds to locate Q′. Note that Alice's velocity according to Bob is $v = \dfrac{\Delta x}{\Delta t} = \dfrac{(-3\text{ ly})}{(5\text{ y})} = -\left(\dfrac{3}{5}\right)c$. (Refer to the last diagram in Figure 7.20.)

***Example 2: Clock effect/twin paradox***

**Question:** After meeting Alice at event O, Bob travels away at velocity $v = \left(\dfrac{3}{5}\right)c$, instantaneously turning around at a distant event Q and returning with velocity $v = -\left(\dfrac{3}{5}\right)c$. (Refer to Figure 7.21.) At the reunion event Z, if Alice has aged 10 y, how much did Bob age?

**Solution:** This is essentially two time dilation problems. As before, $\Delta t_{OQ,\text{ Alice}} = 5$ y, but $\Delta t_{OQ,\text{ Bob}} = 4$ y. Similarly, $\Delta t_{QZ,\text{ Alice}} = 5$ y, but $\Delta t_{QZ,\text{ Bob}} = 4$ y. Thus, Alice has aged $\Delta t_{OZ,\text{ Alice}} = 10$ y, but $\Delta t_{OZ,\text{ Bob}} = \Delta t_{OQ,\text{ Bob}} + \Delta t_{QZ,\text{ Bob}} = 8$ y.

**Interpretation:** When Bob returns at the reunion event Z, he has aged less than Alice who remained inertial the entire time from O to Z. This route-dependence of elapsed time on a clock between separation and reunion events is called the **clock effect**. It is the analogue of the triangle inequality in the Euclidean geometry – that there is a route-dependence of the distance between two points.

A related puzzle known as the **twin paradox** attempts to carry out the analysis from Bob's frame of reference, invoking the relativity principle, and deducing the *mistaken* conclusion that Alice, who moved away and returned, should be the one who has aged less. The flaw in the argument is that *Bob is not an inertial astronaut throughout his trip from O to Q to Z.* By the law of inertia, a ball kept on a frictionless table in Bob's ship will move at the turn-around event Q. A similar ball in Alice's ship will remain motionless throughout her trip from O to P to Z.

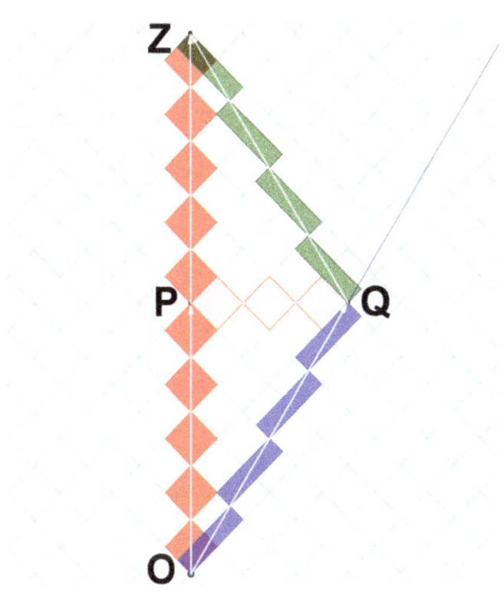


7.21
The clock effect/twin paradox, as consecutive time dilation situations. The elapsed time along the inertial worldline OZ is longer than along any other worldline from O to Z such as Bob's non-inertial worldline OQZ. (*Source: Copyright by Roberto B. Salgado*)

Furthermore, the spacetime diagram for any inertial astronaut (like Alice) will be a complete map of spacetime events, showing OZ as a straight worldline and OQZ as a bent worldline, with more elapsed time along OZ. It turns out that any attempted spacetime diagram drawn by Bob in which OQZ is straight will result in a distorted and incomplete map of spacetime.

## CONCLUSION

We have demonstrated a method to perform calculations in special relativity by visualising the tick marks on a spacetime diagram using light-clock diamonds generated by an astronaut's light clock. By stringing together diamonds, the elapsed times along worldlines and lengths between worldlines can be visualised and counted. This provides immediate physical interpretation. Calculations can be performed by counting and simpler algebra, compared to the traditional textbook formulas. We can develop physical intuition because we can more easily visualise and interpret the results of the calculation, rather than just relying on the numerical or symbolic output of formulas.

For more examples and for connections to the standard formulas, refer to my paper (Salgado, 2016). More examples are in the e-resources. An extension to the uniformly-accelerated astronaut is being developed and will appear elsewhere. Refer to my website: http://visualrelativity.com/ for further updates and related topics. An old set of interactive visualisations is available at https://www.geogebra.org/m/HYD7hB9v.

## FURTHER READING

E-resources: Example 3 – length contraction
E-resources: Graph paper sheets
Chapter 6: Event diagrams – supporting student reasoning in special relativity through thought experiments

## REFERENCES

Bondi, H. (1965). *Relativity and Common Sense: A New Approach to Einstein*. London: Heinemann.

Einstein, A. (1905). "On the electrodynamics of moving bodies" in *The Principle of Relativity*, by A. Einstein, H. A. Lorentz, H. Minkowski, & H. Weyl. New York: Dover.

Galileo, G. (1632). *Dialogue Concerning the Two Chief World Systems. Florence: Giovanni Battista Landini.*

Minkowski, H. (1908). "Space and time" in *The Principle of Relativity*, by H. A. Lorentz, A. Einstein, H. Minkowski, & H. Weyl. New York: Dover.

Moore, T. A. (2016). *Six Ideas that Shaped Physics: Unit R*. New York: McGraw-Hill.

Salgado, R. (2016). "Relativity on rotated graph paper", *Am. J. Phy.*, 84, 344.

Schild, A. (1959). "The clock paradox in relativity theory", *Am. Math. Monthly*, 66, 1.

Taylor, E. F. & Wheeler, J. A. (1966). *Spacetime Physics*. New York: W.H. Freeman.

# 8 Pushing the boundaries of Einsteinian physics education using virtual reality technology

*Jackie Bondell and Mark Myers*

**WHAT?**

This chapter focusses on guiding principles for designing science lessons that incorporate virtual reality into student-centred activities. A specific focus lies on the holistic incorporation of curriculum, pedagogy, content, and technology to highlight how virtual reality technology can be used to teach contemporary physics concepts.

**FOR WHOM?**

This chapter is intended for readers who are interested in emerging new technologies in physics education, and also, for those who wish to further their technological pedagogical content knowledge.

## INTRODUCTION

Effects of modern physics are difficult to imagine because they are not part of our everyday experiences. In Einstein's theory of general relativity, massive objects warp the fabric of spacetime. Colliding black holes and neutron stars send out waves rippling through this fabric, affecting any other masses in their path. As these effects are often hard to comprehend, the incorporation of virtual reality into the classroom can help by allowing for an interactive, immersive exploration of Einsteinian Physics ideas that may be otherwise difficult to conceptualise. Virtual reality offers the potential for students to see and interact with a world where these effects are apparent, directly experiencing Einsteinian effects first-hand. This chapter focusses on guiding principles for designing science lessons and virtual reality interfaces that effectively incorporate this technology into student-centred activities while ensuring pedagogical integrity and respect for the scientific process.

Educators are always looking for the best ways to engage students while remaining true to their content and pedagogy expertise. Using technology without appropriate context, driving curriculum, and thoughtful pedagogy will not only frustrate teachers and students but also waste precious time teachers can have for engaging students. Virtual reality is a technology with promise and pitfalls. Used haphazardly, it can be isolating and distracting for students, simply a new toy with no educational context. Used thoughtfully, however, it can elevate the learning of the students, allowing them to interact with their peers in doing hands-on collaborative learning and in engaging in new learning environments. Approaching students with technology that is engaging to them while utilising it as a tool that reimagines the content and curriculum highlights the promise of virtual reality in the classroom.

Virtual reality is an educational technology that can enhance students' understanding of content by improving their ability to visualise the content (Nooriafshar, Williams, & Maraseni, 2004). Virtual reality is especially promising in topics that are historically difficult for students to visualise and comprehend. In contemporary physics and astrophysics, such topics include scale, warped space, dark matter, and gravitational waves. In virtual reality, parameters such as scale and time can be changed interactively, and the results can be experienced directly. Engaging more senses in the learning process can enhance the understanding and retention of difficult topics (Shapiro & Stolz, 2019).

As a case study, this chapter highlights Mission Gravity, a programme developed by the Education and Outreach Team at the Australian Research Council Centre of Excellence for Gravitational Wave Discovery (OzGrav). To develop this programme, OzGrav focussed on designing science lessons that effectively incorporated virtual reality into student-centred activities while aligning with curriculum standards, ensuring pedagogical integrity, respect for the scientific process, and relevant incorporation of virtual reality technology. Mission Gravity was designed with a focus on the holistic incorporation of curriculum, pedagogy, content, and technology to highlight how virtual reality technology can be used to effectively teach contemporary physics concepts.

## TECHNOLOGICAL PEDAGOGICAL CONTENT KNOWLEDGE

Existing frameworks serve as lenses through which the incorporation of technology in the classroom can be envisioned. One overarching framework is the TPACK: Technological Pedagogical Content Knowledge (Mishra & Koehler, 2006). TPACK is at the overlap of content, pedagogical, and technological knowledge of both the teacher and the structure of the delivery (Figure 8.1). This structure provides guidance not only for the development of technologically relevant lessons but also for the delivery of teacher professional learning related to these overarching themes. TPACK doesn't identify one area as more important than another, but rather acknowledges that the knowledge of the educator and their ability to develop lessons highlighting all three can optimise the effectiveness of incorporating technology in the classroom and impact student learning of difficult topics.

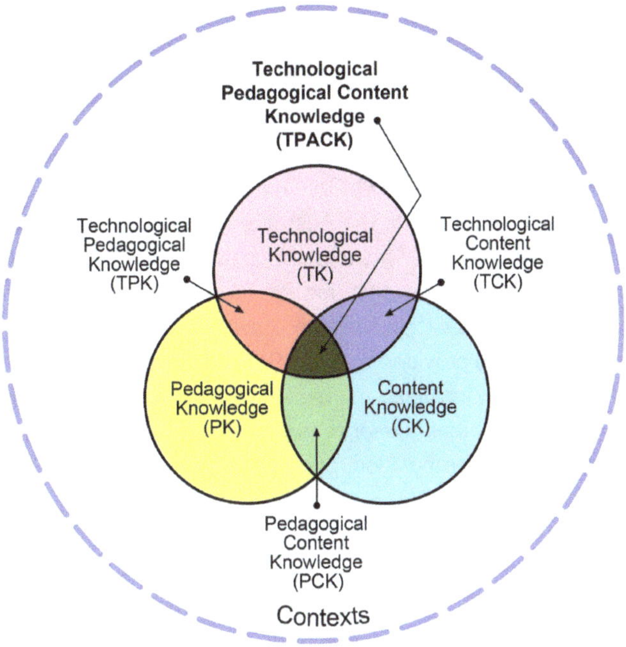


8.1
The TPACK framework shows the overlap of technological, pedagogical, and content knowledge as a model originally designed for effective incorporation of educational technology into the classroom. (*Source: Copyright by Matthew Koehler/ Public domain*)

### *Science content*

OzGrav research dictated the goal to incorporate the science of gravitational waves into Mission Gravity (Figure 8.2). Gravitational waves can align with topics like wave properties while the source of the waves opens the qualitative discussion of curved spacetime due to large masses of black holes and neutron stars.

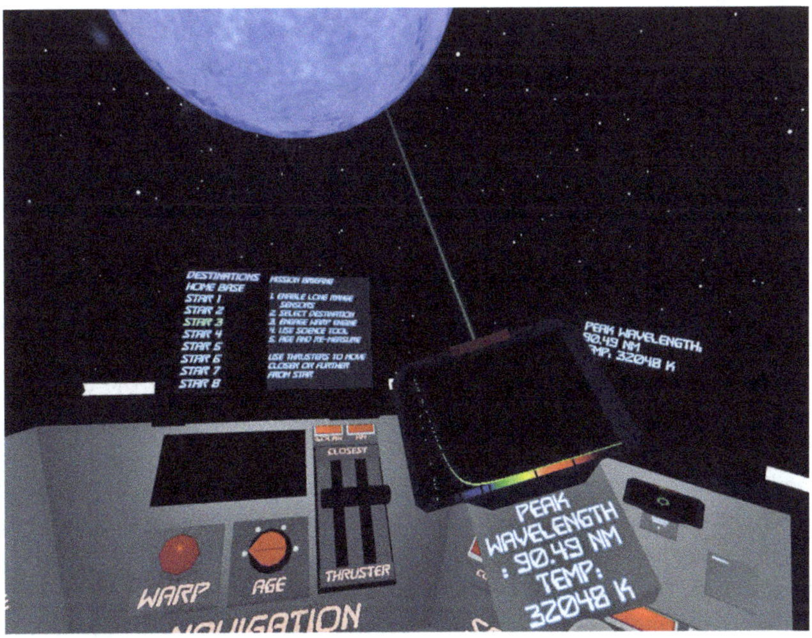


8.2
Screenshot from the Mission Gravity virtual reality environment, showing the temperature and peak wavelength tool being used to investigate the properties of the star. (*Source: Copyright by Mark Myers / Swinburne University of Technology and OzGrav*)

Multimessenger astronomy aligns with the electromagnetic spectrum, atomic spectra analysis, and cosmology topics such as the expansion of the universe. The content in Mission Gravity focusses on answering some questions relevant to understanding the science of gravitational waves such as what are black holes and neutron stars and their warping of the fabric of spacetime. The focus was reached by looking at the major areas of OzGrav research topics and looking for those areas that are best aligned with the Australian Curriculum standards. Specifically, it focusses on allowing students to make observations, both qualitative and quantitative, of stars as they evolve through the major touchpoints of their life cycles. The visualisations within Mission Gravity are driven by stellar evolution data generated by code provided by an OzGrav researcher specialising in stellar and galactic astronomy, including the evolution of stars. This research-validated code allows the virtual reality user to collect realistic data, including temperature, mass, radius, and composition, of multiple stars (Hurley, Pols, & Tout, 2000).

### *Curriculum content*

Mission Gravity was designed focussing not only on content but also on scientific process goals such as scientific inquiry and planning, evaluating, and communicating science discovery, and understanding. For example, curriculum content descriptions focus not only on science understanding but also on science as a human endeavour and science inquiry skills. Mission Gravity's design takes advantage of this guidance and was structured to allow students to engage in the scientific process.

The Australian curriculum aligns well with the Next Generation Science Standards (NGSS) in the United States, and its focus is on 21st century skills such as critical thinking, collaboration, and communication. The NGSS focus on core scientific concepts, science processes and practices, and crosscutting concepts align with the Australian scientific understanding, science inquiry skills, and science as a human endeavour, respectively. Furthermore, there are similar alignments with the European Commission's *Science Education for Responsible Citizenship* which says, "Science education should focus on competences with an emphasis on learning through science and shifting from STEM to STEAM by linking science with other subjects and disciplines".

### *Pedagogy*

The second major theme in TPACK is pedagogical knowledge. The design of Mission Gravity necessitated the contemporary physics concepts, and their incorporation into the virtual reality environment was done with appropriate pedagogical guidance. To ensure this, multiple models were used to develop lesson plans for classroom use and to make the choices about how students engage with the virtual reality environment.

Mission Gravity was developed using the SAMR Model (Puentedura, 2014) for evaluating the role of technology in the classroom (Figure 8.3). This model is

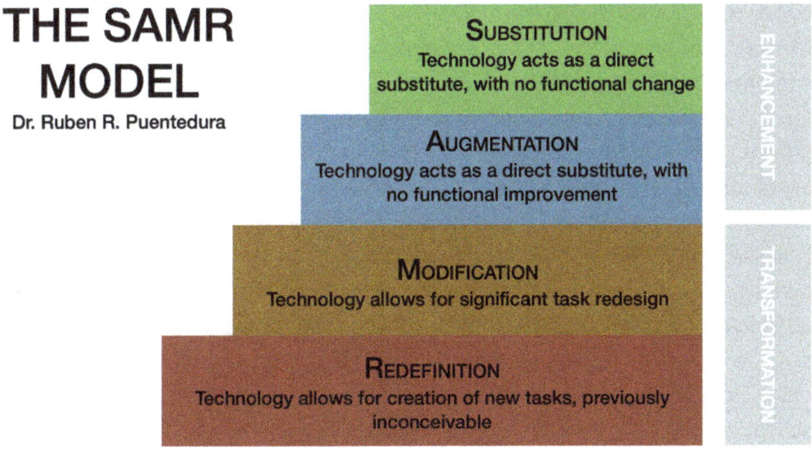


8.3 The SAMR model provides a framework for evaluating the impact and efficacy of technology in the classroom, with the goal to incorporate the technology such that the aim is for the top two tiers. (*Source: Copyright by Lefflerd / CC BY-SA 4.0 (https://creativecommons.org/licenses/by-sa/4.0)*)

a lens through which to evaluate technology and its meaningful use in the classroom, focussing on the relevant incorporation of technology for learning. It looks at how the technology serves to improve the content and accompanies methodology of topics being taught. The questions asked are: does the technology

- Simply serve as a substitution (S) for existing non-technology with no additional benefit
- Provide an augmentation (A) to the lesson and delivery, adding practical benefits for the teacher
- Modify (M) the curriculum allowing for a significant redesign in how the students engage with the material
- Allow for a complete reimagining (R) of the way students can learn?

Using this lens, Mission Gravity was developed with the goal of using virtual reality to modify or reimagine the delivery of the content. If virtual reality only allowed students to see a comparison of flat objects as they would appear on paper, or simply allowed students to passively see a 360-degree video of different types of stars in space, then that would substitute or augment the visualisation of the content. In Mission Gravity, the virtual reality environment allows students to interact with stars and evolve them, use virtual tools to make measurements, manoeuvre in space as stars change in size, and investigate stellar remnants. Students make decisions about how to interact with space and manipulate stars and are provided with virtual resources to make physics measurements and qualitative observations of the stars. The immersive and interactive properties of virtual reality allow students to embody the world of stars and allows teachers to deliver lessons interacting with objects previously impossible in the classroom setting.

The pedagogical structure for the Mission Gravity lesson plans was formatted using the 5 E Lesson Plan Model: Engage, Explore, Explain, Elaborate, and Evaluate. The benefit of virtual reality offers unprecedented exploration by students, while the lessons faithfully incorporate the other Es. The goal of Mission Gravity is to foster discussion, collaboration, and understanding of new content in which virtual reality is one tool in the lesson's arsenal. Student groups start the lesson with a teacher-led discussion of why stars are studied, which provides intention for the activity and the opportunity to show their understanding of stars on a tabletop poster. This engages the students and initially sets the tone that this is a collaborative activity while providing the teacher with a prior assessment of knowledge. With a focus on 21st century skills, the engagement tasks of data collection and explanation tasks involved in the analysis must be done in cooperative groups with role rotation through the lesson. Virtual reality, science, and data guides are provided at appropriate times to allow for elaboration on additional topics and support the students' progression. In addition, school teachers are provided with all of the stellar evolution data for continued use and analysis in subsequent classes.

The 5 E lessons align well with a third structure for the pedagogical design of this program as it was designed to be delivered using a modelling inspired framework. The Modeling Instruction Framework (Jackson, Dukerich, & Hestenes 2008) is a student-centred instructional design focussing on letting students be the scientists, with the teacher prepared to guide, clarify, and address student misconceptions.

With guidance from the teacher to provide the right equipment and engagement, students make their own observations to build their understanding of stellar evolution focussing on the questions regarding the source of different stellar remnants. The groups make conclusion statements regarding their understanding. All of these conclusions must be directly supported by their observations in data. The follow-on class discussion leads to a consensus about major themes and laws related to stellar evolution and remnants.

### *Technology*

The third pillar of the TPACK framework is the Technological Knowledge. Teachers are experts in both content and in pedagogy, though fewer consider themselves experts in understanding the latest technology and its promise and application in the classroom. The expertise from the research visualisation sector was able to fill some of this gap with a goal of accessible, cutting edge, and seamlessly incorporated virtual reality technology for the classroom. In the development of this programme, the focus was to let the curriculum be the driving force in the development of the technology. The team behind Mission Gravity had the benefit of having a curriculum developer and teacher with expertise in education technology working side by side with a virtual reality developer who also has a background in physics and astrophysics. By developing in-house, the team is able to iterate the programmes.

Keeping in mind the SAMR principles above and focussing on letting the technology be the third pillar, no more important than pedagogy and content, the team made thoughtful technology choices. Virtual reality has both hardware and software considerations. While the software is the component with which there is most educational engagement, there need to be choices made regarding hardware. This is true of all hardware incorporated into the classroom: it should be unobtrusive for students to use with an efficient learning curve. It shouldn't hinder the learning with unnecessary barriers for students. It should also streamline engagement with the virtual reality software.

Untethered and standalone virtual reality creates a safer space and ease of sharing the equipment while working collaboratively. The headsets are easily portable and self-contained. Their spatial tracking detects the movement of the wearer and allows for a more realistic immersion with the virtual reality environment. Additionally, a simple remote makes implementation easy for a large group of students so that a short tutorial works for the entire group.

An important factor with virtual reality implementation is to ensure no student spends too much time in the headset so they can remain engaged and on-task with their peers. In addition to building rotating roles into the lesson, the team designed an interface that allows for first-hand observation of fundamental processes and scientific measurement and data collection in the environment. This also ensured students make conscious decisions to manoeuvre in the environment. Software design choices ensured the experience is not gamified. Rather students work in a travelling star lab, using physics tools to investigate, making mistakes and repeating measurements, and otherwise interacting with the environment.

An issue with using virtual reality in the classroom is that often teachers and group members cannot see what students see in their virtual reality environment. The team realised this was a major issue. To resolve this, an in-house Command Centre programme was developed that captures what each student is seeing and screencasts it to the teacher's laptop, which is then projected on a large screen for all to see. This allows students working in groups (2–3 per headset) to continue to collaborate and communicate, and allows more productive engagement of the teacher (Figure 8.4).

### *Reflection*

While Mission Gravity focusses on "how" the universe gets black holes as progenitors of gravitational waves and other Einsteinian effects, it has provided the feedback and structure for the continued development of more educational Einsteinian physics content.

At the time of writing, Mission Gravity has been delivered by our team in over 90 schools (Figure 8.5). Anecdotal feedback from students and teachers has been overwhelmingly positive, with almost all students providing feedback that the virtual reality incorporation helped them to better visualise the changing features of stars as they evolve. One student wrote, "The incursion involved the use of virtual-reality equipment to make interstellar observations and follow the life-cycle of stars (...) Aside from using the virtual reality goggles (...) it was also interesting to see the amazing impact technological advances have had on

8.4
Students collaborate in groups using virtual reality with the Mission Gravity Command Centre in the background, showing the screencasts of what students are viewing in their respective virtual reality headsets. (*Source: Copyright by Jacqueline Musiov / Swinburne University of Technology*)

8.5
Students seamlessly utilise tabletop posters, data tables, and virtual reality in modelling and communicating their understanding of the life cycle of a star. (*Source: Copyright by Jackie Bondell / Swinburne University of Technology and OzGrav*)

our ability to understand spacetime and the implications of Einstein's theories". The team plans to collaborate with education researchers to develop assessment tools and collect evidence to answer: how does a virtual reality incursion programme support the development of students' conceptual understanding in science classrooms?

Furthermore, in order to bring this content to more students, professional development opportunities for teachers will be provided not only to develop content knowledge in contemporary astrophysics but also to do so with thoughtful pedagogy and utilising effective technology. All of the Mission Gravity software and supplemental resources are provided to educators so they can utilise them in their own classrooms. OzGrav is also duplicating the virtual reality resources on multiple platforms (such as computer browser-based) to allow for increased uptake by educators.

## CONCLUSION

The Director of OzGrav, Professor Matthew Bailes states, "The mission of [OzGrav] is to use the historic first detections of gravitational waves to understand the extreme physics of black holes and warped spacetime and to inspire the next generation of Australian scientists and engineers through this new window on the universe". Using visualisations in education is one way to do this. Virtual reality enables students to afford the opportunity to travel to stars and collect data about how the physical properties of stars change over their lifetimes. Students must deliberately move toward and away from stars as they change size during their lifetimes, providing a tangible sense of stellar scales. They use physical tools and need to engage with the environment in order to make the appropriate observations. This utilises the technology to develop a better content understanding. The virtual reality environment which was developed in-house at OzGrav to align with the curriculum goals, using published and continually refined stellar evolution code highlights the benefit of working with a scientific collaboration and provides access to content to incorporate into the technology. Standalone virtual reality headsets that provide an untethered experience and incorporate spatial tracking to increase the experience of immersion and interaction enhance and promote embodied cognition. The laptop-based command centre allows the projection of the point of view of each headset user, and the additional stellar data enhance the collaboration of classmates in virtual reality with those outside of virtual reality. Thus, the technology supports the pedagogy, focussing on 21st century skills such as collaboration and communication for the purpose of completing a goal.

## FURTHER READING

E-resources: lesson plans and data log templates
Web-Based Mission Gravity: http://outreach.ozgrav.org/MissionGravity/
Chapter 19: Designing learning resources and investigating student learning in general relativity and quantum physics in Norway

## REFERENCES

Nooriafshar, M., Williams, R., & Maraseni, T.N. (2004). "The use of virtual reality in education." 7th American Society of Business and Behavioral Sciences International Conference (ASBBS 2004), 6-8 Aug 2004, Cairns, Australia.

Shapiro, L. & Stolz, S. (2019). Embodied cognition and its significance for education. *Theory and Research in Education*, *17*, 19–39.

Mishra, P., & Koehler, M. J. (2006). Technological pedagogical content knowledge: A framework for integrating technology in teachers' knowledge. *Teachers College Record*, *108*(6), 1017–1054.

Hurley, J., Pols, O., & Tout, C. (2000). Comprehensive analytic formulae for stellar evolution as a function of mass and metallicity. *MNRAS*, *315*, 543–569.

Puentedura, R. (2014). Learning, technology, and the SAMR model: Goals, processes, and practice. Available at: http://www.hippasus.com/rrpweblog/archives/2014/06/29/LearningTechnologySAMRModel.pdf (Accessed: 21 July 2020).

Jackson, J., Dukerich, L., & Hestenes, D. (2008). Modeling instruction: An effective model for science education. *Science Educator*, *17*(1), 10–17.

# General relativity

# 9 Standing on the shoulders of giants

## *How historical perspectives on gravity can inform modern physics education*

*Magdalena Kersting*

**WHAT?**

This chapter traces the historical development of scientific models about gravity from antiquity to our modern times. Magdalena Kersting suggests how historical perspectives can help students overcome misconceptions and promote modern intuitions about gravity.

**FOR WHOM?**

This chapter is intended for readers who wish to learn more about historical perspectives of gravity, and also, for those who look for an overview of student conceptions about gravity at all school levels.

## INTRODUCTION

Why do objects always fall down to the Earth and not up to the sky? What keeps us stuck to the ground? And how do planets seem to float in space without falling? The obvious and seemingly simple answer to all of these questions is gravity. However, gravity has puzzled humans for millenniums. Historically, natural philosophers and scientists have held a broad range of different conceptions about gravity. Many of these conceptions resonate with our intuitions about gravitational phenomena in everyday life. Since intuitions can serve as a tool to deepen understanding of the physical world, historical knowledge about different models of gravity entails a twofold potential in physics education. First, ancient conceptions of gravity can offer valuable insights into students' (mis)conceptions about gravity. Second, teachers can draw on historical perspectives to create an awareness of the nature of science that allows challenging students' intuitions about gravity.

As physics teachers and educators, it is our responsibility to re-evaluate instructional practices continually. We should aim to teach students at all levels our current best understanding of both the scientific concepts and their historical developments. In this chapter, we examine different scientific models and theories of gravity, and follow the historical development of these ideas. Neither Galileo nor Newton discovered gravity; scholars from the antiquity and the Indian and Islamic world were the first giants of gravity that made significant contributions in our conceptual quest to understand gravitational phenomena. We supplement the presentation of historical conceptions of gravity with a review of conceptual challenges that students encounter from the primary to the secondary school level. Bringing both historical and educational perspectives together, we draw instructional implications for the teaching of gravity in schools.

## GRAVITY FROM A HISTORICAL PERSPECTIVE

In this section, we trace different conceptions of gravity from the antiquity throughout the middle ages to our modern times. Initially, the name gravity comes from the Latin word "gravitas" meaning "weight". The origin of this name traces back to Plato and Aristotle who conceived of gravity as depended on the weight of the falling object. Today, the Oxford English Dictionary describes gravity as the "force that attracts a body towards the centre of the Earth, or towards any other physical body having mass". This definition reflects the current common sense understanding of gravity as an attractive force; it does not mention Einstein's curved spacetime model that has long superseded the force model.

### *Plato and Aristotle*

> "The tendency of each towards its kindred element makes the body which is moved heavy, and the place towards which the motion tends below, but things which have an opposite tendency we call by an opposite name".
>
> Plato, 360 BC – as cited in (McCloughlin, 2019)

The ancient Greek philosophers Plato and Aristotle believed that all things in the universe have a "proper" or "natural" place, and objects always seek to reach their natural places. According to this world view, the terrestrial part of the world is at the centre of the cosmos and made up of four elements (earth, air, fire, and water); the celestial part consists of a fifth element, ether. Objects that are mostly made out of earth move in the direction that is natural for them, which is to say, towards the ground. If an object was mostly made out of air, it would float away. Consequently, objects rise or fall until they reach their natural place in the universal order (Figure 9.1). Once something gets to its natural place, it remains there unless it is disturbed by something else. The notion of natural movement towards natural places and the geocentric worldview influenced much of the thought in the middle ages (McCloughlin, 2019).

Notably, Aristotle's account of gravity presents an explanation of gravitational phenomena: objects seek their natural places via natural motion. Objects travel "down" because "down" is the direction that is naturally associated with things made of earth and water. Similarly, objects made of air and fire tend to rise. The speed of such motion depends on the density of the element and the weight of the object. Thus, Aristotle thought that more massive bodies fall faster, in proportion to their weight.

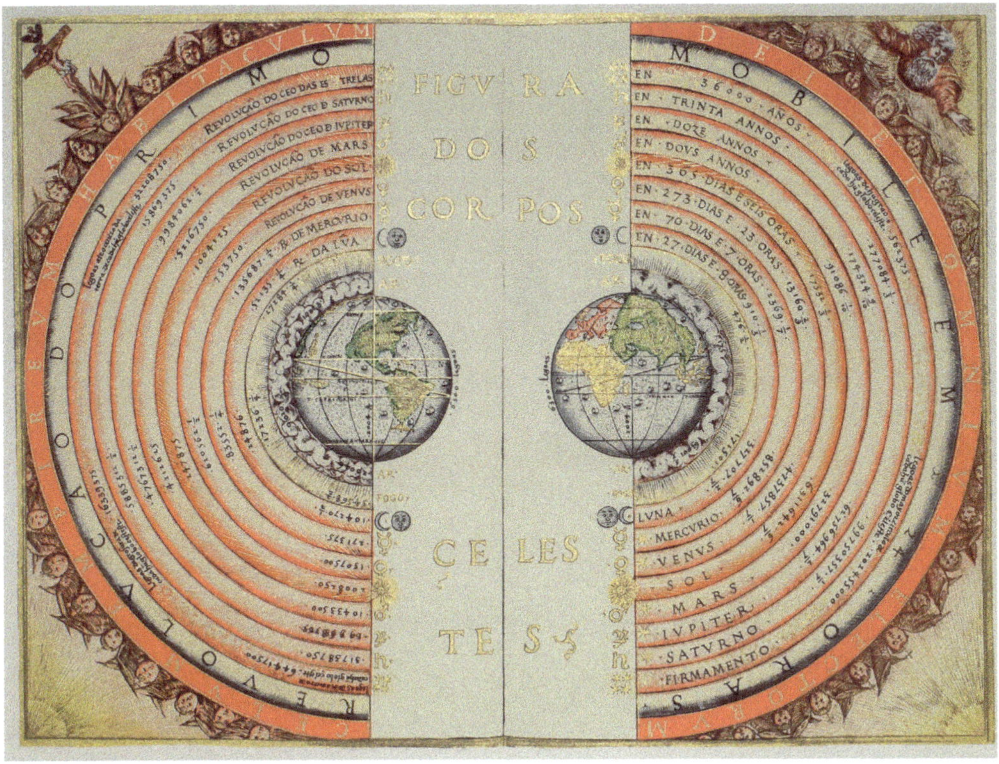


9.1
"Figure of the heavenly bodies", an illuminated illustration of the universe from "Cosmographia", by Bartolomeu Velhomade, 1568. A geocentric model was the predominant description of the universe in the time of Aristotle. The ancient Greek philosopher placed the terrestrial spheres of water and earth at the centre of the universe and described gravity as the natural movement of objects that travel "down" because "down" is the direction that is associated with things made of earth. (*Source: Copyright by Bartolomeu Velho / Public domain*)

### *Philoponus, Al-Biruni, and Galileo*

> "But this [view of Aristotle] is completely erroneous, and our view may be completely corroborated by actual observation more effectively than by any sort of verbal argument. For if you let fall from the same height two weights, one many times heavier than the other you will see that the ratio of the times required for the motion does not depend [solely] on the weights, but that the difference in time is very small..."
>
> Philoponos, ca. 510 – as cited in (McCloughlin, 2019)

Although Galileo Galilei is commonly credited with refuting Aristotle's theory that heavier bodies fall faster than lighter ones, criticism towards this view emerged already in the antiquity. A millennium prior to Galileo, a Byzantine natural philosopher named John Philoponos was one of the earliest thinkers to reject the Aristotelian model of gravity in the 6th century. Philoponos claimed that the difference in the time taken by two bodies with vastly different weights to reach the ground is, in fact, minimal. What makes this claim so remarkable (other than describing the correct behaviour of falling bodies, of course) is its grounding in experimental observation. Philoponos promoted scientific thought based on experiments in the real world. This approach is in stark contrast to the abstract proofs and theoretical reasoning that Plato and Aristotle had established as the core method for gaining knowledge about the world.

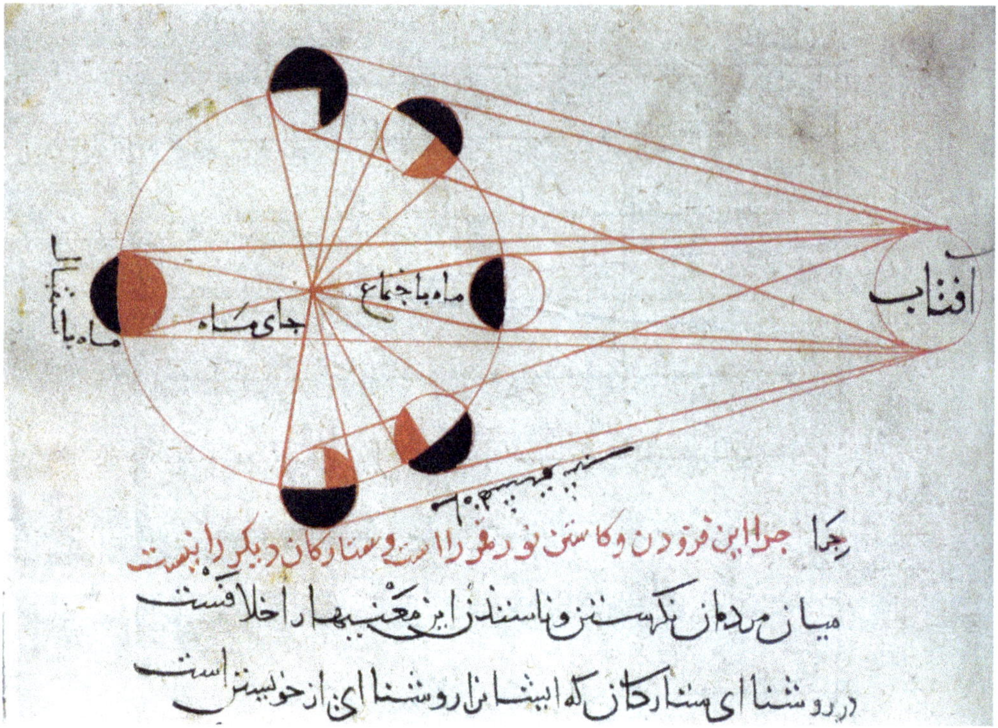


9.2
Abu Rayhan al-Biruni was a Persian scholar who criticised the Aristotelian view that only Earth has mass and gravity. Al-Biruni argued that all heavenly bodies, including the Sun and the moon, should possess these properties. This illustration from al-Biruni's astronomical work explains the different phases of the moon as it orbits the Earth. (*Source: Copyright by Al-Biruni / Public domain*)

With the rise of the Islamic Empire in the 8th century, the world saw tremendous advances in mathematics, astronomy, and the natural sciences. Islamic thinkers held lively discussions about fundamental questions of science and philosophy, thereby often drawing on and refuting Aristotle's work (Starr, 2015). This period of scientific flourishing led to a revival and thorough examination of Greek knowledge and brought forth modern conceptions of gravity. Abu Rayhan al-Biruni was a Persian scholar who criticised Aristotle for believing that celestial bodies are fundamentally different from terrestrial bodies. Al-Biruni put forward the bold idea that the Sun, planets, and stars have mass and are subject to gravitational interactions just like objects on our Earth (Figure 9.2). Similar to Philoponos, al-Biruni based many of his scientific arguments on observable evidence instead of relying on pure logic and abstract thought (Starr, 2015).

Despite the progress made by these early pioneers of gravity, Plato and Aristotle's work continued to influence the way most scholars thought about gravity in the middle ages. The Greek influence reached the very core of how science was conducted: instead of running experiments in the real world, most scholars would study natural philosophy.

Galileo Galilei, however, changed that radically in the early 17th century. Galileo laid the foundation for our modern scientific method, and he was one of the first real scientists as we know them today. Notably, Galileo understood that there was a close link between experimental observations and mathematics, which was a significant contribution to separating science from philosophy:

> "Philosophy is written in that great book which ever lies before our eyes – I mean the Universe – but we cannot understand it if we do not first learn the language and grasp the symbols, in which it is written. This book is written in the mathematical language".
>
> Galilei, 1623 – as cited in (Burtt, 2003)

In contrast to the ancient Greek thinkers who tried to find underlying causes of natural phenomena, Galileo often focussed on purely descriptive accounts of the world. Allegedly, Galileo went to the top of the Leaning Tower of Pisa and dropped two balls of the same size but different weights to find that they landed at the same time. There is no evidence that Galileo indeed conducted those experiments. Still, Galileo is credited with having refuted Aristotle's theory of falling bodies – despite similar scepticism that Byzantine and Persian thinkers had expressed centuries earlier.

### *Newton and Brahmagupta*

> "...that one body may act upon another, at a distance through vacuum, without the mediation of anything else, by and through their action and force may be conveyed from one to another, is to me so great an absurdity, that I believe no man who has in philosophical matters a competent faculty of thinking, can ever fall into it".
>
> Newton, 1693 – as cited in (Ohanian, 1976)

Born just one year after Galileo died, Isaac Newton is probably most famous for having discovered the universal law of gravitation. Newton invented calculus as a new mathematical language to describe gravity as an attractive force between the Earth and nearby objects (Figure 9.3). This force is directly proportional to the mass of the object and falls off at an inverse square rate. What is extraordinary about this force model is that it applies both to an apple falling from a tree and to the movement of the planets in our solar system. This unification of separate physical phenomena is the great achievement of Newton. By describing gravity, Newton truly tried to understand the universe.

While Newton could describe gravity as an attractive force, he had no explanation about the origin of this force. In contrast to the Aristotelian principle of final causation that described objects as always returning to their proper places, the sources of Newton's force of gravity remained mysterious. Newton was aware that his model did not explain but only described gravity through the inverse law relation (Box 9.1).

It was Newton's great scientific feat to unify celestial and terrestrial phenomena by proposing a force model that could describe the mechanical movement of bodies and their gravitational attraction. Nevertheless, Newton was not the first to use the concept of an attractive force to describe gravity. Already a millennium before Newton, Indian astronomer Brahmagupta described gravitational phenomena through attractive forces:

9.3
British painter William Blake depicted Isaac Newton as a divine geometer. Newton developed mathematical tools to describe the movement caused by gravity: planets orbit around the Sun and the moon circles around the Earth in elliptical shapes. By proposing the force model of gravity, Newton tried to understand the universe. (*Source: Copyright by William Blake / Public domain*)

> "The Earth on all its sides is the same; all people on the Earth stand upright, and all heavy things fall down to the Earth by a law of nature, for it is the nature of the Earth to attract and to keep things, as it is the nature of water to flow … If a thing wants to go deeper down than the Earth, let it try. The Earth is the only low thing, and seeds always return to it, in whatever direction you may throw them away, and never rise upwards from the Earth."
>
> Brahmagupta, 7th century – cited in (Bīrūnī, 2006)

**BOX 9.1: INVERSE-SQUARE LAW OF GRAVITY**

Isaac Newton proposed the inverse-square law of gravity to account for the movements of planets in the solar system:

$$F = G\frac{m_1 m_2}{r^2}$$

Here, $F$ is the force of gravity, $m_1$ and $m_2$ are the masses of the two interacting objects, $r$ is the distance between the (centres of the masses of the) objects, and $G$ is the gravitational constant. Newton expressed his mathematical law of gravity in the following way: "I deduced that the forces which keep the planets in their orbs must [be] reciprocally as the squares of their distances from the centres about which they revolve: and thereby compared the force required to keep the Moon in her Orb with the force of gravity at the surface of the Earth; and found them answer pretty nearly" (Chandrasekhar, 2003).

### *Einstein*

> "I was sitting in a chair at the patent office in Bern when all of a sudden, a thought occurred to me: If a person falls freely, he will not feel his own weight. I was startled. This simple thought made a deep impression on me. It impelled me toward a theory of gravitation".
>
> Einstein, 1922 – as cited in (Goldsmith & Bartusiak, 2006)

Undoubtedly, Albert Einstein revolutionised our understanding of the universe and changed the way scientists look at the world. General relativity is Albert Einstein's theory of space, time, and matter and our current best understanding of gravity. According to general relativity, gravity is a manifestation of the curvature of spacetime. Matter tells spacetime how to curve, while curved spacetime tells matter how to move. Thus, gravity is not interpreted as an attractive force, but as the curvature of spacetime.

Einstein's interpretation of gravity is astonishingly different from the Newtonian conception because time and space take active roles in shaping gravitational phenomena. Besides, Einstein succeeded in presenting both a new mathematical description as well as an explanation of the gravitational mechanism – something, that had been missing in Newton's theory of gravity. Quite astonishingly, there is some resemblance between Einstein and Aristotle's explanation of gravity because both approaches rely on the notion of natural movement. In Einstein's theory, objects follow the straightest possible paths through curved spacetime. We can interpret this movement as the "natural" motion of objects. The gravitational pull that we feel when we stand on the ground corresponds to the normal force that prevents us from following "natural" paths through spacetime.

## GRAVITY FROM AN EDUCATIONAL PERSPECTIVE

After having introduced historical models of gravity, we now turn to educational research to unpack students' conceptual challenges related to gravity. Since gravity is an everyday phenomenon that is a part of many physics curricula around the world, we can draw on a vast body of literature to give a comprehensive overview of students' conceptions about gravity. Although many intuitive conceptions that students bring to the classroom are highly context-dependent, it seems that some intuitions have their counterparts in historical models of gravity. For example, properties such as the size or weight of objects, their location relative to the Earth, the presence or absence of air, and the direction of motion of the objects have all shown to influence student conceptions (Palmer, 2001). These findings suggest that students have multiple intuitions about gravity. Let us unpack these intuitions according to different age groups.

### *Primary school students*

Some of the first experiments humans do address gravitational phenomena. There is evidence that infants and toddlers build experiential understanding of a world in which objects always fall downward (Kim & Spelke, 1992). Therefore, it does not come as a surprise that primary school students readily accept the explanation that gravity pulls things downward. For example, being asked "who do things fall?", most 9–13-year-old students answer because of "gravity" or "the Earth's attractive force" (Bar, Zinn, Goldmuntz, & Sneider, 1994). The force of gravity, as an explanation, provides a sense of comfort and obviousness in reaction to the experience of the world (Kapon & diSessa, 2012).

Even though many students take the gravitational pull for granted, the self-evidence of gravitational attraction often comes as misconceptions in disguise. Students seem to struggle when being asked to reason about gravity outside of their everyday experiences. An exploratory case study with Australian primary school students of age 10–11 years showed that students found the concept of gravity in outer space a lot more complicated than gravity on the Earth (Pitts, Venville, Blair, & Zadnik, 2014). Even after instruction, the students displayed misconceptions about gravity, for example, that there is no gravity on the moon or in space. The study concluded that gravity was an elusive concept for many students at the primary level. Other studies have found that students also struggle to understand gravity in everyday situations. For example, students often seem to assume that a ball that is thrown vertically up does not feel the pull of gravity at its highest point before it falls back to the Earth (diSessa, 2006).

### *Middle school students*

While an understanding of gravity as the force that makes things fall towards the centre of the Earth can reasonably be expected of students at the primary and middle school level, many alternative conceptions about gravity seem to be highly resistant, and change little during traditional instruction (Palmer, 2001). In interviews with Australian primary (11–12 years) and middle school students (15–16 years), the majority of students displayed both scientifically acceptable and alternative conceptions when describing whether or not gravity was acting on a series of moving or non-moving objects in everyday situations. Students were more likely to indicate that gravity acted on downward-moving objects than upward-moving or stationary objects. The actual type of object did not seem to be a critical factor for most of the students.

A study with French middle school students of age 15 years looked at the ways students' conceptions of gravity evolved during instructional interventions either based on a Newtonian or Einsteinian perspective (Baldy, 2007). Initially, many students believed that gravity was not universal; students treated the Earth as a special case. From that perspective, the explanation that bodies fall due to attraction is only valid on the Earth – on the moon or in outer space, bodies float because they are in a vacuum (Baldy, 2007). Baldy found that the strength of the Einsteinian perspective was that it changed students' conception of attraction as a property of space rather than a property of bodies.

A study with Australian middle school students (15 years old) found that students readily accepted an Einsteinian model of gravity (Choudhary et al., 2019). Nevertheless, the students struggled to draw on the curved spacetime model to explain the actual mechanism of how gravity makes things fall. Moreover, some students found the idea that gravity is a geometric phenomenon somewhat confusing. After all, at the middle school level, most students think of geometry as basic two-dimensional shapes. These findings suggest that teachers and instructors have to pay particular attention to help students understand the mechanism of gravity in an Einsteinian framework instead of just introducing the spacetime model as a replacement of the force model.

### *Secondary school students*

There are well-documented misconceptions about gravity among students at the upper secondary or introductory university level as well (Williamson, Willoughby, & Prather, 2013). Based on pre-post questionnaires, the Newtonian Gravity Concept Inventory gives insight into such common conceptions. One particularly common misconception links gravity to the atmosphere of the Earth. Many students believe that one must simply leave the Earth (and its atmosphere) to feel zero gravitational force from the Earth. These students may think of floating astronauts, and associate weightlessness with the absence of an atmosphere. Besides, many students confound the force of gravity with effects related to air pressure, rotation, or magnetism (Williamson, Willoughby, & Prather, 2013).

By and large, the force model of gravity seems to be prominent in upper secondary school students' understanding of gravity. Similar to findings among middle school students, secondary school students continue to draw on the force model to explain gravity even when operating within Einstein's curved spacetime model (Kersting & Steier, 2018). Besides, students seem prone to confuse cause and effect. Instead of attributing gravitational phenomena to curved spacetime, some students assume that (the force of) gravity warps spacetime. It is, therefore, essential to point out how massive objects distort spacetime and how physicists, in turn, interpret manifestations of this distortion as gravity. Spacetime is an abstract concept, and it can be challenging for students to fit together the individual pieces of their knowledge of gravity in an Einsteinian framework. Since science educators have only started to explore how students understand gravity in line with an Einsteinian perspective, there remains much to explore.

In summary, students in all age groups display naïve ideas and intuitions about gravity that often remain persistent throughout the instruction. Indeed, educational research suggests that there exist unique reasoning difficulties related to the concept of gravity that go beyond conceptual problems related to forces more generally (Williamson, Prather, & Willoughby, 2016). These persistent difficulties might be related to the fact that many students experience gravity in their everyday lives in an Aristotelian way. Let us see how we can use historical conceptions of gravity to improve instructional practices in the classroom.

## INSTRUCTIONAL IMPLICATIONS

The starting point of this chapter was our assumption that an understanding of historical conceptions of gravity allows us to gain deeper insights into students' conceptions. In this section, we look at the ways students' intuitions about gravity are mirrored in historical models. We suggest how both perspectives can be brought together to improve instruction of gravity at the primary, middle, and secondary school level.

### *Thinking with Aristotle, Philoponus, and Galileo to challenge students' intuitions*

It seems that students' conceptual understanding of gravity is largely based on their intuitions of our world as a place where things fall downwards. Since everyday experience suggests that heavier bodies fall faster than light ones; this framework, albeit flawed, works for students in their daily lives (McCloughlin, 2019). The naïve intuition that the rate of acceleration depends on the weight of the falling body aligns with early historical ideas held by Plato and Aristotle. This view can be a productive starting point from which teachers can guide students to a model of gravity that aligns with our modern understanding. For example, teachers can surprise students when the outcome of a free-fall experiment contradicts students' experiences and expectations. The apparent contradiction between our intuitions and reality might be one of the reasons that falling-body experiments continue to be popular and engaging experiences for students (Crease, 2003; Venville et al., 2012). Besides, instructors have the opportunity to illustrate the nature of the scientific method by emphasising how Philoponus and Galileo keenly observed the world around them instead of just reasoning theoretically about natural phenomena. Above all, science always builds on observations of the world.

### *Thinking with Newton and Al-Biruni to challenge students' ideas of universality*

A persistent misconception among students at all levels is that gravity acts only on the Earth, and that there is no gravity in outer space. Drawing a difference between the Earth and the heavens aligns with ancient and medieval conceptions of gravity. There is much potential to promote conceptual understanding of gravity and create awareness of the role of models in science by challenging students' ideas about the universality of gravity. Al-Biruni was one of the first to posit that the Sun and other celestial objects have mass and, therefore, are subject to gravity just like objects on the Earth. Newton's force model allows unifying the terrestrial and celestial realms based on a simple formula that relates the force of gravity to the masses of and distances between objects (Box 9.1). Hence, when introducing Newton's force model of gravity, it can be instructive to emphasise the unifying power of this theory and its mathematical formalism.

### *Thinking with Einstein to challenge students' ideas of scientific explanations*

In physics classes, most students are familiar with the force concept. For example, students can draw on the force of gravity to describe the movement of objects in free fall. Yet, probing deeper and letting students question the nature of this force of gravity can provide remarkably productive discussions (Kersting,

Henriksen, Bøe, & Angell, 2018). Newton's force of gravity acts at a distance without a mediator. If students understand that Newton was able to describe gravitational phenomena, but could not explain the mechanism of gravity, they may appreciate the explanatory power of Einstein's model. There is a difference between describing and explaining a scientific phenomenon. This insight is a productive starting point to introduce Einstein's model of gravity that is superior to Newton's model because of its greater explanatory scope.

## CONCLUSION

> "If I have seen further it is by standing on the shoulders of Giants".
>
> Newton, 1675 – as cited in (Turnbull, 1959)

In this chapter, we have traced the history of gravitational models to draw parallels between historical and student conceptions about gravity. In contrast to the conventional view that Galileo and Newton were the fathers of gravity, Byzantine, Indian and Persian thinkers had understood and expressed crucial features of gravitational phenomena many centuries earlier. Nonetheless, Galileo and Newton advanced the scientific enterprise significantly by promoting a scientific methodology based on empirical observations and experiments. Knowledge of the historical struggle to find an accurate model of gravity and the methodological progress made along the way can stimulate instructional practices in physics education. For instance, historical conceptions of gravity can offer valuable insights into students' intuitions about gravity. It seems that many students conceptualise gravity in Aristotelian terms. Teachers can tailor their instructional approaches to students' everyday understanding of gravitational phenomena by challenging students' intuitions and their knowledge of the nature of science. In modern physics education, the development of physical intuitions about scientific phenomena is a crucial goal. The history of gravity serves as an excellent example to illustrate the development of scientific models and stimulate reflections about our intuitions of the natural world.

## FURTHER READING

Chapter 1: Intuitions in Einsteinian physics
Chapter 2: Time for changing paradigms in science and in education
Chapter 14: Using the language of gravity to teach about space, time, and matter in general relativity

## REFERENCES

Baldy, E. (2007). A new educational perspective for teaching gravity. *International Journal of Science Education*, *29*(14), 1767–1788.

Bar, V., Zinn, B., Goldmuntz, R., & Sneider, C. (1994). Children's concepts about weight and free fall. *Science Education*, *78*, 149–169.

Bīrūnī, M. ibn A. (2006). Alberuni's India London : Kegan Paul, Trench, Trübner & Co., 1910. Electronic reproduction. Vol. 1 and 2.

Burtt, E. A. (2003). *The metaphysical foundations of modern science.* New York: Courier Corporation.

Chandrasekhar, S. (2003). *Newton's principia for the common reader.* Oxford: Oxford University Press.

Choudhary, R. K., Kraus, U., Kersting, M., Blair, D., Zahn, C., Zadnik, M., & Meagher, R. (2019). Einsteinian physics in the classroom: Integrating physical and digital learning resources in the context of an international research collaboration. *The Physics Educator, 1*(4), 1950016.

Crease, R. P. (2003). *The prism and the pendulum: The ten most beautiful experiments in Science.* London: Random House Inc.

diSessa, A. A. (2006). "A history of conceptual change research: Threads and fault lines." In K. Sawyer (Ed.), *The Cambridge handbook of the learning sciences* (265–282). Cambridge, MA: Cambridge University Press.

Goldsmith, D., & Bartusiak, M. (Eds.). (2006). *E=Einstein, his life, his thought, and his influence on our culture.* New York: Sterling Publishing & Co.

Kapon, S., & diSessa, A. A. (2012). Reasoning through instructional analogies. *Cognition and Instruction, 30*(3), 261–310.

Kersting, M., Henriksen, E. K., Bøe, M. V., & Angell, C. (2018). General relativity in upper secondary school: Design and evaluation of an online learning environment using the model of educational reconstruction. *Physical Review Physics Education Research, 14*(1), 010130-1-010130–18.

Kersting, M., & Steier, R. (2018). Understanding curved spacetime – The role of the rubber sheet analogy in learning general relativity. *Science & Education, 27*(7), 593–623.

Kim, I., & Spelke, E. (1992). Infants' sensitivity to effects of gravity on visible object motion. *Journal of Experimental Psychology: Human Perception and Performance, 18*(2), 385–393.

McCloughlin, T. J. J. (2019). Beware the Greeks: Sources for the history of gravity in science teaching. In M. Matthews (Ed.), *HPS&ST Note Newsletter,* 1–40.

Ohanian, H. (1976). *Gravitation and spacetime.* New York: W.W. Norton.

Palmer, D. (2001). Student's alternative conceptions and scientifically acceptable conceptions about gravity. *International Journal of Science Education, 23*(7), 691–706.

Pitts, M., Venville, G., Blair, D., & Zadnik, M. (2014). An exploratory study to investigate the impact of an enrichment program on aspects of Einsteinian physics on year 6 students. *Research in Science Education, 44*(3), 363–388.

Starr, S. F. (2015). *Lost enlightenment: Central Asia's golden age from the Arab conquest to Tamerlane.* Princeton University Press. Princeton, New Jersey.

Turnbull, H. W. (Ed.). (1959). *The correspondence of Isaac Newton: 1661–1675, Volume 1.* London: Cambridge University Press (Published for the Royal Society).

Venville, G., Blair, D., Coward, D., Deshon, F., Gargano, M., Gondwe, M., & Zadnik, M. (2012). Research informed science enrichment programs at the gravity discovery centre. *Teaching Science, 58*(1), 35–41.

Williamson, K., Prather, E. E., & Willoughby, S. (2016). Applicability of the Newtonian gravity concept inventory to introductory college physics classes. *American Journal of Physics, 84*(6), 458–466.

Williamson, K., Willoughby, S., & Prather, E. E. (2013). Development of the Newtonian gravity concept inventory. *Astronomy Education Review, 12*(1).

# 10 Models and analogies in teaching general relativity

*Markus Pössel*

**WHAT?**

This chapter presents and analyses several teaching models of general relativity and cosmology. Markus Pössel provides a critical discussion of the nature and limitations of each model as well as the concept it is meant to illustrate.

**FOR WHOM?**

This chapter is intended for teachers who look for instructional models to teach concepts of general relativity and cosmology at the upper secondary school level, and also for trainee teachers who wish to learn more about the use of teaching models in physics education.

## INTRODUCTION

"Anyone who has ever tried to present a rather abstract scientific subject in a popular manner knows the great difficulties of such an attempt. Either he succeeds in being intelligible by concealing the core of the problem and by offering the reader only superficial aspects or vague allusions, thus deceiving the reader by arousing in him the deceptive illusion of comprehension; or else he gives an expert account of the problem, but in such a fashion that the untrained reader is unable to follow the exposition and becomes discouraged from reading any further. If these two categories are omitted from today's popular scientific literature, surprisingly little remains. But the little that is left is very valuable indeed".

Popular science has changed considerably since Albert Einstein wrote these words in his foreword to a popular book on relativity (Barnett, 1948). Nevertheless, the core problem is the same for today's science communicators as for Einstein's contemporaries: how to present a complex subject such as the concepts and applications of Einstein's general theory of relativity without becoming too superficial on one hand, or too technical on the other hand. Whoever teaches about general relativity to an audience that is not (yet) in a position to understand and build upon the theory's full mathematical formalism, will need to grapple with that fundamental question. The same is true in the context of teaching undergraduates or high-school students. A central element of any answer is the concept of a teaching model.

## TEACHING MODELS OF GENERAL RELATIVITY

Building on similar definitions in the literature (notably Gilbert, 2004), this chapter adopts the following, deliberately general definition:

> A **(teaching) model of general relativity**, or a part of general relativity, is a concrete or abstract entity that represents some of the structure of general relativity, or structures related to general relativity's concepts and applications, in a simplified way, for the purpose of teaching about general relativity and its applications.

A model can be a physical object, or mathematical in nature, or a narrative that simplifies certain of the arguments by which physical predictions are derived from the fundamentals of Einstein's theory. Mathematical teaching models are closely related to the mathematical models that physicists create to describe specific situations in terms of fundamental physical laws. A key concept in this context is that of **analogy**, that is, the statement that a specific aspect of a teaching model "is like" a corresponding specific aspect of the structure that is being explained (Gilbert, 2004). In the following, when we discuss specific teaching models, a key part of the discussion will always concern the nature and limitations of the analogies between the model and the concept or application it is meant to represent.

Such models fit into various broader theories of teaching. The simplifications they involve are the first step of "pedagogical reduction" or "educational reconstruction" ("didaktische Rekonstruktion"): efforts to transform the complexities of a subject into content suitable for learners (Duit et al., 2012; Grüner, 1967). Specifically, this kind of pedagogical simplification is known as "elementarization" (Kircher, 2015). Given their role in making complex subjects teachable, teaching models are an important part of teachers' pedagogical content knowledge – the "particular form of content knowledge that embodies the aspects of content most germane to its teachability" (Shulman, 1986).

Einstein's two categories of how simplifications can fail, as encapsulated in the quote, apply to models as well. Phrased positively, we are looking for the following qualities in good teaching models:

1. Faithfulness: The model represents all important features of the original.
2. Parsimony: Non-essential, potentially distracting features are kept to a minimum.
3. Confusion avoidance: The model has no (obtrusive) misleading features.
4. Accessibility: The target audience is capable of understanding the structure of the model.
5. Extendability: The model is a stepping stone to additional insight, can be built upon and can be connected with other teaching models.

The criteria are situation-dependent. Accessibility depends on your target group, and the extent to which you aim to avoid confusion will often be a subjective choice. More generally, this is not a checklist to be completed mechanically, but a list of criteria to keep in mind when choosing a teaching model. The resulting decisions will necessarily involve judgement calls by the teacher.

In the following, we will consider some representative examples of general relativity teaching models. A more comprehensive collection can be found in my review "Relatively complicated? Using models to teach general relativity at different levels" (Pössel, 2018). Specific teaching models will also be discussed in Chapters 12, 14 and 24 of this book.

## GENERAL RELATIVITY: TEACHING THE BASICS

Two-dimensional surfaces have a number of applications in teaching relativity. For instance, they can be used to teach about the underlying geometric concepts, such as curvature, or geodesics as straightest possible lines, in an accessible way (Figures 10.1 and 10.2; Effing, 1977; Henderson & Taimina, 2004; Wood et al., 2016).

But arguably the most famous of all relativistic teaching models, the rubber sheet model (Figure 10.3), sets out to do something more. In its simplest incarnation, this is an elastic sheet, suspended in a frame. A heavy sphere is placed on the sheet, pulling the fabric down and creating a warped surface. This corresponds to the second half of John Wheeler's famous two-sentence summary of general relativity (Wheeler, 1990): spacetime tells matter how to move, while matter tells spacetime how to curve. We can reproduce the first half by letting smaller spheres roll on the warped sheet. Their trajectory will be curved around the central heavy sphere, reminiscent of planetary orbits.

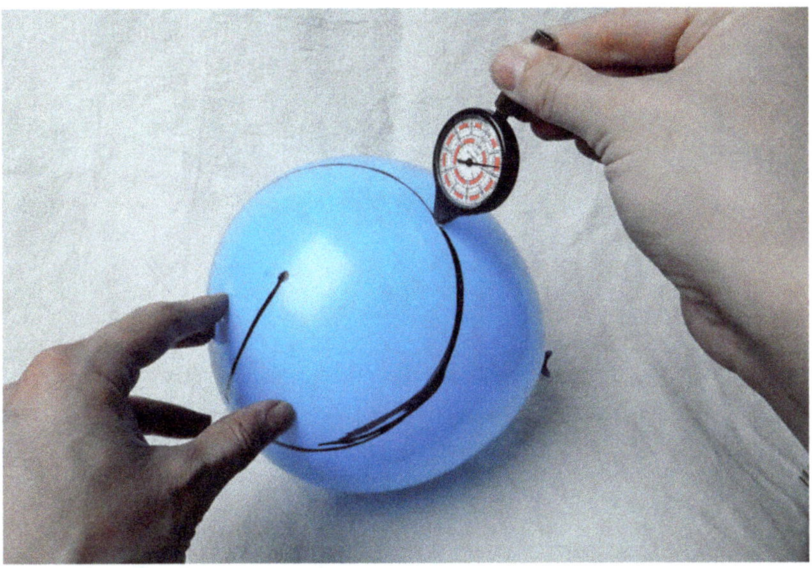

10.1
Measuring the curvature of a balloon surface using a map-measuring device. (*Source: Copyright by Markus Pössel*)

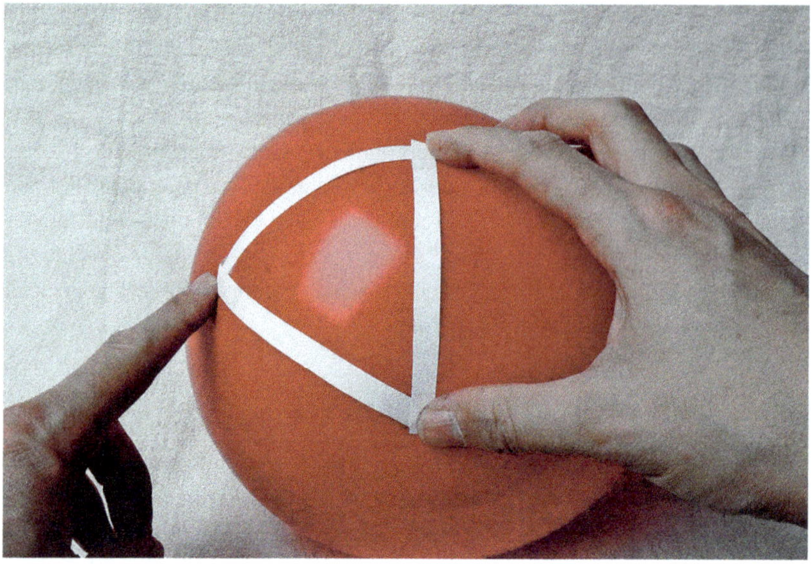

10.2
Measuring the curvature of a balloon surface using paper-strip triangles. (*Source: Copyright by Markus Pössel*)

Consider the faithfulness of this model. As Wheeler's summary shows, it does capture the central features of Einstein's equations: spacetime is represented by the fabric of the sheet, matter by the heavy ball; matter warps spacetime. Warped spacetime, the deformed sheet, influences the way that matter moves – in this case, how the smaller spheres travel across the sheet. The model is certainly accessible: it is readily understood by 15-year-old ninth-graders (Baldy, 2007). It is also parsimonious: sheet, heavier sphere and smaller spheres are all needed for the basic analogy.

The elastic sheet reproduces the field character of gravity: Gravitation is not a force-at-a-distance; instead, a mass will change the properties of surrounding spacetime. Since disturbances of the elastic sheet travel at a finite speed, it is even possible to produce "gravitational waves" – in the simplest case by flicking the sheet with a finger.

10.3
A small wooden sphere, rolling on an elastic sheet whose shape is warped by the heavier sphere in the centre (composite of still images taken 1/5 second apart). (*Source: Copyright by Markus Pössel*)

The model also has potentially confusing features (Janis, 2018; Kersting & Steier, 2018, and references therein). Gravity plays an unusual double role: real-life gravity is needed to pull the heavy sphere downward and create the warped geometry in the first place. The motion of the small spheres is determined by a combination of real gravity (pulling the sphere downwards) and model gravity (the difference between real gravity and the constraint force for an object moving on the curved sheet). As a (two-dimensional) spatial model, the elastic sheet implicitly emphasises the role of space, whereas in reality, at least the effects of classical gravity are encoded in Einsteinian gravity as a distortion of time. Every model that represents spacetime as a material medium singles out a special reference frame, contrary to Einstein's theory, where all the different possibilities of covering spacetime with coordinates are on an equal footing. Other problems are connected with the two-dimensional representation of spacetime: its clash with the three-dimensionality of the spheres, lack of spherical symmetry, and the more general problem of embedded representations of space and of spacetime that we will come back to in the section on cosmology.

Are those drawbacks sufficient reason for discarding the model altogether? That decision is left to the teacher, and depends on the target group and teaching situation. All that general relativity education research can do is to evaluate teaching models systematically – both analytically, for which the qualities listed in the previous section provide a helpful framework, and empirically, by testing a model's usefulness, and its potential for confusion, with selected groups of students. The results provide teachers with guidance for their individual decisions on whether or not to include the model in their teaching.

## BLACK HOLES

Black holes are among the most fascinating phenomena in general relativity. The possibility of regions of space shut off from the rest of the universe, as a cosmic one-way street, is one of the more surprising consequences of Einstein's equations. In the framework of general relativity, the simplest, spherical black hole is described by the so-called Schwarzschild metric. This metric was derived by the German astrophysicist Karl Schwarzschild shortly after Einstein's publication of his theory. It was intended to describe the spacetime geometry around a point particle, and only later recognised to describe the simplest kind of black hole. A crucial quantity in this metric is the Schwarzschild radius, $R_S = \frac{2GM}{c^2}$, which marks the value of the radial coordinate where the black hole's (spherical) horizon can be found – the boundary separating the outside of the black hole from the inside whence, in the classical general-relativistic description, nothing that has fallen in can ever return.

Perhaps the most common teaching model for black holes is the "funnel" model that extends the elastic sheet model from the previous section, and pictures a black hole not as a moderate depression (as that created by a heavy sphere) but as a funnel-shaped depression, ending either in a cusp or in a narrow circular border (e.g., Sagan et al., 1980; Hawking, 1996). Three-dimensional sector models for the spatial and spacetime geometry around a black hole can be found in Chapter 12 of this book. Model activities for the astrophysical context of black holes include hidden magnets to show how invisible objects can be detected by their influence on visible objects, and visualisations of the collapse of a massive star to form a compact object (Tran et al., 2018).

In the counter-intuitive spacetime geometry around a spherical (Schwarzschild) black hole, even the seemingly simple cross-section representation of a black hole by a circle (the horizon) with a dot in the centre (the singularity) is misleading: inside the horizon, the radial coordinate is not a spatial coordinate any more, but a time-like direction. Moving towards the centre of symmetry is as inevitable as moving forward in time; viewed as a snapshot in constant time, the singularity is an infinitely extended boundary where time ends (Droz et al., 1996; Pössel, 2010).

A useful black hole model that is not as widely known as it should be, in this author's opinion, is the *river model of black holes* (Hamilton & Lisle, 2008, on which the rest of the section is based) shown in Figure 10.4.

The top part of the figure shows a sideways cross-section of a river, as seen from an observer at rest on the riverbank, with water flowing at a constant speed (as indicated by the blue arrows) from left to right. In the water, fish are swimming. Fish can either go with the flow or against it, but they cannot go faster than a certain maximal speed-of-fish $c_F$ relative to the surrounding water. Now, consider the bottom part of the figure, where the flow speed varies from

10.4
Basic setup of the river model of black holes. (Top) side-view of fish in a river. The fish can move in either direction; and (Bottom) side-view of fish in a river with variable flow speed. Fish that have moved to the right of H can never return to the left of that boundary. (*Source: Copyright by Markus Pössel*)

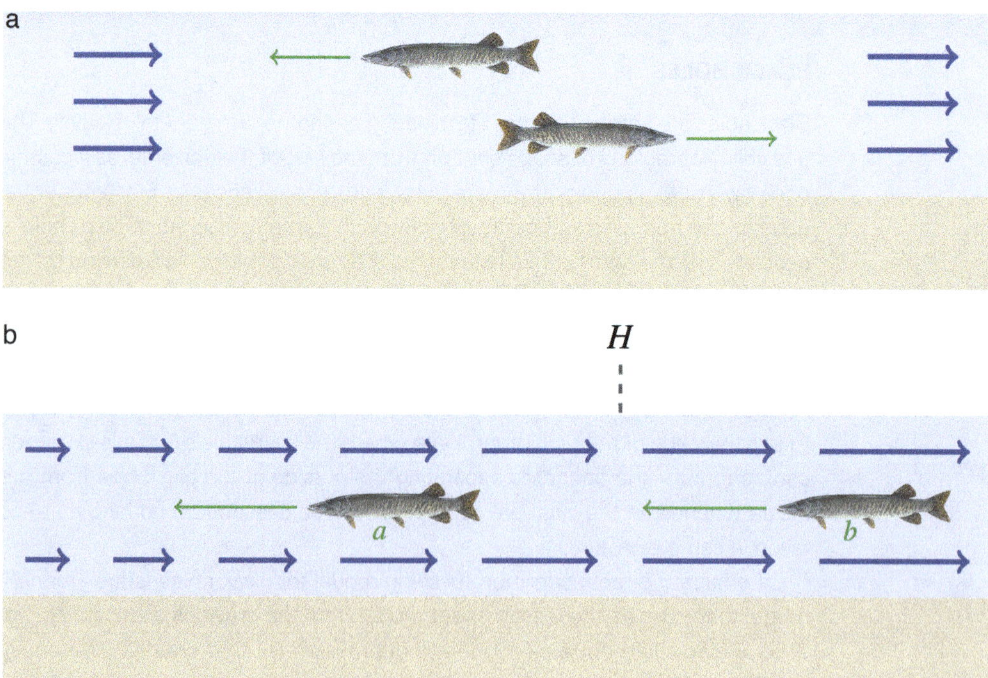

location to location along the river bank. (A natural explanation for this could be that the river becomes progressively narrower.) Flow speeds are once more marked relative to a stationary observer on the river bank, and at the location H, as marked on the river bank, such an observer will find that the flow speed has become larger than $c_F$.

In consequence, H marks a fish boundary: fish will always be able to swim from left to right across H. Our external observer can also find fish to the left of H, making headway to the left. But our observer will find that any fish to the right of H will necessarily move ever further towards the right. In that region, in order to move sufficiently fast against the flow to make progress leftwards, the fish would need to exceed the maximal speed-of-fish $c_F$.

This last situation represents a black hole horizon: the fish correspond to test particles, and the maximal speed-of-fish corresponds to the maximal possible speed of test particles, that is, to the speed of light. The observer on the river bank corresponds to one of the usual choices of a global coordinate system – a way of describing the situation as a whole. Fish swimming in the river correspond to observers or test particles. In this way, the model successfully reproduces the key properties of a black hole horizon: horizons are a global phenomenon. At least in classical general relativity, an observer in free fall towards the black hole will not feel any local effects telling them when they have crossed the horizon – just like a fish swimming in the river has no way of telling if or when they have passed the point of no return to H. Only by applying a global criterion can we chart the different regions involved, and divide the river, or spacetime, into inside and outside. In the case of the river, we make the division by noting that no fish can move from the right of H to the left of H. In the vicinity of a (spherical) black hole of mass $M$, we note that no test particle at a radius value smaller than the Schwarzschild radius can move to a radius value larger than the Schwarzschild radius.

The river model suffers from the same drawbacks as all models do which represent spacetime by a physical medium – in this case, water. For an observer in free fall towards a black hole, any other observer at the same location, but with a constant speed relative to the first, will also be in free fall. In the river, there is a preferred frame of motion: drifting at rest relative to the surrounding water. A fish at constant speed relative to that standard of rest would feel friction and feel a drag. Then, there is the question of symmetry: the river is a one-dimensional model, representing the radial direction in a spherically symmetric black hole spacetime. Making the connection, and seeing how the simplified and the three-dimensional geometry are related, is bound to be a challenge for less advanced students in particular. (A two-dimensional model of water in a basin flowing towards a central drain can be an intermediate step.)

On the plus side, the river model is eminently extendable: with the necessary mathematical tools, it can be formulated more precisely to yield the so-called Gullstrand-Painlevé metric, which is equivalent to the usual Schwarzschild metric. More advanced information about this can be found in the e-resources.

And as if that were not enough, there is an interesting sub-field of experimental physics, known as analogue gravity, dedicated to the analogies between fluids of this kind and relativistic gravity – up to and including attempts to incorporate quantum effects, and obtain measurements analogous to the Hawking radiation of black holes (for an overview, see Barceló et al., 2011).

## COSMOLOGY

General relativity, via the Friedmann-Lemaître-Robertson-Walker (FLRW) solutions of Einstein's equations, provides the foundation for the modern cosmological models of an expanding universe that started in an extremely hot and dense Big Bang phase. Substrate models for representing the systematic changes in distance that constitute cosmic expansion go back to Arthur Eddington, who wrote in 1930, in the same article that proved the instability of Einstein's static universe (Eddington, 1930): "Observationally, galaxies 'at rest' will appear to be receding from one another since the scale of the whole distribution is increasing. It is as though they were embedded in the surface of a rubber balloon which is being steadily inflated".

Eddington's simile has become a staple of teaching cosmology to high-school students or to the general public. At its core is the fact that rubber is stretchy – blow up a rubber balloon, or stretch a rubber band, and the imposed greater area or length will be distributed evenly between all different regions of the balloon or the band. In consequence, distances between marked locations on such a balloon, or rubber band, will grow in direct proportion to one and the same factor. This neatly reproduces the scale factor dependance of cosmic expansion, where all distances between galaxies in the Hubble flow (that is, idealised galaxies whose distance change only in accordance with cosmic expansion, without additional effects) change in direct proportion to the same universal, time-dependent scale factor, which is commonly denoted by a(t).

The one-dimensional version, easily produced by cutting open the loop of a rubber band, is best suited for direct measurements. Having marked the locations, with a pen or by taping small markers to the rubber band, students can document the distances between those galaxies at an earlier and a later moment of cosmic time, with the rubber band stretched less in the first instance and more in the second, and rediscover the Hubble–Lemaître relation: the change in distance between two galaxies over time is proportional to their initial distance (e.g., Price & Grover, 2001; activity H-9 in Fraknoi, 2011). As a downside, the single dimension makes it harder for students to generalise the lesson to three-dimensional space.

In two dimensions, as in the rubber balloon version shown in Figure 10.5, it is easier to see how the pattern of galaxy positions – which depends on distance ratios – does not change, although overall distances do. Quantitative measurements are more difficult, since the balloon surface is not a perfect sphere, but have been described by some authors (e.g., activity H-7 in Fraknoi, 2011).

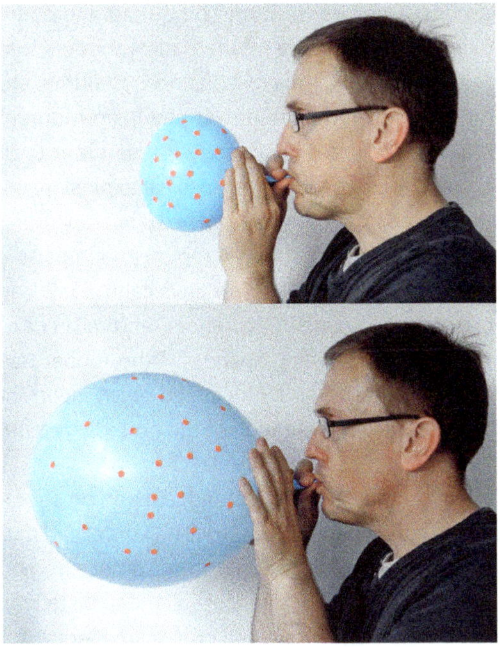

10.5
Rubber balloon representing an expanding universe. The galaxies are little stickers on the balloon surface. As the balloon is inflated, distances between the galaxies change, at least approximately, in proportion to the same scale factor. (*Source: Copyright by Markus Pössel*)

The rubber balloon universe is a handy example for a universe that is finite in size (here: area), yet has no boundaries, no "end of space", being curved into itself to form a closed surface. On the other hand, the fact that the curved balloon surface is necessarily embedded into three-dimensional space is a common source of confusion – ask students where the centre of the balloon universe is, and most will identify the centre of the three-dimensional balloon. In employing this model, it is crucial for the teacher to clarify that "the universe" in this model corresponds to the surface of the balloon only – a non-trivial step of abstraction. While less easy to stretch homogeneously, a plain rubber sheet has a better chance of avoiding the embedding-related confusion (e.g., Chapter 14 in Harrison, 2000). It is also closer to the flat geometry of the three-dimensional space in our best current cosmological models.

The three-dimensional expanding-substrate model is typically presented as a thought experiment: picture a raisin cake in the oven, rising (and thus increasing its volume) with the raisins moving apart from each other in something approximating scale-factor expansion.

Astronomy education researchers have found a number of misconceptions among students that are relevant to the expanding substrate models. One is the notion of the Big Bang as a classical explosion, with galaxies starting out from a single well-defined point in space (taken to be "the location of the Big Bang") and flying outward into the surrounding empty space. In this incorrect image of the Big Bang as a *classical explosion*, the galaxy-filled part of the universe has a boundary, and it increases in volume as the boundary moves ever further out into space (Trouille et al., 2013; Wallace et al., 2012, and references therein).

The physical models provide opportunities for both cementing and counteracting this misconception. A key property of the mathematical models that form the foundations of our understanding of cosmic expansion is that they are based on a universe that is homogeneous and isotropic: at any fixed moment in cosmic time, properties such as the local energy density are same at each location in space, and there is no preferred direction. The classical explosion violates both assumptions: it creates a fundamental inhomogeneity by separating space into an empty outer region and a galaxy-filled inner region and imparts a preferred radial direction onto the expansion process. The physical models can support that misconception: rubber band, expanding flat sheet, and raisin cake have outer boundaries moving into the surrounding space; for the rubber balloon, misunderstanding the role of embedding space can lead to an incorrect view of the expanding rubber balloon as a representation of the boundary of the classical explosion.

That is why it is important to stress that the rubber band, expanding sheet, and raisin cake each represent no more than part of the overall universe, and that the reason why cosmic expansion can be seen even within such a smaller region is because expansion is a feature of *all* inter-galaxy distances: all those distances grow larger over time and the number density of galaxies grows smaller.

Part of the initial Eddington quote foreshadows a debate about the interpretation of cosmic expansion that continues to this day. Interested readers can find the details in the literature on the "expanding space" interpretation which sees the cosmic expansion as distinct from ordinary motion through space, and thus is particularly close to the physical models for an expanding substrate (e.g., Davis & Lineweaver, 2004; Francis et al., 2007) versus the "relativistic motion" or "relativistic explosion" interpretation that shows how, using appropriate tools from differential geometry, the cosmological redshift can be interpreted as a relativistic Doppler effect (e.g., Bunn & Hogg, 2009; Narlikar, 1994; Pössel, 2019, and references therein).

## GRAVITATIONAL WAVES

Ever since the February 2016 press conference reported the first direct detection of gravitational waves (on September 14, 2015, by the LIGO collaboration), there has been intense public interest in the subject and corresponding interest on the part of educators.

For propagation speed and transversal nature of gravitational waves, electromagnetic waves provide a simple analogy (e.g., Price et al., 2012), which has been extended to more advanced topics such as the generation of gravitational waves (e.g., Kalckar & Ulfbeck, 1974; Hilborn, 2018). Sound waves, while not transversal, can be used as models for gravitational waves as carriers of information about large-scale phenomena, in contrast with the resolved small-scale information provided by typical electromagnetic observations in astronomy (Kwon, 2017).

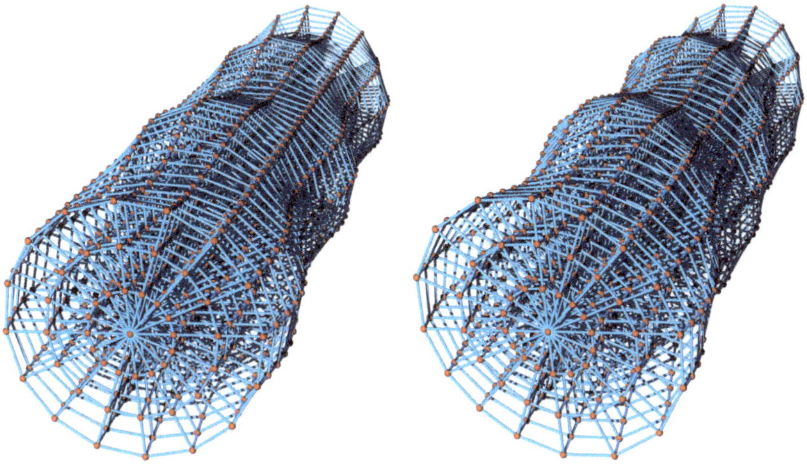

10.6
Distortion pattern of two gravitational waves in space. *Left*: linearly polarised gravitational wave. *Right*: circularly polarised wave. (*Source: Copyright by Markus Pössel*)

Most fundamental teaching models in this context are those meant to visualise the action of gravitational waves on free-falling particles. The most common model involves test particles floating in free space, forming a ring in, say, the *x*-*y* plane. A gravitational wave is assumed to propagate through the particle ensemble, its direction of propagation at right angles to the ring plane, e.g. in the positive *z*-direction. As a simple, sine-shaped gravitational wave passes through, the inter-particle distances shrink alternately in the *x*- and the *y*-direction. Whenever distances shrink in the *x*-direction, they are stretched in the *y*-direction, and vice versa, making for such waves' characteristic "quadrupole pattern". Series of snapshots of this characteristic change over time, often for a ring but sometimes for a rectangular pattern of particles, can be found in most of the general relativity textbooks.

Figure 10.6 shows the three-dimensional generalisation: nested versions of such a ring, placed at different z locations which show the wave nature of the gravitational waves and the associated particle distortion: stable spatial patterns propagating over time, which are seen as oscillatory changes if we observe not the wave as a whole but its action at a certain fixed location in space. An animated version can be found on http://www.haus-der-astronomie.de/gw-animations, together with animations of the production and the detection of gravitational waves.

Hands-on teaching models of the characteristic sequence of patterns include applying the pattern to an elastic rectangular sheet which is then alternately stretched and relaxed in two orthogonal directions, and "snapshots" of the test particle configurations at different times, printed on transparent plates, which students are then asked to bring to the correct order (Kraus & Zahn, 2016).

There are several aspects to keep in mind when using these models. Once more, space is represented by a physical medium, implying a standard of rest that is just not there in general relativity. The test particles are free of external forces; the presence of such forces will alter their behaviour (e.g., in so-called resonant gravitational wave detectors). Last but not least, it is tempting to look at all those elementary, sine-shaped gravitational waves and conclude that this simple oscillatory nature is somehow a property of gravitational waves themselves. It is not – in a field satisfying the wave equation, any pattern will propagate through space. That certain important gravitational wave patterns are nearly oscillatory is a property of the gravitational wave sources (e.g. orbiting bodies), not of gravitational waves in general or of spacetime as a medium for wave propagation. This can be reproduced using the elastic sheet model, with two spheres made to orbit each other artificially to produce ripples that propagate outwards in a spiral (e.g., Mould, 2016; Overduin et al., 2018).

Michelson interferometers have been used as teaching models for modern interferometric detection, either as part of interactive exhibitions or in affordable versions suitable for the use in a high school setting (e.g., Choudhary et al., 2018; Farr et al., 2012). While there are various complications, the simple layout of a Michelson interferometer does capture the essential layout of modern interferometric detectors. There is, however, a twist in that, in general, a gravitational wave will influence not only the distance between the detector's optical elements but also light propagation and the wavelength of the light inside the detector, in a manner analogous to the cosmological redshift in an expanding universe (Faraoni, 2007; Saulson, 1997).

## CONCLUSIONS

Teaching models are a key part of teaching about general relativity and its applications, in particular on all those occasions where teaching does not include the formalism and the exact mathematical formulation of the theory. But each model comes with its characteristic limitations, of which the teacher must be aware in order to minimise confusion and deal with possible misunderstandings. Limitations of teaching models can be utilised to learn about general relativity itself – understanding the limitations of the rubber balloon model for an expanding universe, for instance, can foster understanding of the intrinsic nature of geometry and the role of embeddings. More generally, such limitations can be used to initiate a discussion about the role of models and simplifications in science. "Due diligence" in selecting and using appropriate teaching models is the responsibility of the teacher; that of the general relativity education community in this context is to analyse the various teaching models and provide teachers with the information they need to use the models responsibly.

## ACKNOWLEDGEMENTS

I would like to thank the participants of the 690th WE Heraeus seminar for helpful discussions – in particular Bernard F. Schutz (who pointed out the "sine waves are not elementary" confusion to me, in the section on gravitational waves above), and Martin Hendry.

## FURTHER READING

E-resources: River model and Gullstrand-Painlevé metric.
Chapter 14: Using the language of gravity to teach about space, time, and matter in general relativity.
Chapter 24: Introducing general relativity without special relativity – Classroom experience from the Czech Republic.

## REFERENCES

Baldy, E. (2007). "A new educational perspective for teaching gravity." *International Journal of Science Education*, 29, 1767–1788.

Barceló, C., Liberati, S., & Visser, M. (2011). "Analogue gravity." *Living Reviews of Relativity* 14(1), 3.

Barnett, L. (1948). *The Universe and Dr. Einstein*. Victor Gollancz. London.

Bunn, E.F. and Hogg, D.W. (2009). The Kinematic Origin of the Cosmological Redshift. *American Journal of Physics*, 77, 688–694. http://dx.doi.org/10.1119/1.3129103

Choudhary, R. K., Foppoli, A., Kaur, T., Blair, D. G., Zadnik M. & Meagher R. (2018). "Can a short intervention focused on gravitational waves and quantum physics improve students' understanding and attitude?" *Physics Education*, 53(6).

Davis, T. M. & Lineweaver C. H. (2004). "Expanding confusion." *Publications of the Astronomical Society of the Pacific*, 21, 97–109.

Droz, S., Israel, W. & Morsink, S. M. (1996). "Black Holes: The Inside Story." *Physics World*, 9(1), 34–37.

Duit, R., Gropengießer, H., Kattmann, U., Komorek, M. & Parchmann, I. (2012). "The Model of Educational Reconstruction – A Framework for Improving Teaching and Learning Science." D. Jorde & J. Dillon (Eds.). *Science Education Research and Practice in Europe: Retrospective and Prospective*. 13–37. Rotterdam: Sense Publishers.

Eddington, A. S. (1930). On the Instability of Einstein's Spherical World, *Monthly Notices of the Royal Astronomical Society*, Volume 90, Issue 7, May 1930, Pages 668–678.

Effing, M. (1977). "Squashed space: an activity concerning general relativity." *The Physics Teacher*, 15, 298–299.

Ellis, G. F. R. & Williams, R. M. (2000). *Flat and Curved Spacetimes*. 2nd ed. Oxford: Oxford University Press.

Faraoni, V. (2007). "A common misconception about LIGO detectors of gravitational waves." *General Relativity and Gravitation*, 39, 677–684.

Farr, B., Schelbert, G. & Trouille, L. (2012). "Gravitational wave science in the high school classroom." *American Journal of Physics*, 80, 898–904.

Fraknoi, A. (Ed.) (2011). "The Universe at Your Fingertips 2.0." *A DVD-ROM of Astronomy Activities and Resources*. Astronomical Society of the Pacific. https://myasp.astrosociety.org/product/DV122/the-universe-at-your-fingertips-20-dvd-rom (last accessed 21 September 2019)

Francis, M. J., Barnes, L. A., James, J. B. & Lewis, G. F. (2007). "Expanding space: The root of all evil?" *Publications of the Astronomical Society of Australia*, 24, 95–102.

Gilbert, J. (2004). "Models and modelling: Routes to more authentic science education." *International Journal of Science and Mathematics Education*, 2(2), 115–130.

Grüner, G. (1967). "Die didaktische Reduktion als Kernstück der Didaktik." *Die Deutsche Schule* 59, 414–430. http://www.digizeitschriften.de/dms/resolveppn/?PID=PPN509092632_0059%7Clog100 (last accessed December 21, 2019).

Hamilton, A. J. S. & Lisle, J. P. (2008). "The river model of black holes." *American Journal of Physics*, 76, 519–532.

Harrison, E. (2000). *Cosmology. The Science of the Universe.* 2nd ed. Cambridge University Press. New York.

Hawking, S. (1996). *The Illustrated: A Brief History of Time.* Updated and Extended Version. Random House.

Henderson, D. W. & Taimina, D. (2004). *Experiencing Geometry: Euclidean and Non-Euclidean with History.* Pearson.

Hilborn, R. C. (2018). "Gravitational waves from orbiting binaries without general relativity." *American Journal of Physics*, 86, 186–197.

Janis, A. I. (2018). "On mass, spacetime curvature, and gravity." *The Physics Teacher*, 56, 12–13.

Kalckar, J. & Ulfbeck, O. (1974). "The problem of gravitational radiation." *Det Kongelige Danske Videnskabernes Selskab Matematisk-fysiske Meddelser*, 39, 6.

Kersting, M. & Steier, R. (2018). "Understanding curved spacetime. The role of the rubber sheet analogy in learning general relativity." *Science & Education*, 27(7–8), 593–623.

Kircher, E. (2015). "Elementarisierung und didaktische Reduktion." Kircher, E., R. Girwidz, and P. Häußler (Eds.): *Physikdidaktik. Theorie und Praxis*, Chapter 4. Springer Spektrum.

Kraus, U. & Zahn, C. (2016). "Gravitational Waves: Models and Experiments on Waveforms, Effects and Detection." *Proceedings of the Heraeus Summer School 'Astronomy from 4 Perspectives: Gravitational Wave Astronomy'*". Ed. by Lotze, K.-H. and S. Völker. University of Jena, URL: http://www.spacetimetravel.org/gwave16/gwave16.html.

Kwon, D. (2017). "Turning on the cosmic microphone." *Science in School*, 39, 8–11. URL: https://www.scienceinschool.org/content/turning-cosmic-microphone.

Mould, S. (2016). "Visualising gravitational waves." YouTube Video posted 9 August 2016, URL: https://youtu.be/dw7U3BYMs4U.

Narlikar, J. V. (1994). "Spectral shifts in general relativity." *American Journal of Physics*, 92, 903–907.

Overduin, J., Perry, J., Huxford, R. & Selway, J. (2018). "Classroom simulation of gravitational waves from orbiting binaries." *The Physics Teacher*, 56, 586–590.

Pössel, M. (2010). "Changing places – space and time inside a black hole." Einstein Online 4, article id 1009. url: https://www.einstein-online.info/en/spotlight/changing_places/

Pössel, M. (2018). "Relatively complicated? Using models to teach general relativity at different levels." arXiv:1812.11589.

Pössel, M. (2019). "Cosmic event horizons and the light-speed limit for relative radial motion." arXiv:1912.11677.

Price, R. H., & Grover, E. (2001). "Cosmological expansion in the classroom." *American Journal of Physics*, 69, 125–128.

Price, R. H., Belcher, J. W. & Nichols, D. A. (2012). "Comparison of electromagnetic and gravitational radiation: What we can learn about each from the other." *American Journal of Physics*, 81, 575–584.

Sagan, C., Druyan, A. & Soter, S. (1980). Cosmos. *Episode 9: "The Lives of the Stars".* PBS.

Saulson, P. R. (1997). "If light waves are stretched by gravitational waves, how can we use light as a ruler to detect gravitational waves?" *American Journal of Physics*, 65, 501–505.

Shulman, L. S. (1986). "Those who understand: knowledge growth in teaching." *Educational Researcher*, 15(2), 4–14.

Tran, H., Russo, P. & Russell, T. (2018). Black Hole Activities – A Quick Reference Guide. Online at https://www.unawe.org/static/archives/education/pdf/Black_Hole_booklet.pdf.

Trouille, L. E. et al. (2013). "Investigating student ideas about cosmology iii: Big Bang theory, expansion, age, and history of the universe." *Astronomy Education Review*, 12, article id. 010110.

Wallace, C. S., Prather, E. E. & Duncan, D. K. (2012). "A study of general education astronomy students' understandings of cosmology. Part IV. Common difficulties students experience with cosmology." *Astronomy Education Review*, 11, article ID 010104.

Wheeler, J. A. (1990). *A Journey into Gravity and Spacetime*. W. H. Freeman & Co Ltd. New York.

Wood, M., Smith, W. & Jackson, M. (2016). "Curvature of spacetime: A simple student activity." *The Physics Teacher*, 54, 572–573.

# 11 Gravitational lensing as a focal point for teaching general relativity

*Karl-Heinz Lotze and Silvia Simionato*

**WHAT?**

This chapter presents pedagogical approaches to light deflection by gravitational fields. Karl-Heinz Lotze and Silvia Simionato put particular emphasis on historical perspectives to contextualise gravitational lensing in light of the larger development of Einsteinian physics.

**FOR WHOM?**

This chapter is intended for readers who look for instructional models of the gravitational-lens effect in general relativity suitable at the secondary school level.

## INTRODUCTION

In this chapter, we treat the deflection, or curvature, of the trajectories of light propagating in weak gravitational fields. We follow two guidelines:

- maintaining the broadest possible perspective and
- making the selected topics accessible to secondary school students and teachers.

Since "Einsteinian physics" does not mean that there is a "modern" Einsteinian body of knowledge disconnected from the rest of physics, our perspective takes a historical note. In fact, we start tracing the ideas about light deflection that go back to Newton's time, and discuss the pre-Einsteinian understanding of this phenomenon, the so-called ballistic light deflection. The next step deals with Einstein's conceptually different path towards the deflection of light, which starts from the principle of equivalence and ends with his theory of general relativity. It was nobody else but him who meditated on this phenomenon, as early as in 1912, under the broader intuition that there are other light-deflecting celestial bodies besides the Sun. Consequently, we start discussing the gravitational lens effect, in which we find quasars, galaxies, and clusters of galaxies as light sources and gravitational lenses. To allow readers to investigate this effect on their own, we describe an optical gravitational lens experiment together with instructions on how to perform it using a piece of equipment at everyone's reach.

This chapter is mainly written in a narrative style. However, we believe that, when teaching physics at any level, we should never miss the opportunity to demonstrate the power of mathematics as a method to create knowledge in physics. In fact, it is not possible in the framework of this chapter to explicitly show the various derivations of light deflection by the gravitational field. However, we are going to discuss the results in terms of simple formulas, most of them enclosed in separate boxes, and ask questions like: what are the underlying model assumptions and approximations? Does the result seem to be structurally "reasonable" even though we are not able to "guess" every numerical value in advance?

The audience we primarily want to address includes secondary school teachers who want to create a deeper understanding of the gravitational lens effect in the minds of their students, rather than merely reporting on it. We want to provide them with the background information as well as some suggestions on how lessons concerning this aspect of Einsteinian physics, including demonstrations with simulated gravitational lenses, could be organised.

## LIGHTS ALL ASKEW IN THE HEAVENS

"Lights all askew in the heavens" was written on the title page of the New York Times of 10 November 1919. It was in May 1919 when two British solar-eclipse expeditions confirmed one of the results of Einstein's general relativity theory: light rays, when passing the limb of the Sun, are deflected from their straight path under the influence of the gravitational field of the Sun by an amount of 1.75 seconds of arc.

Thus, the question naturally arises whether we observe the universe at all only in the light of gravitation since the sky is full of sun-like stars. Indeed, suppose that a star relatively close to us is located on the line connecting us with another, more distant star. Then the closer star, by its gravitational field, should give rise to a deflection of the light of the more distant one. Today, this effect, called gravitational lens effect, shows up in numerous observations with galaxies and clusters of galaxies instead of single stars as light-deflecting gravitational lenses.

Thinking about light deflection in a gravitational field has a long tradition that goes back to Newton's time. Even Einstein, a few years before completing his theory of general relativity, predicted only half of the above-mentioned deflection angle like all of his predecessors did in the framework of Newtonian physics. From the point of view of the "final" theory, and from the point of view of observations as well, this result is wrong. Is it then allowed to teach what is wrong?

We are thoroughly convinced that the agreement between theoretical predictions on one hand and observations or experiments on the other hand, is not the only standard for the educational relevance of a theory or a phenomenon. The historical development of scientific thoughts is of equal importance for real understanding. In the case at hand, we learn a lot about how the concepts of light developed in the course of time from Newtonian celestial mechanics of corpuscles of light to the light-like geodesics of a curved spacetime.

## BALLISTIC LIGHT DEFLECTION

On 26 May 1783, Reverend John Michell (1724–1793) wrote a letter to Henry Cavendish (1731–1810) and asked for support in publishing its scientific content in the Philosophical Transactions of the Royal Society. In this letter, Michell arrived, among other things, at the conclusion that stars with a mean density equal to that of the Sun and a size of about 500 times that of the Sun remain invisible for observers in the distant universe. This is because, for such stars, the escape velocity from their surface exceeds the speed of light.

Whereas Michell thought about the deceleration of light after leaving a star in radial direction, Cavendish, perhaps inspired by Michell's letter, was interested in whether and how, if at all, light rays approaching a celestial body "from infinity" are deflected by its gravitational attraction when closely passing by its surface.

Isaac Newton (1642–1726), in the appendix to his opus "Opticks", published in 1704, handed over to the next generations a whole series of queries for future research. The first of these queries reads as follows:

> "Do not Bodies act upon Light at a distance, and by their action bend its Rays, and is not this action (caeteris paribus) strongest at the least distance"?

Another query gives us even an idea of how Newton visualised light itself:

> "Are not the Rays of Light very small Bodies emitted from shining Substances"?

Accordingly, we can imagine a light ray being a cascade of tiny bodies or corpuscles. Consequently, optics was, at that time, nothing but the application of Newtonian mechanics to these corpuscles of light.

This is indeed the method employed by Michell and Cavendish as well. However, Cavendish's calculations from 1783/84 of the deflection of light have never been published. They are part of a manuscript on comets which is not astonishing because if the corpuscles of light approach "from infinity" with the initial velocity $v_\infty = c$, they do so on hyperbolic orbits in the same way as many comets do. In a short report (Thorpe, 1921) on Cavendish's astronomical manuscripts, at least his result is communicated:

"Let the velocity of body revolving in a circle at a distance [of the radius of a celestial] body be to the velocity of light as 1 : *u*, then will the sine of half bending of the ray be equal to $\frac{1}{1+u^2}$".

According to this result, the deflection angle amounts to $\delta_{N\odot} = 0.87''$ for the limb of the Sun. Moreover, it confirms Newton's expectations (also consider equation (11.2) of Box 11.1): the light deflection is strongest at the least distance, and this is the radius of the celestial body itself. The strength of gravity,

### BOX 11.1: CAVENDISH'S BALLISTIC LIGHT DEFLECTION

We want to translate Cavendish's result into our modern terminology without deriving it. The velocity of circular motion around a celestial body of mass *M* at a distance of the radius *R* from the centre of that body is, according to Newtonian mechanics,

$$\boldsymbol{v}_{\mathbf{circ}} = \sqrt{\boldsymbol{G}\frac{\boldsymbol{M}}{\boldsymbol{R}}}.$$

Here, *G* is Newton's gravitational constant. According to Cavendish's definition, we have $\boldsymbol{u} \equiv \frac{\boldsymbol{c}}{\boldsymbol{v}_{\mathbf{circ}}}$, and the sinus of one half of the deflection angle $\delta$ explicitly reads (see Figure 11.1)

$$\sin\frac{\delta}{2} = \frac{1}{1+\frac{Rc^2}{GM}} = \frac{GM}{Rc^2}\cdot\frac{1}{1+\frac{GM}{Rc^2}}.$$

In all cases considered in this chapter, the condition

$$\frac{GM}{Rc^2} \ll 1 \tag{11.1}$$

will be fulfilled, whose meaning will become clearer in due course. For the mass and radius of the Sun, we get, for instance,

$$\frac{GM_\odot}{R_\odot c^2} = 2.1\cdot 10^{-6}.$$

In view of such a small number, we are allowed to set the sinus of one half of the deflection angle approximately equal to one half of the deflection angle itself. Thus, we obtain in very good approximation

$$\delta_N = 2\frac{GM}{Rc^2}. \tag{11.2}$$

(Here, the subscript N stands for Newtonian mechanics and the Newtonian conception of light.)

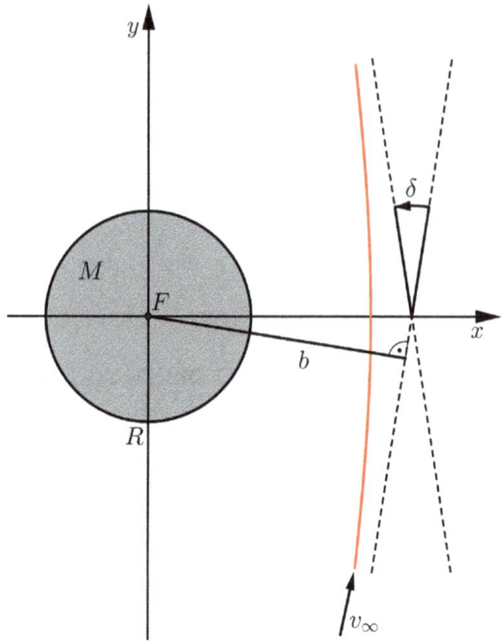


11.1
Ballistic light deflection. The light path is the hyperbola in the focus, *F*, of which the light-deflecting celestial body (mass *M*, radius *R*, e.g. the Sun) is located. The dashed lines are the asymptotes of the hyperbola (deflection angle δ). The impact parameter *b* is the shortest distance at which the light ray would pass by the celestial body if there were no deflection. (*Source: Copyright by Karl-Heinz Lotze, Silvia Simionato*)

expressed by *G*, together with the mass *M* of the celestial body, determines its light-deflecting power. The reason why the resulting deflection angle is still so small is the – according to human standards – incredibly high speed of light. From now on, we will call the result shown in equation (11.2) "ballistic light deflection".

The first who, in 1801, published his calculations of the deflection angle was Johann Georg von Soldner (1776–1833). He also transferred the laws of free fall of mechanics to optics. However, the way Soldner posed the problem of light deflection was not the same as Cavendish's. He first calculated the path of a light ray emanating with tangential velocity *c* from a star "towards infinity", and afterwards he reversed the course of the light ray. As was pointed out by Will (Will, 1988), Soldner's result is, within the scope of equation (11.1) and in leading order of approximation, the same as that of Cavendish (Soldner, 1801): When light passes by the solar limb, the "angle of perturbation" amounts to 0.84 seconds of arc.

Moreover, Soldner reflected on the observability of that angle of perturbation or deflection and arrived at the conclusion (Soldner, 1801) (English translation adopted from (Jaki, 1978)):

> "If one could observe the fixed stars very close to the Sun, then one would have to take this very much into account. But since this is not known to happen, the perturbation caused by the Sun can also be neglected".

After Soldner, for one century apparently, nobody was interested in the problem of light deflection. It was Einstein (1879–1955) who resumed the quest for an answer to the problem of how the propagation of light was influenced by gravity. In doing so, he had no need any more to fall back to Newton's conception of light corpuscles.

## LIGHT DEFLECTION AND EQUIVALENCE PRINCIPLE

In his famous paper (Einstein, 1911) "Über den Einfluss der Schwerkraft auf die Ausbreitung des Lichtes", published in 1911 in the "Annalen der Physik", Einstein presented an idea that reminds us of his ingenuity of 1905 when he postulated the validity of Galileo's relativity principle of mechanics for the whole physics.

Einstein, in a manuscript in 1920 that was unpublished in those days but has been included in (Janssen et al., 2002), considered the following idea, which we quote in a slightly modified version, the happiest of his life: How a roofer, when falling down from the roof, would describe the motion of the hammer he, being frightened, released?

Since the acceleration exerted by the gravitational field of the Earth is constant and the same for all bodies, the hammer will stay at rest with respect to the roofer as seen by himself if it was initially at rest. (It will be in free-fall motion at the same speed as the roofer, as seen by an observer on the ground.) Once given a kick, the hammer will move in a straight line with constant speed. Everything is like in a frame of reference where Galileo's law of inertia holds and that we call an inertial frame of reference.

According to Einstein,

> "this experience of the identical falling of all bodies in the gravitational field is one of the most universal experiences that the observation of nature has yielded to us; nevertheless, this law has not been granted a place in the foundations of our physical edifice".

Einstein assigns such a place to this law by his principle of equivalence:

> In a small laboratory that is in free-fall motion in a gravitational field, the laws of physics (all of them!) are the same as those valid in the absence of a gravitational field in a Newtonian inertial frame of reference.

This principle "*possesses great heuristic significance, provided that it is really correct*" (English translations adopted from (Klein et al., 1994)). The principle of equivalence is a local principle. However, it is not possible to precisely state what size the laboratory is allowed to have in order that it can be considered "small". In the case at hand, we claim that the laboratory is small in the sense that the gravitational field inside may be considered homogeneous for the whole duration of an experiment in free fall. In other words, the field lines of the gravitational field, actually converging towards the Earth's centre, may be replaced by parallel ones.

11.2
Gedanken experiment about the application of the principle of equivalence to light deflection inside an elevator cage that is freely falling in a homogeneous gravitational field. (*Source: Copyright by J. Hookham, G. Gamov. With permission from Cambridge University Press adapted from Gamov [1965] where this laboratory is located inside a spaceship.*)

Now, we want to demonstrate the significance of this principle when applied to the problem of light deflection.

Imagine that in a small, freely-falling elevator cage, a projector shines light on a screen at the opposite side of the cage (see Figure 11.2). According to Einstein's principle of equivalence, the light, as seen by the freely-falling observer inside the cage, propagates in the same way as it would in an inertial frame of reference, that is, in a straight line. Now, imagine replacing the projector by a laser cannon such that a light ray, propagating parallel to the floor of the cage, has the power of burning a hole into the screen at the same height at which it was released.

An observer on the ground notices that the cage, during the interval of time needed by the laser beam to cross the cage, together with all things inside, is falling through a certain distance towards the ground. When after a hopefully soft landing, both observers debate their observations, there will be one thing they will agree upon: there is one and only one hole in the screen. We have already predicted the place on the screen where this hole will be from the perspective of the freely-falling observer. In order that it is located at the very same place as viewed by the observer on the ground, the light ray, for that observer, has to have bent towards the ground because the target point falls down while the light

### BOX 11.2: LIGHT DEFLECTION IN THE GRAVITATIONAL FIELD NEAR THE EARTH'S SURFACE

Let's consider, for instance, a cage that is freely falling near the surface of the Earth with acceleration $\boldsymbol{g} = \mathbf{9.81}\frac{\mathbf{m}}{\mathbf{s^2}}$ and a light ray that has to cover the distance $\ell$. For light, this takes an interval $\boldsymbol{t} = \frac{\ell}{\boldsymbol{c}}$ of time, whereas the cage is falling towards the ground through the distance $\boldsymbol{s} = \frac{\boldsymbol{g}}{\mathbf{2}}\boldsymbol{t}^2$. In order that the light ray hits the predicted position of the target spot, it has to be deflected by the angle

$$\delta = \frac{g\ell}{c^2} = \frac{GM_E}{R_E c^2} \cdot \frac{\ell}{R_E},$$

where, $M_E$ and $R_E$ now stand for the mass and the radius of the Earth. If $\ell = 10$ m, the light ray crossing the cage will be deflected merely by unmeasurable 0.000 000 000 2 seconds of arc. This corresponds to a free-fall distance of about the diameter of an atomic nucleus.

is on its way from the laser cannon to the screen. (For a quantitative example, see Box 11.2.)

If the two observers wonder why a light ray propagates along a straight line for one observer and along a curved path for the other, they come to the conclusion that gravity must be the cause. Indeed, the observer inside the cage could follow the influence of gravity freely, thus feeling weightless. The other observer, however, could not because the ground he was standing on prevented him from doing so.

When applying the basic idea of the principle of equivalence to light that, approaching from infinity and receding to infinity, is deflected by the Sun, we have to bear in mind that we are no longer allowed to treat the gravitational field of the Sun as being homogeneous when viewed over the entire light path. Therefore, we have to implement the idea of lining up a lot of freely falling cages, all of them fulfilling the conditions of the principle of equivalence. For each of these cages, we have to calculate, at its respective position, the light deflection with respect to a locally straight line which is parallel to the straight lines of all the other cages. In the end, we have to add up all the results.

According to Einstein (Einstein, 1911), (English translation adopted from (Klein et al., 1994))

> "it turns out that ... rays of light passing near the Sun experience a deflection by its gravitational field, so that a fixed star appearing near the Sun displays an apparent increase of its angular distance from the latter, which amounts to almost one second of arc".

To be more precise, Einstein obtained a deflection angle of 0.83 seconds of arc, which is the ballistic result again, but now based on a completely different idea. When this correspondence between Newton's theory and the older relativistic theory of gravitation comes to the problem of the deflection of light, it must not mislead us to think that both theories agree in all their predictions about the propagation of light in gravitational fields. The contrary is true (Treder, 1970): according to Newton's theory, for a light corpuscle approaching the light-deflecting celestial body from infinity, the speed increases, whereas it decreases by the same amount according to Einstein's theory of 1911. Nevertheless, the deflection of light turns out to be the same in both the theories.

In order to confirm these predictions, many expeditions set out to photograph the positions of fixed stars that became visible in the vicinity of the Sun during total solar eclipses. Their aim was to determine whether the star positions were different from those when the Sun was not located in the region of the sky in question. An expedition sent from Argentina to Brazil in 1912 was not successful because of bad weather. Another one set out to Russia in 1914 was unfortunate as well because after the outbreak of World War I, the members of the expedition were temporarily arrested.

Since total solar eclipses are rare events and success is by no means guaranteed, Einstein, on 14 October 1913, wrote a letter to the American astronomer, George Ellery Hale asking, *"if it was possible to image stars close to the Sun in daylight (without an eclipse)"* (Kennefick, 2019). But Hale could only suggest waiting for the next solar eclipses, and this was – as we shall see – fortunate for Einstein, at least from a historical perspective.

## LIGHT DEFLECTION AND GENERAL RELATIVITY

In the years between 1911 and 1915, Einstein completely elaborated the fundamentals of his general theory of relativity. He presented them in November 1915, and published them in the "Annalen der Physik" (Einstein, 1916) in 1916. At the end of a short calculation of the light deflection at the limb of the Sun, Einstein, without referring to his earlier result, let the reader know that (English translation adopted from (Klein et al., 1996))

> "... a ray of light going past the Sun undergoes a deflexion of 1.7″; and a ray going past the planet Jupiter a deflexion of about .02″".

How can we understand this doubling of the deflection angle as compared to its value obtained on the basis of the principle of equivalence?

In Einstein's general theory of relativity, gravitation is no longer a force in the sense of Newtonian dynamics. It manifests itself as the curvature of the spacetime, caused by the presence of matter and energy, where material bodies and light move. Even mass can be expressed "more geometrico", namely as the length

$$R_S = 2\frac{GM}{c^2}, \tag{11.3}$$

that is called Schwarzschild radius, after Karl Schwarzschild (1873–1916), who found an exact solution of Einstein's field equations where this length scale naturally occurs. For instance, the Schwarzschild radius of the Sun is $R_{S\odot} \approx 3$ km, and that of the Earth is $R_{S(E)} \approx 0.89$ cm.

Curvature of spacetime is the curvature of time and space. The curvature of space cannot, however, be taken into account by the principle of equivalence being a local principle. In fact, light deflection was described by a deviation from a Euclidean straight line that was traversed by a light ray in an inertial frame of reference. What if space is curved? In that case, the bodies, light as well, follow trajectories, called geodesic lines or geodesics, that are as straight as possible, but different from the straight lines of Euclidean geometry.

In order to illustrate this, it is very instructive not to study light deflection from the outset but instead, from the deflection of a massive, albeit ultra-relativistic particle that passes by a celestial body of mass $M$.

The result consists of two additive contributions. Only one of them (the second term of equation (11.4) in Box 11.3) depends on a characteristic of the particle, namely, the velocity $v_\infty$ it has "at infinity". This is the ballistic part of the result. For $v_\infty = c$, this term alone corresponds to the ballistic deflection angle shown in equation (11.2). The other contribution describes the deviation of the straightest possible line of the curved space with respect to a Euclidean straight line. The amount of this deviation depends, according to equation (11.3), exclusively on the mass $M$ that produces the curvature of that space. As long as equation (11.1) is fulfilled, it is just the same as the Newtonian deflection angle $\delta_N$ for light.

For light rays grazing the limb of the Sun ($M = M_\odot$ and $b = R_\odot$), both contributions to the deflection angle add up to the predicted value $\delta_{E\odot} = 1.75''$. That

### BOX 11.3: GENERAL-RELATIVISTIC DEFLECTION OF AN ULTRA-RELATIVISTIC PARTICLE

Suppose that an ultra-relativistic particle, having the velocity $v_\infty$ "at infinity", would pass by the celestial body at a distance $b$ in case that no attractive power emanated from that body (see Figure 11.1). Obviously, the radius $R$ of the celestial body is the smallest possible impact parameter $b$. For such a particle, the deflection angle in a weak gravitational field, where equation (11.1) holds, turns out to be

$$\delta = \frac{R_S}{b} + \frac{R_S}{b} \cdot \frac{c^2}{v_\infty^2}. \tag{11.4}$$

Applied to the deflection of light ($v_\infty = c$), we end up with twice the Newtonian deflection angle (E: Einstein)

$$\delta_E = 2 \cdot \frac{R_S}{b} = 4 \cdot \frac{GM}{c^2 b} = 2 \cdot \delta_N. \tag{11.5}$$

can be described, in a catchy way, as twice the ratio between the Schwarzschild radius of the Sun and its real size.

We realize in retrospect that, irrespective of the underlying theory, the deflection angle $\delta$ is determined by the dimensionless ratio of the only two available length scales of our problem: the Schwarzschild radius $R_S$ of the light-deflecting celestial body, and the impact parameter $b$ whose minimum value is given by the size $R$ of that body. This we could have anticipated without any calculation from a dimensional analysis. The only matter of debate is the occurrence or non-occurrence of the factor 2. It turns out that the ratio $\frac{R_S}{R}$ is of crucial importance for estimating the order of magnitude of other relativistic effects as well.

The general theory of relativity allows for yet another interpretation of the deflection of light that is very instructive because of its affinity to the phenomenon of fata morgana. Instead of thinking about the propagation of light in a space that is curved by the mass of the Sun, we can imagine a flat space endowed with a medium whose refractive index $n$ varies with the distance $r$ from the Sun as

$$n(r) = 1 + \frac{R_S}{r}.$$

For light propagating in that space from a fixed initial position to a fixed final position, Einstein's deflection angle shown in equation (11.5) results from Fermat's principle, according to which the light trajectory realised in nature is of minimum duration (at least in most cases).

Had the early expeditions been successful, the precision of measurement at that time would have been high enough to provide deflection angles, and error estimates not grouping around 0.83 seconds of arc, but more likely around twice that value. Einstein would have had to explain this discrepancy and to revise his theory.

The next opportunity to test this new prediction was the solar eclipse of 29 May 1919. Two British expeditions to Sobral (Brazil) and to the island of Principe (São Tomé and Príncipe, West Africa) yielded the results $\delta = 1.98'' \pm 0.16''$ and $\delta = 1.61'' \pm 0.40''$, respectively. These were presented at a joint meeting of the Royal Society and the Royal Astronomical Society on 6 November 1919 in London (Kennefick, 2019). The Astronomer Royal, F.W. Dyson, stated:

> "After a careful study of the plates, I am prepared to say that there can be no doubt that they confirm Einstein's prediction. ... Light is deflected in accordance with Einstein's law of gravitation".

The president of the Royal Society, J. J. Thomson, who was at the same time the chairman of that meeting, put this result into a historical context:

> "This is the most important result obtained in connection with the theory of gravitation since Newton's day.
>
> ...

If it is sustained that Einstein's reasoning holds good [...] then it is the result of one of the highest achievements of human thought".

The observation of a total solar eclipse on 21 September 1922 at Wallal (Australia) yielded the result for the deflection angle (Campbell & Trumpler, 1923), $\delta = 1.72'' \pm 0.11''$ (see Figure 11.3).

The results of 1919 already entailed the "canonization" of Einstein. In fact, how different his position would be if a light deflection of 0.87 seconds of arc were to be proven!

## THE GRAVITATIONAL LENS EFFECT

In modern terminology, we call the Sun a gravitational lens because it deflects light from a straight trajectory due to gravitational attraction like a "real" lens does by its refractive power.

As the curved light path of Figure 11.1 indicates, the deflection of light is a continuous process. However, the distance $D_L$ between observer and lens and, likewise, the distance $D_{LS}$ between lens and source are very large as compared

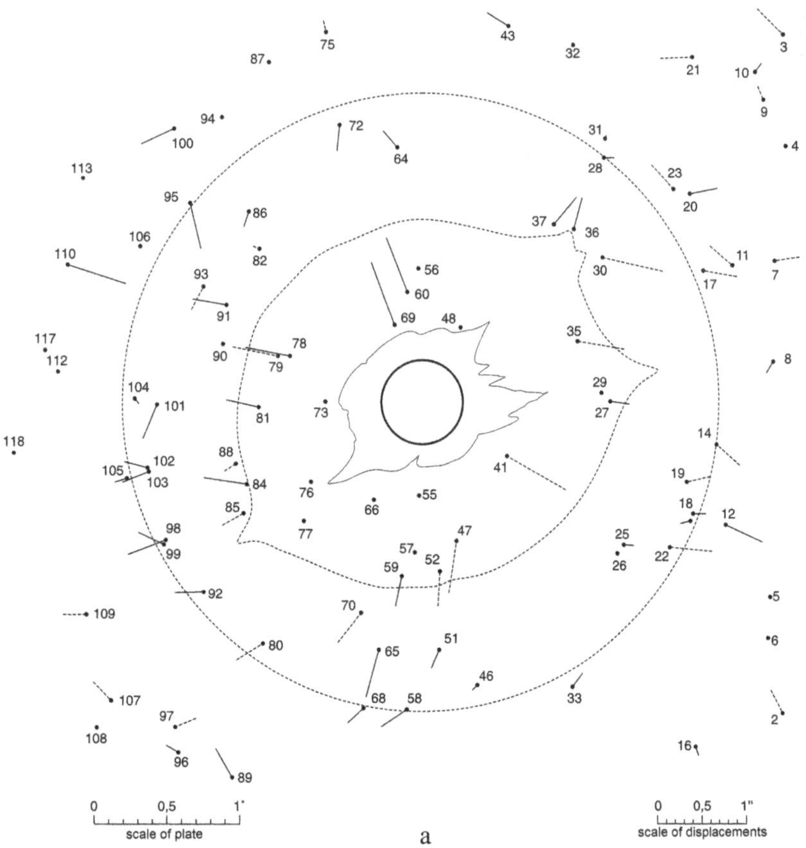


11.3 The solar eclipse of 21 September 1922: (a) the lines, enlarged by a factor 2250, indicate the displacement of the positions of the stars and (b) radial lengths of the displacement lines from part (a) of the figure, plotted as a function of their distance from the centre of the Sun. The upper graph (designated as $\gamma = 1$) is the prediction of the general theory of relativity, and the lower graph corresponds to ballistic light deflection. (*Source: Copyright by Campbell & Trumpler (Campbell & Trumpler 1923), Courtesy Special Collections, University Library, University of California Santa Cruz. Lick Observatory Records.*)

11.3 *(continued)*

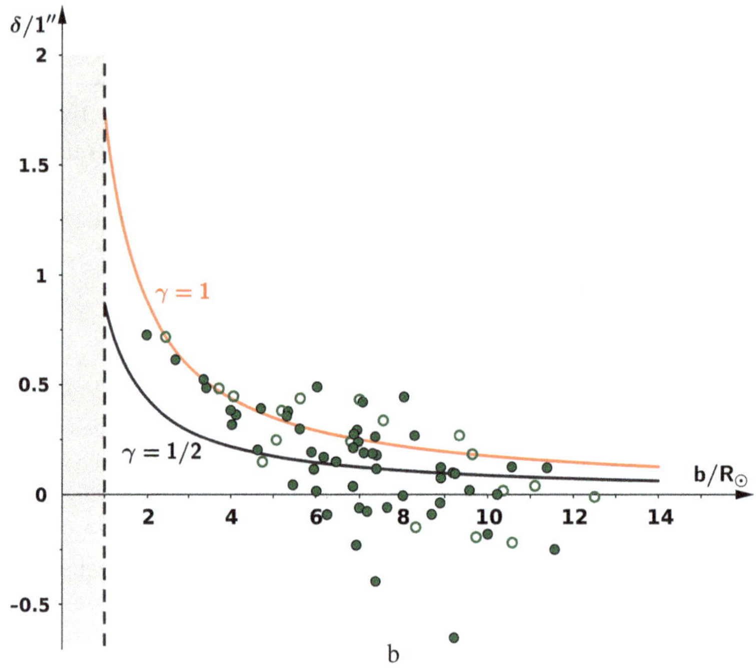


to the size of the lens itself. Therefore, we may assume that the process of deflection does not proceed continuously, but only in a single act at a point P close to the lens. This point, a distance *b* away from the gravitational lens, defines, together with that lens, the lens plane. In optics, this approach corresponds to the approximation of a thin lens.

The general situation of the gravitational lens effect is depicted in Figure 11.4. Without the light-deflecting gravitational lens L, the observer O would see the light source at its original position S at the angle $\theta_S$ against the line $\overline{OL}$. However, since the light is deflected by the angle $\delta$, the image I of the source appears, instead, at the angle $\theta$ against this line.

Even though we started with one image I in Figure 11.4, it turns out that already in the simplest case of a point-mass gravitational lens, there are two images. The second image is located on the opposite side of the line $\overline{OL}$ as seen from the first one. If the alignment is perfect like $\overline{OLS_0}$, the two images, for reasons of symmetry, merge into what is called an Einstein ring (see Figure 11.4). In fact, the two image positions $\theta_1$ and $\theta_2$ and the angular radius $\theta_E$ of the Einstein ring as well are solutions of a quadratic equation.

If, however, the gravitational lens is extended in size and transparent (as is the case for galaxies and galaxy clusters but not for stars), the impact parameter *b* may be smaller than the outer dimensions of the lens. Then, the mass *M* occurring in the deflection angle shown in equation (11.5) is the projected mass

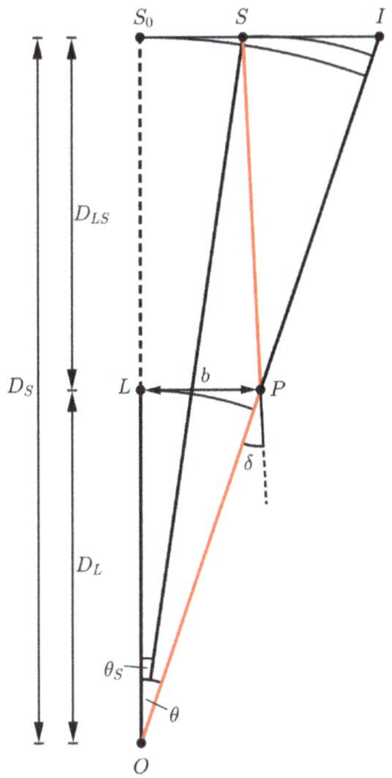


11.4 (a)
The geometry of the gravitational lens effect. The meaning of the symbols is as follows: O: observer, L: gravitational lens, S: source, I: image, $D_L$, $D_{LS}$, $D_S$: the corresponding distances, $b$: impact parameter, $\delta$: deflection angle. (*Source: Copyright by Karl-Heinz Lotze, Silvia Simionato*)

11.4 (b)
The Einstein ring B1359+154, known as the "Cosmic Horseshoe". In the centre, the lensing galaxy can be seen. For that image to arise, the light source has to have the position $S_0$ in the left part of the figure. (*Source: Copyright by ESA/Hubble, NASA/Public domain*)

inside an imaginative circle of radius $b$ in the lens plane, thus becoming a function of $\theta$. Therefore, in general, equation (11.6) of Box 11.4, that is called lens equation, is more complicated than the quadratic equation (11.7). Accordingly, in general, there are more than two images.

Einstein investigated the point-mass case as early as 1912 (Renn et al., 1997), that is, three years before he completed and published his general theory of relativity where he calculated the correct value of light deflection by the Sun. The results of 1912 were published only in 1936 in a short essay in the journal Science (Einstein, 1936). In this essay, he calculated the angular radius $\beta$ (here called $\theta_E$) of that luminous ring that would form if both stars and the observer were perfectly aligned. The ring then decays into a double image if the observer is not aligned with the two stars. After estimating the order of magnitude of the effects, Einstein arrived at the conclusion:

> "Of course, there is no hope of observing this phenomenon directly. First, we shall scarcely ever approach closely enough to such a central line. Second, the angle $\beta$ will defy the resolving power of our instruments".

### BOX 11.4: LENS EQUATION AND EINSTEIN RING

Because the angles $\delta$, $\theta_S$, and $\theta$ are small, we may, in good approximation, identify the distance *b* of Figure 11.4 with the impact parameter *b* defined in Figure 11.1.

For the same reason, it is allowed to interchangeably replace the lines $\overline{S_0I}$, $\overline{S_0S}$, and $\overline{LP}$ by arcs centred at O, and $\overline{S_0I}$ by an arc centred at P, as shown in Figure 11.4. Then, we can read off from that figure that, within the scope of our approximations, the image position $\theta$ of a source obeys the equation

$$\theta \boldsymbol{D}_\mathbf{S} = \theta_\mathbf{S} \boldsymbol{D}_\mathbf{S} + \delta \boldsymbol{D}_\mathbf{LS}. \tag{11.6}$$

This equation adopts its simplest form for a point-mass gravitational lens because in that case, all impact parameters *b* we can conceive are bigger than the lens itself. Consequently – and as always so far – *M* in equation (11.5) is the total mass of the lens. Inserting the deflection angle shown in equation (11.5) and $b \approx \theta \cdot D_L$, equation (11.6) turns into the quadratic equation

$$\theta^2 - \theta_\mathbf{S} \cdot \theta - 2\boldsymbol{R}_{\mathbf{S(L)}} \frac{\boldsymbol{D}_\mathbf{LS}}{\boldsymbol{D}_\mathbf{L} \boldsymbol{D}_\mathbf{S}} = \mathbf{0}, \tag{11.7}$$

where $R_{S(L)}$ is the Schwarzschild radius of the lens.

If, in particular, $\theta_S = 0$ which means that the light source, now at position $S_0$, is arranged precisely on the line $\overline{OL}$, the image, because of symmetry, is a circle called Einstein ring. Its angular radius

$$\theta_\mathbf{E} = \sqrt{2\boldsymbol{R}_{\mathbf{S(L)}} \frac{\boldsymbol{D}_\mathbf{LS}}{\boldsymbol{D}_\mathbf{L} \boldsymbol{D}_\mathbf{S}}} \tag{11.8}$$

is a measure of the size of the gravitational lens effect (see Table 11.1).

One year after this pessimistic assessment about the observability of the gravitational lens effect, the imaginative Swiss astronomer, Fritz Zwicky presented a new proposal (Zwicky, 1937). He made

> "some calculations which show that extragalactic nebulae offer a much better chance than stars for the observation of gravitational lens effects".

Indeed, nebulae, whose extragalactic nature has been established only less than a decade before by Hubble, have two advantages over stars as gravitational lenses: they are much more massive because they can be composed of a few hundred billion stars and, moreover, they are not point-like so that perfect alignment of light source, gravitational lens, and observer is not essential.

**_Table 11.1_ Examples of Einstein radii**

| *Lens* | $\frac{M}{M_\odot}$ | $D_L$ | $\frac{D_S}{D_{LS}}$ | $\theta_E$ |
|---|---|---|---|---|
| Sun | 1 | 1 AU | ≈ 1 | 41.3″ |
| Sirius | 2.12 | 2.67 pc | ≈ 1 | $8.1 \cdot 10^{-2}$″ |
| Massive Compact Halo Object (MACHO) in front of Large Magellanic Cloud (LMC) | 0.08 | 8 kpc | 1.18 | $2.7 \cdot 10^{-4}$″ |
| Galaxy | $10^{12}$ | 1 Gpc | 2 | 2″ |
| Galaxy cluster | $10^{14}$ | 1 Gpc | 2 | 20″ |

(1 AU = $1.5 \cdot 10^8$ km; 1 pc = $3.086 \cdot 10^{13}$ km)

In Table 11.1, a few examples of angular radii of Einstein rings are listed. In what follows, we comment on them in more detail.

- In this discussion, the Sun occupies an exceptional position because of its close proximity to us as the observers. Therefore, the condition $D_S \approx D_{LS}$ is fulfilled exceptionally well. If we compare the angular radius $\theta_E = 41.3''$ with the one of the solar disk in the sky that, on average, amounts to 960″, it clearly becomes obvious that the Sun obscures its Einstein ring. The same is true for the second image of a star that is not perfectly aligned with the observer and the Sun's centre. What we do observe is only one of the two solutions of equation (11.7), namely, the displacement $\theta - \theta_S = \delta_E$ of the stars like in Figure 11.3.

The invisibility of the Einstein ring raises the question of how far we would have to fly away from the Sun in order that the ring becomes visible. It is at least $D_L = 540$ AU or, equivalently, almost 18 times the distance of planet Neptune from the Sun. At that distance, for the first time, the Sun and its Einstein ring have the same angular radius in the sky, that is, 1.77″. Erroneously, this distance is sometimes called the focal length of the Sun as a gravitational lens. However, gravitational lenses don't have a focus but a focal line instead, as Sir Oliver Lodge noticed as early as in 1919 (Lodge, 1919).

- Sirius, the star in the cosmic neighbourhood of the Sun, acting as a gravitational lens for a much more distant star, would give rise to an effect in the range of hundredths of seconds of arc.
- An object in the halo of the Milky-Way galaxy called Massive Compact Halo Object (MACHO), that we can envisage as a Jupiter-like planet or a "brown dwarf", this time acting as a gravitational lens for one of the many stars in the Large Magellanic Cloud (LMC), would produce a micro gravitational lens effect of the order of magnitude of $10^{-4}$ seconds of arc.

These orders of magnitude seem to be in favour of Einstein's pessimism. Nowadays, however, the precision of measurement has become so advanced that, for instance, the GAIA satellite can measure the light deflection by the Sun not only for stars near the solar limb but also for those that are more than 90° away. The deflection caused by the big planets has to be taken into account as well if we want to derive from observations the "genuine" positions of the stars.

- If galaxies and clusters of galaxies act as gravitational lenses for light sources twice as distant from the observer as the lens itself, they make the gravitational lens effect strong, thus confirming Zwicky's predictions. To give an example, the Einstein ring of Figure 11.4 has the angular radius $\theta_E = 5.1''$, which is in good agreement with the order of magnitude given in Table 11.1.

## THE OPTICAL GRAVITATIONAL LENS EXPERIMENT

Finally, we want to tackle the problem of how to construct a lens made of plexiglass in order that its light deflection due to refraction mimics the deflection shown in equation (11.5) that is due to gravitation. This problem that is particularly useful from the point of view of teaching has been treated repeatedly in the literature (Refsdal & Surdej, 1994). However, we can add new results.

As it is realised in nature for the examples we want to simulate, we are looking for lenses with rotational symmetry. This enables us to restrict the discussion of the trajectory of light to a problem in a plane.

We choose the *x*-axis of Figure 11.5 to be the axis of symmetry. Aiming at a lens that has, in the end, a simple shape, we assume that the surface facing the source is plane. Therefore, light emanating from the source S enters the plexiglass block (the shaded region of Figure 11.5) through this surface at the distance $b = y_P$ from, and parallel to, the axis of symmetry. The opposite surface of the lens is the one to be determined by means of Snell's law and the deflection angle given in equation (11.5). An observer would not see the light source at position S; instead, tracing the refracted light ray backwards, at position S′.

As is to be expected, the problem to find the shape of the curved surface of the plexiglass lens is most easily solved for a point-mass gravitational lens because in that case, the light-deflecting mass is independent of the impact parameter $y$ (see Box 11.5). Consequently, the precise shape of the resulting exponential function is determined by the Schwarzschild radius $R_S$ of the gravitational lens.

In Figure 11.6, all "gravitational" lenses we manufactured from plexiglass are shown.[1] The lenses in nature and their respective counterparts made of plexiglass are from left to right:

1 For information regarding numerical solutions, calculations, and simulations with Geogebra for the plexiglass lenses, please contact the authors.

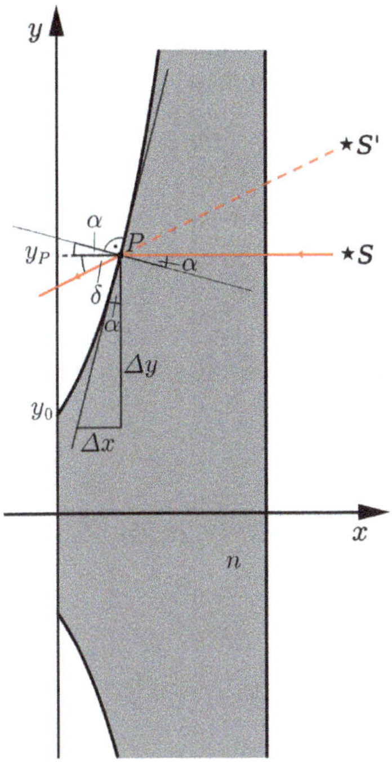


11.5 (a)
Construction of a point-mass "gravitational" lens made of plexiglass. (*Source: Copyright by Karl-Heinz Lotze, Silvia Simionato*)

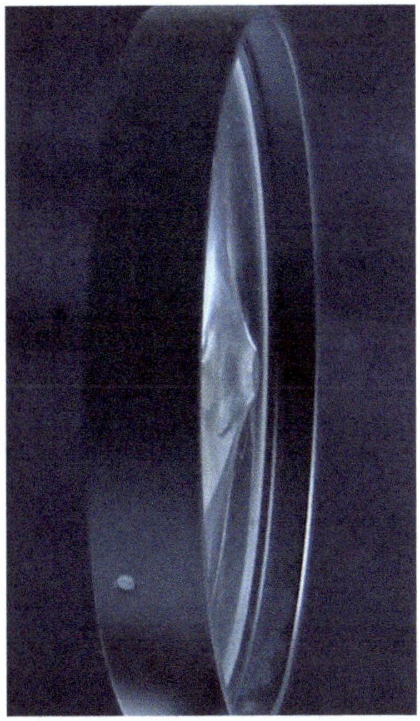

11.5 (b)
A lens made of plexiglass. It has a diameter of 15 cm and is made of plexiglass with the refractive index $n$ = 1.5 against air. Its equivalent Schwarzschild radius $R_S$ = 0.92 cm corresponds to that of a celestial body of about the mass of the Earth. (*Source: Copyright by Th. Schott*)

- The point mass and the exponential function of Figure 11.5.
- The singular isothermal sphere that results if stars inside a galaxy move under the influence of their own gravitational field, and the lens called axicon, a straight cone with rotational symmetry.
- The uniform disk and the corresponding paraboloid with rotational symmetry.
- The Plummer sphere and the Kuzmin disk, closely related to each other. Both of them have more complicated mass distributions inside the gravitational lens. Accordingly, the plexiglass lenses, despite their rotational symmetry, cannot be described by simple geometrical figures.

Equation (11.9) is the basis of the calculation of the shape of the plexiglass simulator, not only for the point-mass gravitational lens. However, in all the other cases, the light-deflecting mass $M$ is a function of $y$. This renders the solution for $y$ of equation (11.9) more demanding.

## GRAVITATIONAL LENSES IN THE CLASSROOM

Gravitational lenses are a powerful educational tool. Light deflection by gravitational fields is the key mechanism behind the gravitational lens effect. In

## BOX 11.5: THE OPTICAL "GRAVITATIONAL LENS" SHAPED LIKE THE FOOT OF A WINEGLASS

At point P of Figure 11.5 where the light ray enters the surface yet to be determined, we replace the function $y(x)$ that describes the lens shape by its tangent, and draw the perpendicular through P. Measured against this perpendicular, $\alpha$ is the angle of incidence of the light that gets refracted away from the perpendicular when it leaves the plexiglass block of refractive index $n$. The refraction angle is the angle $\alpha$ of incidence, enlarged by the small angle $\delta$ of deflection. In terms of these angles, Snell's law reads

$$n = \frac{\sin(\alpha + \delta)}{\sin\alpha} = \cos\delta + \cot\alpha \cdot \sin\delta.$$

According to Figure 11.5, the slope of the tangent line through P is

$$\frac{dy}{dx} = \cot\alpha = \frac{n - \cos\delta}{\sin\delta}.$$

Because the deflection angle shown in equation (11.5) is small, this expression simplifies considerably so that we have to solve the equation

$$\frac{dy}{dx} = \frac{n-1}{\delta} = \frac{c^2(n-1)}{4GM} \cdot y. \tag{11.9}$$

The result is the exponential function

$$y = y_0 \cdot \exp\left[\frac{c^2(n-1)}{4GM} \cdot x\right] = y_0 \cdot \exp\left[\frac{(n-1)}{2} \cdot \frac{x}{R_S}\right]. \tag{11.10}$$

Anticipating this result, we have already plotted it correctly in Figure 11.5. If we rotate the shaded region of that figure clockwise by 90° and keep in mind the rotational symmetry, our lens looks like the foot of a wineglass.

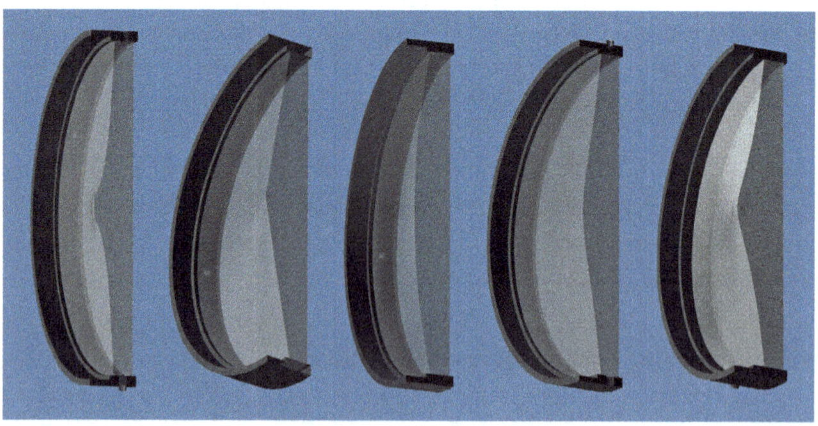

11.6
The "gravitational" lenses manufactured in the workshop of the Faculty of Physics and Astronomy of the Friedrich Schiller University Jena, Germany. Cross sections are displayed, from left to right, of lenses corresponding to a point mass, a singular isothermal sphere, a uniform disk, a Plummer sphere, and a Kuzmin disk. (*Source: Copyright by Computer graphics: Th. Köhler*)

11.7
Upper panel: experimental setup of the optical gravitational lens experiment using an optical bench; lower panel: a second setup that, including a video camera, is well-suited for projection in a lecture hall. The component parts of both setups are described in the text. (*Source: Copyright by Upper photo: the authors Lower photo: J. Schneider*)

particular, the simulation of this effect by suitably shaped lenses made of plexiglass possesses an integrating power in bringing together general relativity and classical optics.

In the upper panel of Figure 11.7, a setup of the optical gravitational lens experiment is shown. The component parts are from right to left: the light source; a halogen lamp (12 V,100 W); a collimator lens; the gravitational lens simulator, in this case, the point-mass lens shown in Figure 11.5; a screen with a pinhole (pinhole camera); and a screen on which to project the image of the source as it is produced by the lens simulator.

In Figure 11.7, the lens is placed perpendicular to the line connecting the light source and the pinhole, with its centre on that line. The resulting image is the Einstein ring on the screen. (The bright spot in the centre of the ring is not an image produced by "gravitational" lensing. It is merely the undisturbed view at the light source through the plane central part, with radius $y_0$, of the lens, see Figure 11.5. This part is blocked in a real gravitational lens event.)

We can break the perfect alignment of source, lens, and pinhole by moving the lens aside and/or tilting it in such a way that the parallel light coming from the source does not enter the plane surface of the lens perpendicularly. In doing so, we can create multiple images and arcs instead of an Einstein ring.

In the bottom panel of Figure 11.7, we have shown a second setup that has a few advantages over the first one. The component parts are (again from right to left): the light source, here the black spot on the white screen; a lamp, mounted at the table, to illuminate the source if necessary; the gravitational lens simulator, here again, the point-mass simulator of Figure 11.5 that can easily be replaced by one of the other lenses placed on the edge of the table; and a video camera.

The advantages of this setup are:

- It is possible to project the images on the wall of a lecture hall in real-time. Thus, a broader audience can follow and discuss the experiment.
- We can easily replace the point-like source by other patterns like grids of horizontal and vertical lines and even coloured images of real astronomical objects. In this way, it is possible to study the images of extended sources and their parity (inversion).
- If we move the lens perpendicular to the optical axis of the setup, that is, across a field of light sources in the background, we can see, in motion pictures, distorted images and short-lived Einstein rings.

Using exactly this last setup, we have produced the left panel of Figure 11.8, which is just a screenshot of such a motion picture. We chose a picture of the Hubble Ultra Deep Field in inverted colours as the light source. Its undisturbed view can be seen in the leftmost part of that panel. The broad outer ring is nothing but the frame of the lens. Inside this ring, highly distorted galaxies as well as an Einstein ring as an image of a galaxy that happened to be perfectly aligned with the lens and the camera can be seen. (The feature in the very centre of the ring has nothing to do with the gravitational lens effect, see comment above.)

For the sake of comparison, the photo on the right side of Figure 11.8 shows a real gravitational lens event. Here, the gravitational lens is the galaxy cluster Abell S1063. We can easily see arclets and other distorted images of background galaxies, very similar to the left-hand panel of Figure 11.8.

However, we should bear in mind that one of those images is produced by a whole cluster of galaxies, the other one by a point-mass gravitational lens

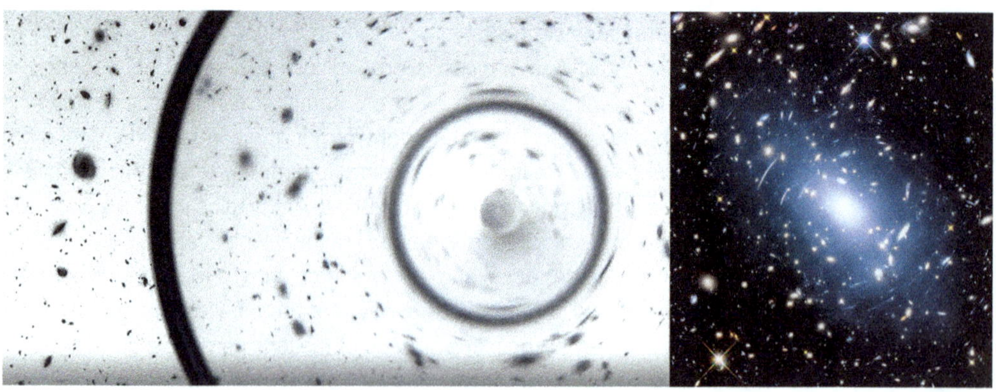

11.8
Left: screenshot of a "gravitational" lens event produced using the equipment shown in the lower panel of Figure 11.8. Background (light source): Hubble Ultra Deep Field; right: the galaxy cluster Abell S1063 acting as a real gravitational lens. (*Source: Copyright by Photo: Th. Schott, Background and right figure: ESA/ Hubble, NASA/ Public domain*)

corresponding to approximately the mass of the Earth. The primary criterion of the construction of our plexiglass lenses was the practicability in the mechanical workshop that produced them instead of the correspondence to real astronomical lenses.

At this point, the reader could argue that none of these equipment, in particular the gravitational lens simulator, is at his disposal. This is, however, only partially true. The foot of a wineglass, with the goblet broken away, is available to everybody. In this case, one side of the wineglass foot is not perfectly plane, and the shape of the other one does not precisely follow an exponential function. Nevertheless, the wineglass foot is close enough to the lens shown in Figure 11.5 in order to reproduce gravitational lens effects at least qualitatively. In order to do so, we simply should use this lens as a magnifying glass, move it in front of the source, tilt it, and vary the distances between source and lens as well as between lens and observer [Huwe & Field, 2015].

## CONCLUSION

In this chapter, we discussed the gravitational lens effect. As a matter of fact, the subject of light deflection in a gravitational field covers many more topics than those we selected from the point of view of teaching general relativity. In the narrow sense of these words, "gravitational lens effect" denotes the deflection of light in weak gravitational fields, as described by the inequality shown in equation (11.1).

In weak gravitational fields, we discriminate among three effects (see also (Congdon & Keeton, 2018)), namely:

- The **strong** effect, which produces spatially resolved multiple images; among them are giant luminous arcs. The light sources can be quasars or galaxies, and the gravitational lenses can be again galaxies or clusters of galaxies. In cosmology, this effect is used as an instrument to study the distribution of (dark) matter, and to determine cosmological parameters, e.g. the Hubble constant.
- The **micro** effect, where the images cannot be resolved. However, it is possible to measure the total brightness amplification of all images that emerges when source, lens, and observer are in relative motion. In this case, so-called MACHOs (see Table 11.1) serve as gravitational lenses. Light sources are stars in the LMC. Moreover, this micro effect is used as a method to detect exoplanets around stars other than the Sun.
- The **weak** effect, which is characterised by galaxy clusters as gravitational lenses that influence the shape and orientation of background galaxies. In contrast to multiple images, these perturbations ("cosmic shear") can only be detected by statistical investigations of many sources. Applications of this effect are in the field of cosmology, regarding, in particular, the large-scale structure of the universe.

Within this classification scheme, the deflection of light at the solar limb, as it is treated in this chapter, belongs to the strong effect in the weak gravitational field of the Sun. In particular, the comparatively small distance between the gravitational lens and the observer gives rise to some peculiarities we described above. Moreover, the stars, which are the sources of these images produced by the Sun as a gravitational lens, can normally be observed as non-deflected original light sources in the night sky, that is, without a solar eclipse.

In this chapter, among all these effects, we selected the strong gravitational lens effect in weak gravitational fields for educational reasons and linked it up with the deflection of light at the solar limb. These reasons are:

- Together with the perihelion shift of the orbit of the planet Mercury and the gravitational redshift, the light deflection at the solar limb is one of the three classical tests of general relativity in weak gravitational fields.
- The principle of equivalence, as well as the Schwarzschild radius, belongs to the very fundamentals of general relativity. The consequences drawn from this principle and the explanation why this principle, local by its very nature, can yield only one half of the amount of light deflection, as compared to the prediction of the full theory, can hardly be overestimated for an understanding of general relativity. It is a classic example for the gedanken experiments which are so characteristic of Einstein's way of thinking.
- We can learn a lot about the "nature of science" from the long history of light deflection. This concerns, in particular, the relationship between theory and experiment that has changed with this test. While the main task of theoretical physicists used to consist of evaluating the results of experiments, this changed over time: now an experiment (an observation) is often *predicted*, and experimentalists are challenged to prove or disprove this prediction. This, by the way, has had a considerable influence on the intellectual development of philosophers and historians of science like Sir Karl Popper [Kennefick, 2019].
- The strong gravitational lens effect is imaginable because of the impressive images it produces like the Einstein ring of Figure 11.4. This can give rise to a strong motivation to study this effect.
- Given Einstein's expression in equation (11.5) of the deflection angle, the lens equation (11.7), describing the gravitational lensing by a point mass, can be derived even in upper high school. Being a quadratic equation, its solutions can be discussed analytically or graphically in detail. In this chapter, we did not demonstrate this only because of lack of space. Instead, we refer to the literature (Congdon & Keeton, 2018; Dodelson, 2017).
- The instructive combination "general relativity meets classical optics" is an added bonus that allows for hands-on simulations of gravitational lens effects by optical instead of real gravitational lenses.

Apart from gravitational lens effects in weak gravitational fields, there exist, of course, strong field effects for which $\frac{R_S}{R} \lesssim 1$ holds instead of equation (11.1).

These effects are related to neutron stars and black holes. Specifically, they refer to the deflection of light that emanates from the environment, for instance from accretion disks, of those celestial bodies. In these cases, the observer not only can experience regions that lie, as seen from his position, behind a black hole but also he can observe multiple images and "shadows", which are formed after light orbited the black hole partially or even several times.

The idea of light deflection in strong gravitational fields has a long history as well and can be traced back to the late 18th century (Michell, Laplace (Montgomery et al., 2009)). But it was only in 2019, that the "shadow" of a black hole was observed for the first time. It was the black hole in the centre of the galaxy Messier 87 (EHT Collaboration, 2019), seen against its surrounding luminous accretion disk.

Light deflection in strong gravitational fields might be as much motivating. However, the image of the "shadow" of a black hole is the result of a complicated interplay between observation and theoretical modelling, and thus, much harder to understand.

## ACKNOWLEDGEMENT

The authors gratefully acknowledge the always pleasing and fruitful collaboration with the editors of this book, Magdalena Kersting and David Blair. Their comments and suggestions helped very much to give shape to this chapter. Our special thanks go to David who was kind enough to read our text carefully from the point of view of English wording and style.

## REFERENCES

Campbell, W.W., Trumpler, R. (1923). "Observations of the Deflection of Light in Passing Through the Sun's Gravitational Field." *Lick Observatory Bulletin XI*, 346, 41–54.

Congdon, A. B., Keeton, Ch. R. (2018): *Principles of Gravitational Lensing*. Springer Nature Switzerland AG.

Dodelson, S. (2017). *Gravitational Lensing*. Cambridge University Press.

Einstein, A. (1911). "Über die Ausbreitung der Schwerkraft auf die Ausbreitung des Lichtes." *Annalen der Physik*, 35, 898–908.

Einstein, A. (1916). "Die Grundlage der allgemeinen Relativitätstheorie." *Annalen der Physik*, 49, 769–822

Einstein, A. (1936). "Lens-Like Action of a Star by the Deviation of Light in the Gravitational Field." *Science*, 84, 506–507.

Gamov, G. (1965). *Mr. Tompkins in Wonderland*. Cambridge University Press.

Huwe, P., Field, S. (2015). "Modern Gravitational Lens Cosmology for Introductory Physics and Astronomy Students." *The Physics Teacher*, 53, 266–270

Jaki, S.L. (1978). "Johann Georg von Soldner and the Gravitational Bending of Light, with an English Translation of His Essay on It Published in 1801." *Foundations of Physics*, 8 (11/12), 927–950.

Janssen, M., Schulmann, R., Illy, J. et al. (Eds.) (2002)." Einstein, A., *Grundgedanken und Methoden der Relativitätstheorie, in ihrer Entwicklung dargestellt*. In: Janssen, M.,

Schulmann, R., Illy, J. et al., "The collected papers of Albert Einstein, vol. 7, the Berlin years: Writings 1918–1921." Princeton University Press.

Kennefick, D. (2019). *No Shadow of a Doubt – The 1919 Eclipse That Confirmed Einstein's Theory of Relativity.* Princeton University Press.

Klein, M. J., Kox, A. J., Renn, J. et al. (Eds.) (1994). *The Collected Papers of Albert Einstein, Vol. 3, The Swiss Years: Writings 1909–1911.* Princeton University Press (English Translation Supplement).

Klein, M. J., Kox, A. J., Schulmann, R. et al. (Eds.) (1996). *The Collected Papers of Albert Einstein, Vol. 6, The Berlin Years: Writings 1914–1917.* Princeton University Press (English Translation Supplement).

Lodge, O. J. (1919). "Gravitation and Light." *Nature*, 104, 2614, 354.

Montgomery, C., Orchiston, W., Whittingham, I. (2009). "Michell, Laplace and the Origin of the Black Hole Concept." *Journal of Astronomical History and Heritage*, 12(2), 90–96.

Refsdal, S., Surdej, J. (1994). "Gravitational Lenses." *Reports on Progress in Physics*, 56, 117–185.

Renn, J., Sauer, T., Stachel, J. (1997). "The Origin of Gravitational Lensing: A Postscript to Einstein's 1936 Science Paper." *Science*, 275, 184–186.

Soldner, J. (1801). "Über die Ablenkung eines Lichtstrals von seiner geradlinigen Bewegung durch die Attraktion eines Weltkörpers, an welchem er nahe vorbeigeht." In: Bode, J. (Ed.), *Astronomisches Jahrbuch für das Jahr* 1804, 161–172.

Thorpe, E. (Ed.). (1921). *The Scientific Papers of the Honourable Henry Cavendish, F.R.S., Volume II: Chemical and Dynamical.* Cambridge University Press

Treder, H.-J. (1970). "Über die Schwere des Lichtes in der Newtonschen und in der Einsteinschen Gravitationstheorie." *Annalen der Physik*, 7.Folge, 25 (3), 315–320

Will, C. (1988). "Henry Cavendish, Johann von Soldner, and The Deflection of Light." *American Journal of Physics*, 56, 413–415.

Zwicky, F. (1937). "Nebulae as Gravitational Lenses." *Physical Review*, 51, 290. The Event Horizon Telescope Collaboration. First M87 Event Horizon Telescope Results. I. Shadow of the Supermassive Black Hole. Astrophysical Journal Letters 875 (2019) L1, (April 10).

# 12 Introducing the geometric concepts of general relativity with sector models

*Ute Kraus, Corvin Zahn, and Sven Weissenborn*

**WHAT?**

This chapter introduces sector models as a geometric tool to visualise the geometry of curved spaces. Ute Kraus, Corvin Zahn, and Sven Weissenborn describe an introductory teaching unit for secondary school students that aims to foster an intuitive geometric understanding of general relativity.

**FOR WHOM?**

This chapter is intended for readers who wish to teach the basics of general relativity at the secondary school level.

## INTRODUCTION

Gravity is geometry, and geometry lends itself to the use of models and graphic methods. We propose a class of models that we call sector models, which facilitate studying curved spaces and spacetimes of general relativity. In this chapter, we show how sector models can visualise three-dimensional curved spaces, and how they allow constructing geodesics with pencil and ruler. More generally, sector models can provide a clear and physically sound visualisation of many general relativistic concepts.

## THE TIME HAS COME TO TEACH GENERAL RELATIVITY AT SCHOOL

Relativity has a formidable reputation for being difficult. Nevertheless, we propose to teach it in school as a standard topic, and we argue for teaching the core concepts as early as grade ten (age 16 years). This chapter is written keeping teachers in mind who want to teach the basics of general relativity at school. It does not presume that the reader has studied relativity, either special or general or that the reader is familiar with non-Euclidean geometry. We describe a teaching unit that is part of a novel approach to teaching general relativity that we have been developing for some time and testing it in many school classes over the years.

General relativity was developed a hundred years ago and at the outset was considered particularly abstract and hardly comprehensible. Over time, the theory evolved into a topic that is routinely taught at universities. At present, it is making its way into school curricula. Roman Sexl analysed this process by which a new theory becomes accessible to teaching (Sexl, 1980). He described three important factors that explain the transformation of a new theory into regular learning matter: increased faith in the theory arising from theoretical and experimental work, the achievement of verbal explanations in non-mathematical language, and the development of new audiovisual aids. By all three accounts, the time has come to teach general relativity at all levels.

Progress in the development of audiovisual learning aids in the form of visualisations and interactive simulations is the focus of this chapter. We present models and computer simulations that are designed for student activities so that these activities can take the place of student experimental work as used in teaching classical physics.

## GRAVITY IS GEOMETRY

### *... And geometry is intuitive*

The use of models for visualisation is exceptionally well-suited to general relativity because this is a geometrical theory by nature. In teaching geometry, much can be done without equations, by hands-on work with ruler and protractor. Such work builds geometric intuition, an ability to mentally see an answer without doing the mathematics or the construction. Working with models makes geometry accessible, even in the earlier stages of education. And it means that the core concepts of general relativity are accessible on a non-technical level and at an early stage.

For general relativity, we need to extend geometric intuition beyond the familiar Euclidean case. General relativity tells us that the geometry of the world we live in is not, in general, Euclidean. To a very good approximation, it is Euclidean on the Earth (so much so that the difference is only measurable with great effort). However, it is quite different in high-gravity environments, e.g., close to a black hole.

So, we need to develop intuition about non-Euclidean geometries, and much can be done by working with models, by using ruler, protractor, and compass in hands-on work with models in order to study geometric rules and properties. While the study of non-Euclidean geometries is not a standard topic in school, the realisation that there are different types of geometries provides a broader view, which also fosters a deeper understanding of Euclidean geometry.

### *The geometry of curved surfaces – Go intrinsic!*

To tell apart curved surfaces from plane surfaces is something school students can do well at an early age. The surface of a globe, an egg, or a potato crisp is obviously curved as opposed to a piece of paper lying flat on a desk.

To explore the geometry of curved surfaces is the established starting point in teaching general relativity. Hands-on activities have been described for many different types of models: from specific teaching material like the Lénárt sphere (Lénárt, 1996) via everyday objects like globes and inflatable beachballs or swim rings to objects that can easily be crafted for the specific purpose like 3D–printed surfaces and crocheted surfaces (Taimina, 2018).

One way to imagine a curved surface is to see it as something bulging into space, that is, as a part of the ambient three-dimensional space. We call this the extrinsic view, and our intuition generally works well with this view. The intrinsic view of a surface, on the other hand, does not refer to the ambient space. It can be described as the point of view of a small bug crawling on the surface and never looking up. Or it can be imagined as the point of view of a "flatlander", an idealised being that is itself two-dimensional, lives in the surface, and has no notion of a direction "up" from the surface into space. This point of view is elaborated in Edwin Abbott's tale "Flatland" (Abbott, 1884) and in a movie based on this narrative (Flatland: The Movie, 2007). We can understand surfaces either way, extrinsic or intrinsic. However, in order to study gravitation, the intrinsic view is the one we have to adopt.

It appears that the extrinsic view is the natural one for many of us. For example, some people say that a sphere is a three-dimensional object, arguing that you need *x*-, *y*-, and *z*-axes to assign the points coordinates. Also, many would say that the surface of a cylinder is curved since it is bulging into space. The intrinsic point of view leads to different conclusions. The small bug moving on the sphere finds that it can move in two independent directions: forwards-backwards or right-left, and concludes that a sphere is a two-dimensional object. Think of another bug moving on a (large) piece of paper. The paper is first laid out flat on the table; later, it is picked up and held to form a cylinder. There is no difference in geometry that the bug could detect (that is, could detect locally – it could discover being able to come back to a certain point on the cylinder after a long roundtrip). The geometry is the same; therefore, the cylinder, like the plane, is not curved. When Einsteinian gravity states that the presence of matter is linked to the curvature of spacetime, this means curvature from the intrinsic point of view. The intrinsic viewpoint is, therefore, necessary to grasp the concepts used in Einsteinian gravity.

The intrinsic view is also essential for our purpose of understanding gravity in a visual way since, in this regard, the extrinsic view has severe limitations. When we take the extrinsic view, and we want to talk about three-dimensional space being curved, we are led to think that space has to bulge into something; this something is four-dimensional (at least) and we cannot picture this higher dimensional space. Thus, we may conclude that we cannot picture three–dimensional curved space at all, hence cannot apply intuitive reasoning to it, and thus have to resort to working with the equations. Fortunately, this conclusion is false and geometric intuition can be extended to curved three-dimensional space and, incidentally, also to curved spacetime.

## SECTOR MODELS AS A TOOL FOR TEACHING GENERAL RELATIVITY

### *Sector models of curved surfaces*

When curved surfaces are introduced, e.g., with such models as listed above, then switching to sector models provides easy-to-use models for studying arbitrary surfaces, and also, paves the way towards models of curved spaces and spacetimes.

In order to show how a sector model is defined, we consider the sphere as an example (Kraus and Zahn, 2014): the sphere is divided up into small pieces of area (Figure 12.1a). Each small piece of area is approximated by a small *flat* piece of area. A piece on the sphere and its approximating flat piece are related by their edges: the lengths of the four edges on the sphere are measured (or computed), and a flat trapezium is constructed that has the same edge lengths. To build the sector model of the sphere, the small flat pieces, the sectors, are laid out in the plane (Figure 12.1b). When the flat pieces are glued together, they form a faceted sphere (Figure 12.1c). Obviously, the faceted sphere approximates the sphere, and the smaller the pieces, the better the approximation. This procedure can be carried out for any curved surface in the way described here. The choice of how to subdivide the surface is free; we could, e.g. have chosen to use triangles instead of quadrangles.

12.1
Sector model of a sphere: (a) part of a sphere is divided up into small pieces of area, (b) each small piece is approximated by a flat sector. Together, the sectors form the sector model of the respective part of the sphere. (c): if the sectors are glued together, they form part of a faceted sphere. (*Source: Copyright by Ute Kraus, Corvin Zahn*)

a

12.1
*(continued)*

The sector model is useful because it permits to explore the geometry of the surface as will be shown in the teaching unit described in the following section. One of its strengths as a model is being true-to-scale. Consider the surface of the Earth and imagine that you want to compare the sizes of Greenland and the Arabian peninsula. Greenland is the smaller of the two, as can clearly be seen on a globe. On a typical world map, in contrast, Greenland may well appear much larger than the Arabian peninsula (Figure 12.2a). This is a consequence of the projection used to create the map. A given distance appears larger on the map, when it is closer to one of the poles, meaning that the scale of the map is different at different places. So, to correctly use a world map, we need some expert knowledge about the projection. No expert knowledge is required to use a sector model that represents the surface of the Earth. The whole model is on the same scale, that is, what you see is what there is. Greenland looks smaller than the Arabian peninsula (Figure 12.2b), and therefore is smaller; the intuitive interpretation gives the correct result.

12.2
(a) World map (Gall stereographic projection), (b) Sector model of the surface of the Earth. Earth texture adapted from NASA. (*Source: Copyright by Ute Kraus, Corvin Zahn*)

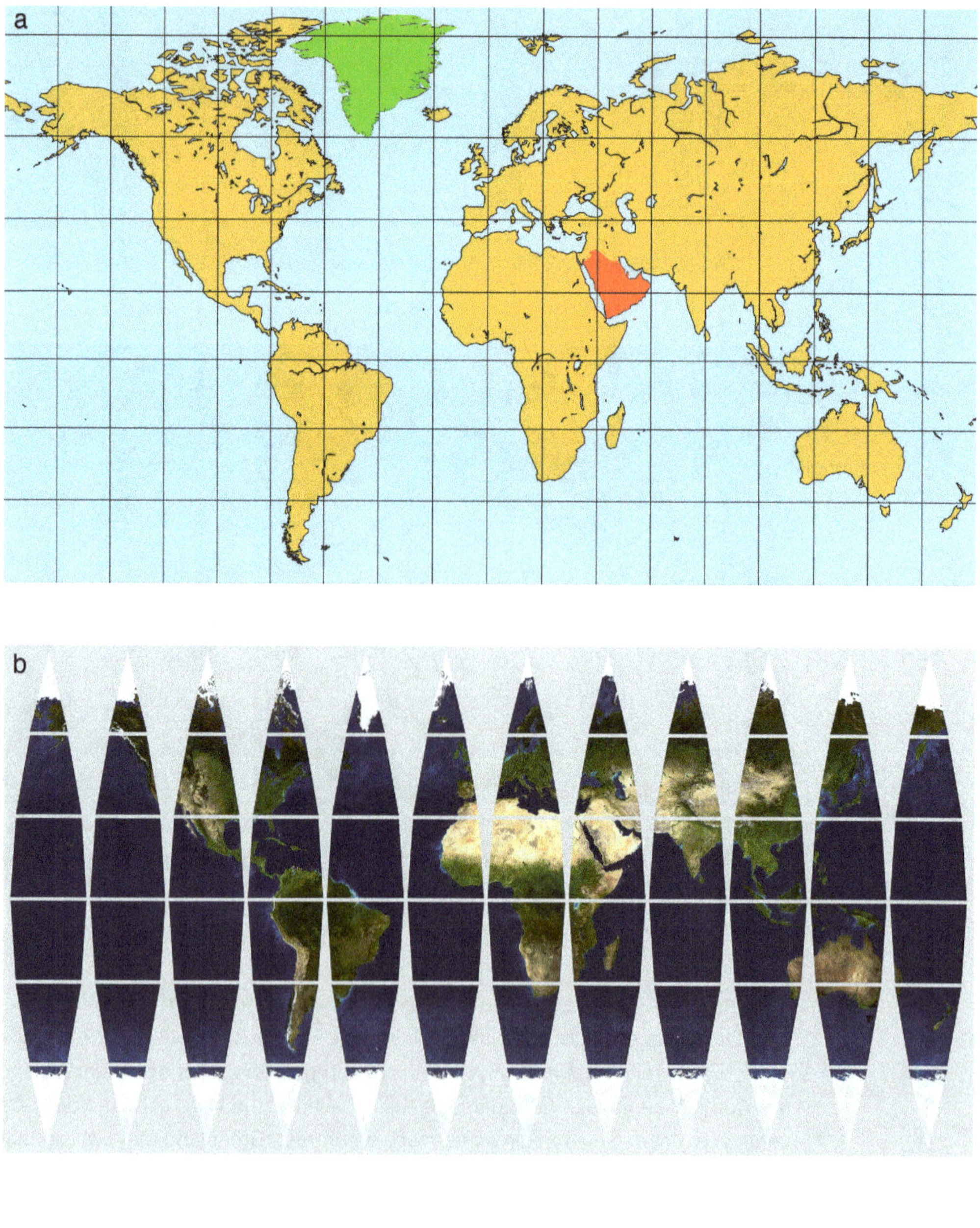

A sector model is also useful from a practical point of view. It can easily be cut out of paper or cardboard. This involves less cost and effort than, e.g., providing each student with a writable globe. It is still easier to use a sector model in the form of an interactive computer model as described in the next section.

Finally, a sector model is innately intrinsic: in the prescription of how to create the model, there is no reference to the embedding three-dimensional space. The sector model depends only on the measurements of the edge lengths, and this is within the scope of what a flatlander with a purely two-dimensional world view can do. Thus, using sector models automatically means adopting the intrinsic point of view, which is the point of view taken in Einsteinian gravity.

### *Go three-dimensional and beyond*

Being, in essence, an intrinsic description is one of the main strengths of sector models. If a flatlander can handle models of curved surfaces, then we, as spacelanders, can handle models of curved space. We can, therefore, take the step from two to three dimensions: subdivide space into small regions, and approximate these by small blocks of Euclidean space (Kraus and Zahn, 2014). The collection of blocks is a faithful model of the curved space (Figure 12.3). It permits to study the geometry of curved space and to develop a geometric intuition for it. In terms of teaching general relativity, the study of curved spaces leads to a first appreciation of the field equations, that is, the equations that relate the curvature of a spacetime to its matter content (Kraus and Zahn, 2019b).

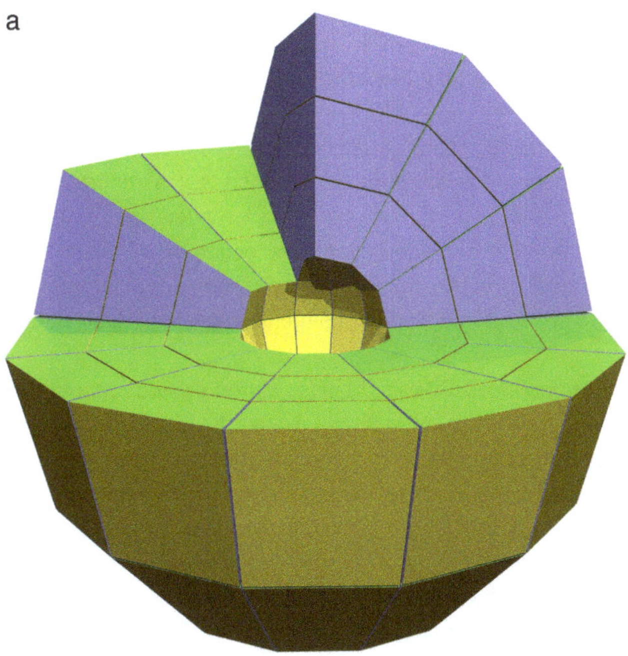


12.3
Sector models of three-dimensional space: (a) a sector model of Euclidean space. Space is divided up into small blocks, (b) a sector model of curved space around a black hole. Space is divided up into small pieces of volume, and each small piece is approximated by a block of Euclidean space. (*Source: Copyright by Ute Kraus, Corvin Zahn*)

b

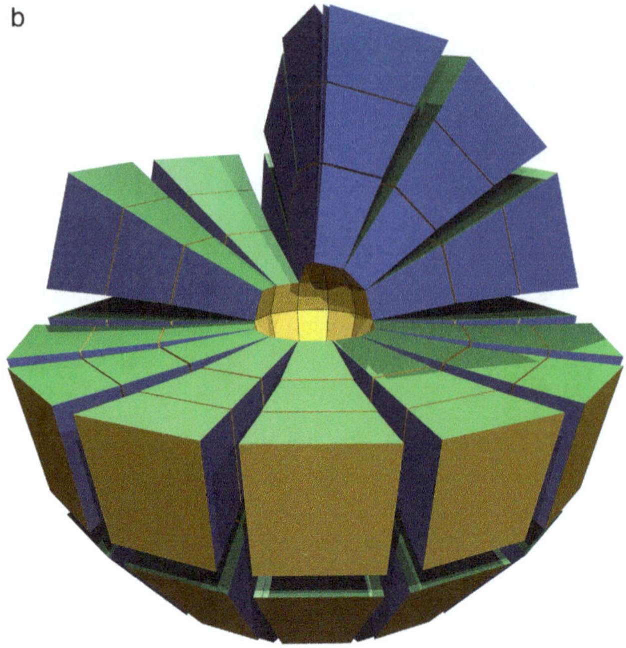

12.3
*(continued)*

The approach can be extended to sector models of spacetimes when spacetimes are restricted to one or two spatial dimensions. This offers additional possibilities for the visualisation of general relativistic concepts (Kraus and Zahn, 2019a).

### *General features of an introductory teaching unit on light deflection*

We present an introductory teaching unit on general relativity that introduces two of the core concepts: curved space and geodesics. Its astrophysical context is gravitational light deflection, a relativistic phenomenon that is important both for its role in current research (e.g., the search for dark matter) and from a historical point of view, with the observation of light deflection near the sun as an early confirmation of the theory. This teaching unit provides a first introduction to general relativity. No previous knowledge of either special or general relativity is assumed.

The unit was developed for a 90-minute class aimed at students of grade 10 to 13 (ages 16 to 19 years). It has been taught more than 50 times to physics classes who visit the "Spacetime-Travel" School Lab of Hildesheim University.

The basis of the teaching unit is the visualisation of curved space near a black hole by a sector model. Due to the spherical symmetry of the black hole (which we assume to be non-rotating), the paths of light and particles are planar, that is, a path is always contained within a symmetry plane of the black hole space. We can, therefore, confine our study to such a symmetry plane. A two-dimensional sector model is used to describe a symmetry plane, and students use the model to construct geodesics. As a result of the hands-on activities with the sector model, students discover typical properties of the propagation of light close to a spherical mass: change of direction, double images, and the formation of an Einstein ring. A detailed theoretical background of this approach is given by Kraus and Zahn (2014; 2019), together with online teaching resources.

## TEACHING UNIT: LIGHT DEFLECTION NEAR A BLACK HOLE

### *Astrophysical context: Light deflection and black holes*

As an introduction to the unit, we first give examples of how gravitational light deflection is observed. The choice depends on the previous knowledge of the participants and their interest, and may include the deflection of light near the sun, astronomical images showing lensed galaxies, or computer simulations showing what an astronaut would see when getting close to a black hole on a fictitious spaceflight. In the introduction, we also provide some information about black holes. This includes the facts that black holes form when massive stars collapse, and that a high number of such stellar-mass black holes is estimated for the Milky Way. We mention observations, e.g. black holes that accrete matter from a companion star and appear as X-ray sources, or black holes that coalesce with a black hole or neutron star companion, emitting gravitational waves. We give an elementary description of the event horizon of a black hole as a boundary that can only be crossed inwards.

### *Basics: Geodesics and the geometry of curved surfaces*

The first part of the unit is about the geometry of curved surfaces. It aims to introduce the concept of a geodesic. To motivate this part, we explain that general relativity is formulated in geometric terms, and that we need to know about geodesics in order to understand what the theory predicts about light.

We introduce a geodesic as a line that keeps its direction at each point. By way of illustration, this means that a pedestrian follows a geodesic if they "always follow their nose", never turning to either side.

Students then explore this definition in order to become familiar with geodesics and their properties. From the definition, they conclude that, in the plane, a geodesic is a straight line. On a curved surface, a geodesic is a line that lies on the surface, again turning neither left nor right at any point. One can think of a car going over a hill with its steering wheel held fixed. The car keeps its forward direction at each point, and so follows a geodesic.

Geodesics on a sphere are then studied in a student activity with writable globes. Students draw geodesics on the globe using narrow flexible rulers. These are flexible enough to lie flat on the globe, but in the transverse direction too stiff to be bent sideways, that is, the lines drawn with these rulers are geodesics (Figure 12.4a). Students explore geodesics on the sphere by working through several tasks designed to build up understanding step-by-step.

1. Draw a geodesic starting from the equator due north.
   Result: The line runs along a line of longitude. Lines of longitude are thus identified as geodesics.
2. Are circles of latitude geodesics, too?
   Result: Tests with the flexible ruler show that the equator is a geodesic, but other circles of latitude are not. Here, students must argue using the definition of the geodesic as a locally straight line. This is to reinforce their understanding of the definition.
3. Draw a second geodesic from the equator, starting close to the first, and again starting due north.
   Result: The two geodesics are parallel at first, then converge, and meet at the North Pole (Figure 12.4b). Thus, students discover that on the sphere, initially parallel geodesics converge. The discussion of this result includes a comparison with geodesics in the plane and leads to the realisation that spherical geometry is different from plane geometry.
4. Draw a geodesic starting from the equator towards the northeast and continue on and on.
   Result: The line closes; it is a great circle. We conclude that the geodesics on the sphere are the great circles, and thus reach the general description.
5. Construct a triangle by drawing a geodesic from the equator to the North Pole, a second one starting at the North Pole at right angles to the first

12.4
Geodesics on the sphere: (a) drawing a geodesic on a writable globe using a narrow flexible ruler, (b) on the globe – geodesics starting parallel to each other converge, (c) the construction of a geodesic on a sector model, and (d) on the sector model – geodesics starting parallel to each other converge. (*Source: Copyright by Ute Kraus, Corvin Zahn*)

a

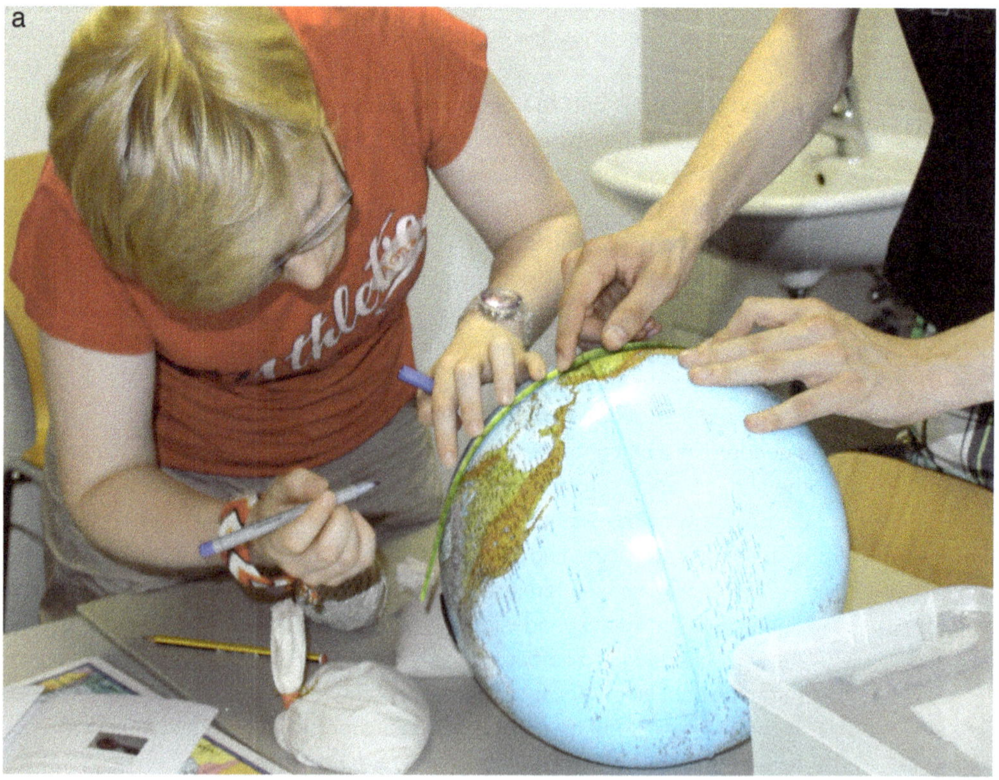

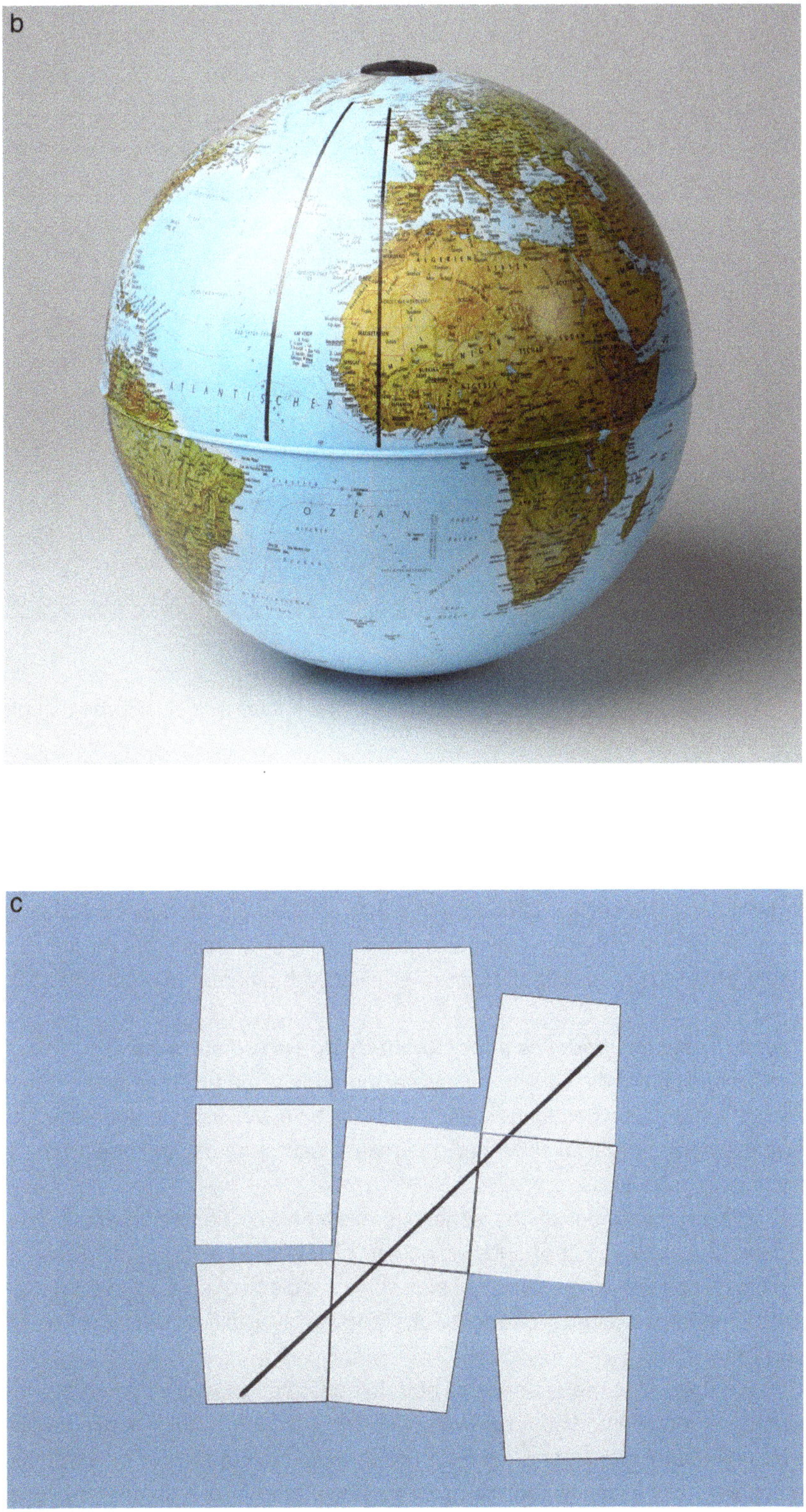


12.4
*(continued)*

12.4
*(continued)*

geodesic and going back to the equator, and a third that closes the triangle along the equator.
Result: We obtain a triangle with three right angles. This is a second example of the fact that spherical geometry has laws that are different from plane geometry.

In summary, this activity aims at two main results: first, students practise to work with the definition of a geodesic, and second, they realise that besides the geometry of the plane, other geometries exist with different geometric laws.

***Tools: The sector model as a tool for studying curved surfaces***

The second part of the unit introduces sector models, the tool that will be instrumental for studying geodesics close to a black hole. We use the sphere as the first example in order to show how to create a sector model and how to use it for finding geodesics.

The sector model of the sphere is explained as described above (see Figure 12.1), as a result of subdividing part of the sphere into small pieces of area and approximating each small piece by a *flat* piece of area. We provide the sector model as a paper model (cut-out sheets are available for download (Kraus and Zahn, 2019)).

We then raise the question of how to construct geodesics when a given surface is represented as a sector model. The answer is developed in discussion with the students, starting from the definition of a geodesic as a line that maintains its direction at each point. In parallel, a teacher or a student executes a step-by-step construction under a visualiser (Figure 12.4c).

1. Within a sector, a geodesic is a straight line.
2. When a geodesic reaches the border of a sector, it crosses into the neighbouring sector. The point of entry is determined by joining the two sectors.
3. It remains for us to find the direction in the new sector. Since a geodesic always keeps its direction, it must cross the border running straight ahead.

Next, students construct a second geodesic on their own. This second geodesic starts in parallel with the first (Figure 12.4d). The two geodesics converge and then intersect. This is exactly what students found earlier, when working on the writable globes. The aims of this activity are twofold: it delivers the rules for constructing geodesics on sector models. And, by comparison with the results on the globes, it provides confidence in the use of sector models for constructing geodesics.

This graphic construction approximates a geodesic since it uses a sector model that is itself an approximation of the sphere. Both the model and the geodesics can be made more precise by using more and smaller sectors (Kraus and Zahn, 2019).

## LIGHT DEFLECTION: GEODESICS CLOSE TO A BLACK HOLE

This is the main part of the teaching unit. It shows how general relativity describes light deflection and lets students explore the predictions. In this part, we build on the understanding that non-Euclidean geometries exist, and we use the tools developed in the previous part.

We describe the relevant statements of the theory of general relativity concerning the propagation of light in the vicinity of a black hole:

- Light follows geodesics in spacetime. That is, to study the propagation of light, we need to find geodesics.
- The geometry of spacetime is related to the presence of mass. So, spacetime close to a black hole has a different geometry from spacetime far away from all masses.

In order to study geodesics, we introduce a sector model that visualises the geometry close to the black hole. The sector model covers a ring around the black hole, lying in the black hole equatorial plane. This is a plane that divides the space around the black hole into two equal halves. The geometry within the ring is different from the geometry of a Euclidean plane, due to the presence of the black hole.

To define the sector model, we describe its construction in a thought experiment: astronaut Karl S. is sent to a black hole in order to survey space in its vicinity. His task is to erect a grid around the black hole according to the pattern shown in Figure 12.5a. The inner edge of this grid is a little outside the horizon[1]. The grid subdivides part of the equatorial plane into quadrangles. Karl S.

1 At a radial coordinate of 1.25 Schwarzschild radii, the outer rim is at 5 Schwarzschild radii.

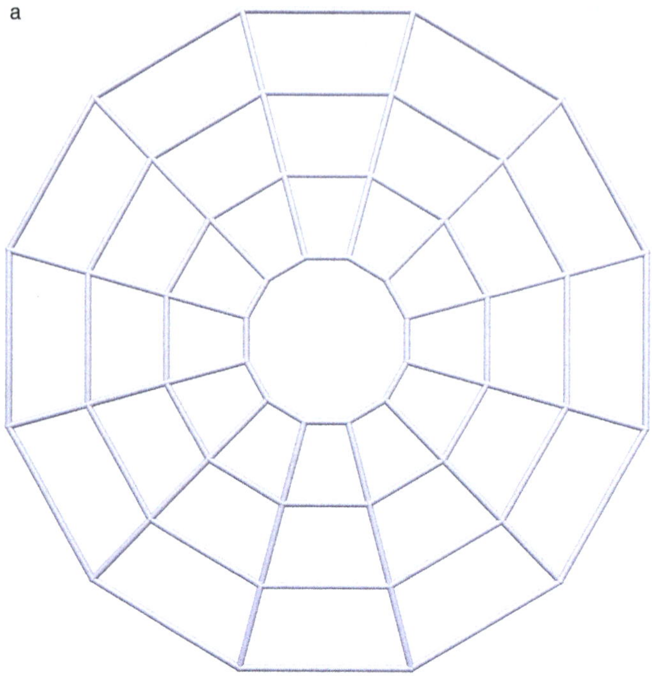


12.5
Geodesics near a black hole: (a) pattern for subdivision of the equatorial plane, (b) resulting sector model, (c)–(e) two geodesics on the sector model, forming a digon. A light source on one of the crossing points will present a double image to an observer on the other crossing point. Images (b)–(e) are screenshots of the web application ViSeMo. (*Source: Copyright by Ute Kraus, Corvin Zahn*)

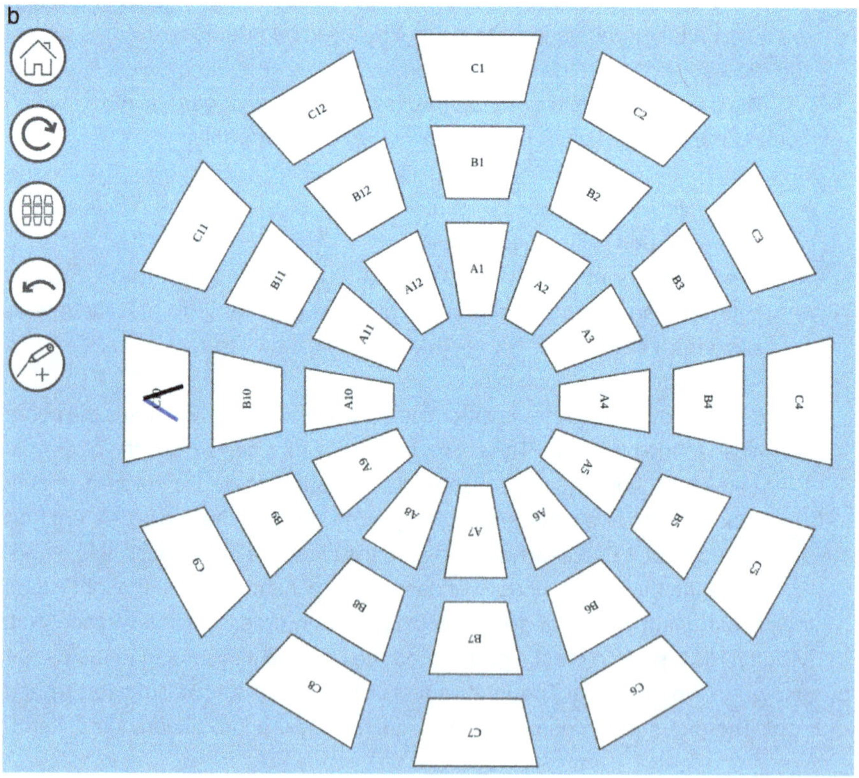

12.5
*(continued)*

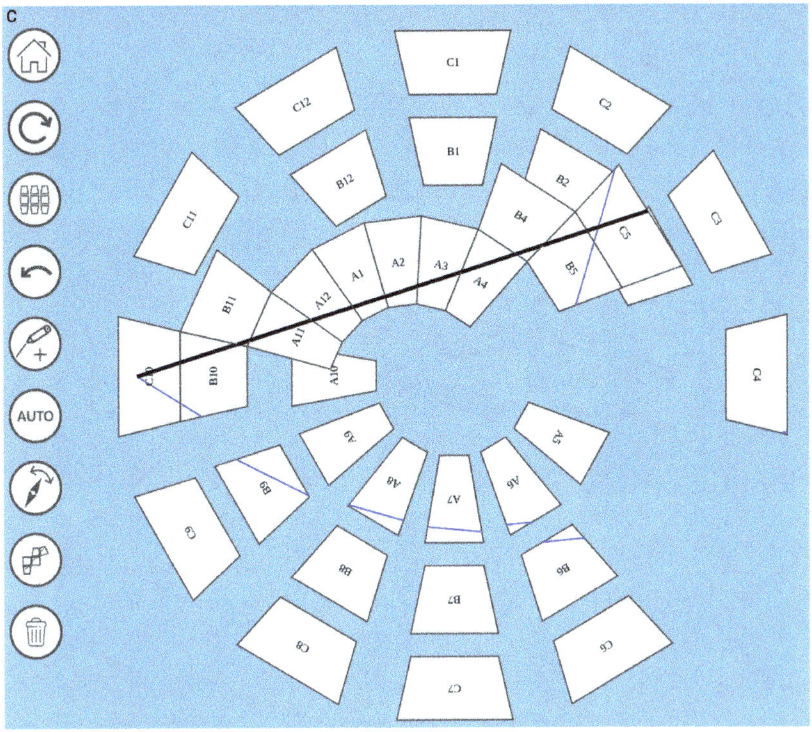


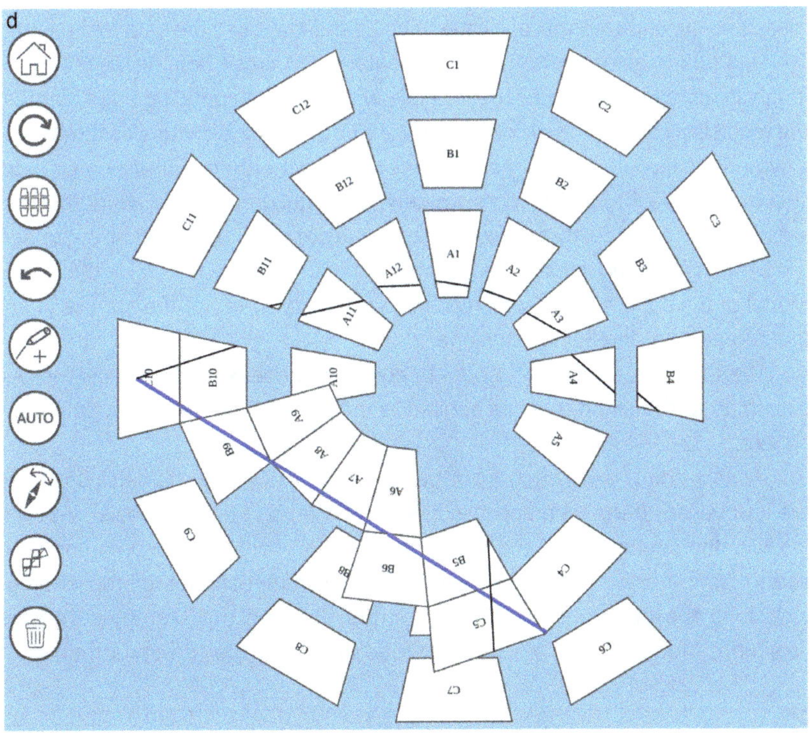

e

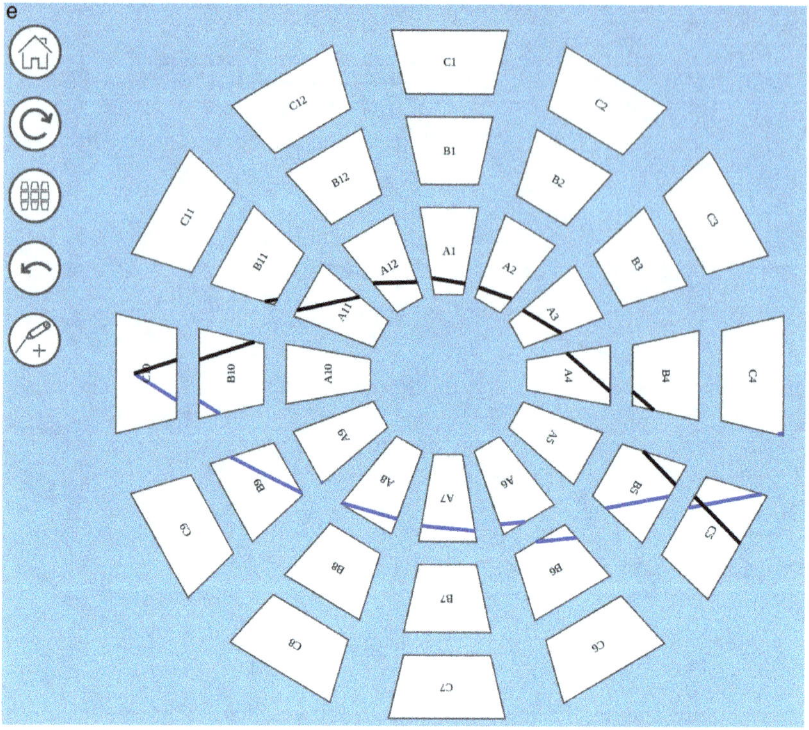


12.5
*(continued)*

measures the edge lengths of all quadrangles and transmits the data to the Earth. There, a true-to-scale sector model is created from these data: each grid quadrangle is represented by a flat quadrangle with the corresponding edge lengths; the result is a sector of the model. Figure 12.5b shows the sectors assembled in a symmetric layout. (This model is a slice through the three-dimensional model shown in Figure 12.3b. The approximately horizontal green sides in Figure 12.3b are identical to the sectors in Figure 12.5). It is not possible to join all sectors at the same time without leaving gaps. Incidentally, the same is true for the sector model of the sphere. This fact indicates that the geometry of the surface is not the geometry of the Euclidean plane.

We then use the sector model to construct geodesics. This is carried out according to the same rules as applied in the case of the sector model of the sphere.

At this point, we switch from paper models to virtual models. Students are introduced to the web application ViSeMo (Virtual Sector Models). ViSeMo displays the sector model and permits to position sectors with drag and drop and to draw lines onto the sectors. Geodesics can be completed step-by-step, similar to the procedure on paper sectors. Beyond the step-by-step construction, though, there are support functions to make the construction of geodesics quick and easy. These include the automatic completion of a geodesic defined by an initial line segment and the automatic positioning of sectors along a geodesic that permits to show that the geodesic is a locally straight line (Figure 12.5c). The interactive simulation significantly reduces time and effort per geodesic, and students can study a larger number of geodesics. Using ViSeMo, students work through tasks defined in several worksheets that step-by-step show up the relevant properties of geodesics near a black hole:

1. Continue an initial line segment to form a geodesic that runs across the whole model (Figure 12.5c, top line).
   Result: When the geodesic is drawn, and the sectors are then arranged symmetrically, it is clear that the directions "far ahead" and "far behind" the black hole are different (Figure 12.5e, top line).
   The construction clearly shows how the change in direction comes about: the line itself is straight at each point. The change in direction occurs because the geodesic passes a region of curved space. This is the principle of general relativistic light deflection.
   One must note, however, that this geodesic is purely spatial, whereas light follows geodesics through space and time. The spatial geodesics studied here share the properties of light paths in a qualitative way, but exhibit smaller deflections.
   By varying the direction of the initial line segment, students seek to create large and small deflections. They find that a large deflection requires a close approach to the black hole.
2. Continue two parallel line segments each as a geodesic.
   Result: The two geodesics are initially parallel, then diverge. This characterises the geometry of the black hole space.
3. Complete two geodesics that start at the same point in different (predefined) directions (Figure 12.5c–e).
   Result: The two geodesics pass the black hole on opposite sides, and then intersect. They form a digon. This illustrates the formation of a double image: when a light source is located at the common starting point and an observer at the point of intersection, the observer sees the light source twice.
4. Complete two geodesics like in task 3, but in a symmetric setup.
   Result: The point of intersection is aligned with the black hole and the starting point. This is the case that produces an Einstein ring.

In summary, this activity permits students to explore geodesics close to a black hole. In particular, they study setups that explain the observational phenomena of displacement, double images, and Einstein ring. An important goal of this teaching unit is the understanding that there is a change in direction of a geodesic in passing the black hole, and at the same time, the geodesic is locally straight: light is deflected, yet light rays are straight.

***Applications: A black hole as a gravitational lens***

At the end of the unit, we come full circle and again look at the phenomena that are observed when gravitational light deflection occurs. An interactive simulation (Kraus and Zahn, 2006) lets students position a black hole in front of a background image showing a cluster of galaxies. Light deflection makes the galaxies appear distorted. Students can vary black hole mass and black hole position. They can create distortions showing the apparent displacement of a background object, a double image, or an Einstein ring, and they can then link the visual phenomena to the theory they explored before.

## ASSESSMENT OF THE TEACHING UNIT

At the end of each event, we invite students' comments. Their overall rating is that they find the unit interesting, can follow and understand well, and appreciate the activities that make up a large part of the class. The level of difficulty generally suits grade 10 students well, whereas students in grade 11 or 12, especially if they take advanced level physics, tend to say that the class is fairly easy and that they wouldn't mind it to be somewhat more challenging, e.g. involving some calculations. We also discuss the students' work with their physics teachers after they have observed their classes working through the programme. They generally confirm that their students are well able to cope with the unit, and that they show motivation and activity. Most teachers say that they would be able to teach this unit at school themselves.

## THE USE OF SECTOR MODELS FOR TEACHING GENERAL RELATIVITY

***Visualisation and the intrinsic point of view***

We have outlined the potential of sector models as tools for teaching general relativity. This approach strongly relies on visualisation: curved surfaces are visualised by means of sector models, as is the equatorial plane of a black hole, and geodesics are visualised as straight lines on sector models. Sector models are true-to-scale representations and geometric conclusions drawn from them are correct results, in agreement with results obtained by setting up and solving the relevant equations. Thus, this unit aims both at conceptual understanding and at providing tools that permit to solve problems and answer questions in a quantitative way. Working with a sector model necessarily means taking up an intrinsic point of view since the embedding space plays no role in the construction and use of the model. Activities with sector models, therefore, also provide practice in adopting the intrinsic viewpoint.

### *Virtual sector models vs. paper models*

The web application ViSeMo represents a digital realisation of sector models. It offers possibilities to construct geodesics simple and fast. Where in the paper version, scissors, pen, and ruler are the tools of choice, mouse clicks, drag and drop mechanisms, and simple gesture controls come to the fore. Purpose-built methods, such as the automatic maintenance of the appropriate direction while continuing a geodesic, support the user in the construction process.

The advantage of the virtual sector model is the fast construction of many different geodesics. Whereas it can become tedious to construct several geodesics on a big paper model, the software provides possibilities to track and compare entire geodesic clusters with just a few clicks.

On the other hand, the haptic handling of pen and ruler emphasises the defining characteristic of the geodesics in that they always maintain their direction. We, therefore, chose to introduce the sector model using the paper version.

### *Beyond the introduction*

The unit described here can serve as a starting point for more detailed studies of curved surfaces, spaces, and spacetimes.

On one hand, sector models have other uses besides the construction of geodesics. In particular, they can be used to study curvature. Also, a more advanced unit could introduce the concept of the metric so that students can then construct sector models on their own (for a teaching unit with this approach, see Kraus and Zahn, 2016).

On the other hand, crucial for teaching general relativity, sector models can be extended far beyond the case of curved surfaces. They afford a description of curved three-dimensional space, and thereby lead to phenomena that are beyond our intuition based on everyday experience. They furthermore describe curved spacetimes. To build a sector model, a curved spacetime is subdivided into small regions, each of which is approximated by a small region of Minkowski spacetime – a spacetime sector. The construction of spacetime sector models and their use in teaching is described by Zahn and Kraus (2014) and Kraus and Zahn (2019a).

## CONCLUSION

We have discussed how to teach the geometric concepts of general relativity at the secondary school level. In doing so, we have emphasised the benefits of adopting an intrinsic point of view and the need to strengthen geometric intuition. Both aspects are inherent to sector models, a purpose-built set of tools for the study of non-Euclidean geometries. This chapter introduces sector models and illustrates their use for teaching general relativity with a teaching unit on geodesics near a black hole. Teaching material described in this chapter is available online at spacetimetravel.org (see online resources for Kraus and Zahn, 2014,

2019). This website presents a wide range of open educational resources on special and general relativity.

## FURTHER READING

Chapter 1: Intuitions in Einsteinian physics
Chapter 10: Models and analogies in teaching general relativity
Chapter 20: Towards a comprehensive general relativity course for secondary school

## REFERENCES


Abbott, E. (1884). *Flatland, A Romance in Many Dimensions*. Seely & Co, London, UK.

Flatland: The Movie (2007), Flat World Productions.

Kraus, U. & Zahn, C. (2016). "Teaching Gravitational Light Deflection: A Short Path from The Metric to The Geodesic." spacetimetravel.org/aur16.

Kraus, U. & Zahn, C. (2019a). "Sector Models – A Toolkit for Teaching General Relativity: III. Spacetime Geodesics." *European Journal of Physics*, 40, 015602, online version on spacetimetravel.org/sectormodels3.

Kraus, U. & Zahn, C. (2019b). "Teaching General Relativity with Sector Models: The Field Equations." *IOP Conf. Series: Journal of Physics: Conf. Series*, 1286, 012025.

Lénárt, I. (1996). *Non-Euclidean Adventures on the Lénárt Sphere*, Key Curriculum Press, online information: lenartsphere.com, viewed 7 November 2019. Budapest.

Sexl, R. U. (1980). "Relativitätstheorie als didaktische Herausforderung" [Theory of Relativity as didactical challenge]. *Die Naturwissenschaften, 67*, 209.

Taimina, D. (2018). *Crocheting Adventures with Hyperbolic Planes*, CRC Press, Boca Raton, USA.

Zahn, C. & Kraus, U. (2006). "Interactive Black Hole." spacetimetravel.org/isl.

Zahn, C. & Kraus, U. (2014). "Sector Models – A Toolkit for Teaching General Relativity: I. Curved Spaces and Spacetimes." *European Journal of Physics*, 35, 055020, online version on spacetimetravel.org/sectormodels1.

Zahn, C. & Kraus, U. (2019). "Sector Models – A Toolkit for Teaching General Relativity: II. Geodesics." *European Journal of Physics*, 40, 015601, online version on spacetimetravel.org/sectormodels2.

# 13 Where do gravitational waves come from and how can we detect more?

*Joris van Heijningen*

**WHAT?**

This chapter explores the origin and detection of gravitational waves. Joris van Heijningen suggests classroom exercises to link one of the most spectacular predictions of Einstein's general theory of relativity to instructional practices in classrooms.

**FOR WHOM?**

This chapter is intended for readers who wish to learn more about the physics of gravitational wave detection, and also, for those who wish to incorporate topics of Einsteinian physics into their astronomy lessons.

## INTRODUCTION

More than one billion years ago, in a galaxy *far, far away*, two black holes smashed into each other at more than half the speed of light. No flashes of light occurred, nothing that would signal to the rest of the Universe something happened. Nothing? Two *pure* vacuum objects that were described solely by their dynamic curvature in space and time became one, and this merger rippled the fabric of that very same spacetime. The magnitude of those ripples was gigantic at nearby distances, but here on the Earth, more than a billion lightyears away, the ripples were minuscule. What is described here is the first-ever detected gravitational wave – it was named GW150914 (in "GWYYMMDD" format) – and was published on 11 February 2016 (Abbott et al., 2016).

Black holes had never been directly observed before, let alone a merger of two of these spacetime objects. The gravitational wave measurements are the most precise ever performed. This accuracy is comparable to measuring the distance from here to the nearest star outside the solar system (Alpha Centauri, 4.32 lightyears away) with the precision of the width of a human hair! What is truly monumental about these first detections is that we can see the Universe in an entirely new way. The ability of a human to directly detect gravitational waves is comparable to being *deaf* and suddenly gaining the ability to *hear*. An entirely new realm of information is now available both for physicists to study the Universe and for physics teachers to excite and motivate their students with the most recent topics in Einsteinian physics.

The detected binary systems have collided hundreds of millions to several billion years ago after a very long cosmic dance around each other. When some of the measured waves began their journey, multi-cellular life began developing here on the Earth. This life evolved to the human species. One century ago, a genius emerged within humanity and predicted the existence of gravitational waves. This genius, Albert Einstein, thought that the minuscule effect of gravitational waves on the Earth would never be measured. Even if gravitational waves were real, Einstein deemed it technically impossible to measure their tiny effects. About 50 years ago, scientists began trying to measure gravitational waves anyway. Ultimately, we built detectors sufficiently sensitive to observe these waves at the end of their cosmic journey. One could say that we have proved Einstein wrong and right by our first detection.

How did the gravitational wave scientists and engineers pull it off? The gravitational waves travelled for more than one billion years to get to us, spreading out and getting weaker along the way. In fact, when GW150914 passed through all of us, *you* probably didn't feel that your body was stretched and squeezed by a thousandth of a billionth of a billionth of a metre. In this chapter, we explore the physics of gravitational wave detection to give teachers a comprehensive resource, which they can draw on when planning their lessons. Our focus is on all the new possibilities that gravitational wave astronomy makes possible - often in synergy with traditional astronomy. We first look at the sources of gravitational waves and what we can learn from detecting them. We then look at gravitational wave detectors and make suggestions for classroom exercises that link the physics of gravitational waves to the instructional practices of classrooms.

## WHERE DO GRAVITATIONAL WAVES COME FROM, AND WHAT CAN WE LEARN FROM THEM?

The first discovery of gravitational waves, GW150914, showed us that three solar masses of energy could be converted into spacetime ripples travelling at the speed of light. This event was the single most powerful astronomical event ever measured, clocking in at 50 times the radiative power of the entire visible Universe at peak luminosity (the total amount of radiation per unit time emitted by a star, galaxy or another radiating astronomical object) The black holes involved, having masses of about 29 and 36 solar mass each, are among the most massive stellar-mass black holes detected. The existence of such massive black hole binary systems was new to astronomy before gravitational waves were part of our scientific toolbox.

The first few detections of gravitational waves were done with two Laser Interferometer Gravitational-wave Observatory (LIGO) detectors in the USA. Since gravitational waves travel at the speed of light, they arrive at different detectors at slightly different times. In Figure 13.1 we see their output during the GW150914 compared to what Einstein's theory would predict such a signal should look like. These timing differences allow two detectors to position the

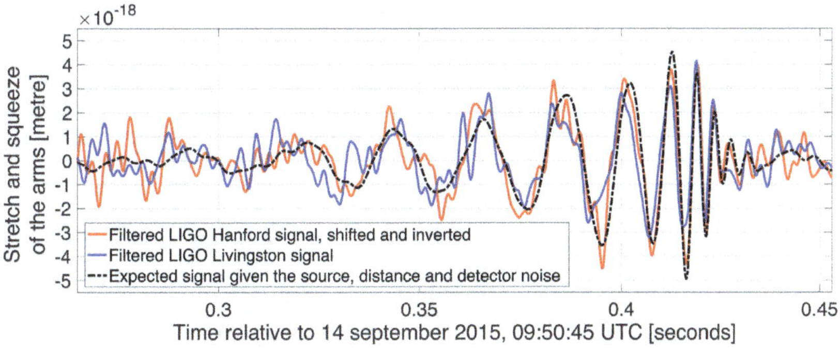


13.1
The distinctive, and by now iconic, chirp signals as recorded by the two LIGO laser-interferometers of GW150914 on September 14, 2015. (*Source: Copyright by Joris van Heijningen*)

source of the gravitational waves in a ring-shaped area in the sky. However, with only two detectors, this area is quite large, typically an area hundreds of times larger than the Moon in the night sky. It is only with the addition of the European Virgo detector that we have seen this area shrink dramatically. Moreover, the combination of different gravitational wave detectors and traditional telescopes opens up possibilities for so-called multi-messenger astronomy, a new type of astronomy that uses complementary measurements of electromagnetic and gravitational waves. Multi-messenger astronomy is an exciting topic to bring to classrooms, not least because many gravitational wave sources link to topics that often are part of national curricula, e.g. what matter is and how it is formed, what stars are, what light is, what gravity is, etc.

The detection of GW170817 provides one example of how we can link cutting-edge discoveries in gravitational wave astronomy to a standard classroom topic: the formation of matter. GW170817 was detected and determined to be coming from a merger of two neutron stars (Figure 13.2). The precise localisation of GW170817 by three gravitational wave detectors helped conventional astronomers to find an afterglow of this massive collision. In this afterglow, we were able to see for the first time the theorised process, now known to be responsible for the abundance of elements heavier than iron in the Universe such as gold, platinum, and uranium.

Besides neutron stars and black holes, there are other potential sources of gravitational waves. For instance, we should *hear* the so-called stochastic background radiation of cosmological or astrophysical origin. Analogous to the cosmic microwave background, the stochastic background radiation is omnipresent in the Universe and could entail a snapshot of the Universe just after the Big Bang! Another source could be small deformations in rapidly rotating neutron stars causing disturbances in spacetime. In realistic neutron stars, such distortions may be supported by strong magnetic fields misaligned to the axis of rotation. If the magnetic field poles are in the line of sight, polarised electromagnetic emission causes the star to appear pulsed; you may have hear of those stars before pulsars. Radio telescopes here on the Earth can observe these radio pulses to gain insight into the nature of these spinning and highly magnetised neutron stars. Again, the combination of gravitational waves and electromagnetic astronomy allows us observing the Universe in new and better ways.

13.2
Artist's illustration of two merging neutron stars. (*Source: Copyright by NSF/LIGO/ Sonoma State University/A. Simonnet / Public domain*)

The future of gravitational wave astronomy is promising. It seems that gravitational wave astronomy and its necessity for multi-messenger astronomy are already living up to the expectations the (scientific) world had up until September 2015. The entire scientific community is counting on us to combine our results with other fields to explain what we see in the Universe. Some already call the first detection of gravitational waves the discovery of this century.

**CLASS EXERCISE: THE STRENGTH OF A GRAVITATIONAL WAVE (A CRUDE ESTIMATE)**

Imagine you jump into a pool, what is the biggest wave you might make? Surely not more than your own size. The same is true for black holes and the fabric of spacetime. Now, a black hole's size is its Schwarzschild radius, $r_s = 2GM/c^2$, where $G$ denotes the Newtonian gravitational constant, $M$ the mass of the black hole, and $c$ denotes the speed of light. For instance, a black hole of 33 solar masses has a radius of abouty 100 km = $10^5$ m.

When two black holes collide, let's say they make a wave in spacetime of that size. Those gravitational waves spread out like ripples in a pond, and their magnitude falls with $1/d$, where $d$ is the distance from the source. At the collision, the strain $h = \Delta L/\Delta L$, the relative length change of space) is roughly equal to $h_{coll}$ = 1. In other words, the wave you make is of your own size. At distances far from the collision $r_{obs}$, the strength of the wave is equal to $h_{coll}(r_s/r_{obs})$, therefore at 1 billion lightyears ($r_{obs} = 10^{25}$ m), the strength of the wave arriving on the Earth is $h = 1 \times (10^5/10^{25}) = 10^{-20}$. This is roughly what we observed in our first detection.

*You* also stretched and squeezed about $10^{-20}$ m as the wave passed by. Of course, no one felt anything. How close would two black holes need to crash into each other for us to feel it? How about within our galaxy, the Milky Way? The typical size of our galaxy is about ten thousand lightyears, a factor $10^5$ less than 1 billion lightyears. The stretch and squeeze would, thus, be $10^{-15}$ m, or about the size of an atomic nucleus. The crash should thus be much closer; suppose we will feel a 1 cm stretch and squeeze, how close should black holes collide?

## HOW DO WE MEASURE GRAVITATIONAL WAVES?

Because there is a correlation between spacetime curvature and gravity, a gravitational wave changes the way objects move and fall with respect to each other. When a gravitational wave passes two objects, a measurable effect occurs. The physical distance between the two objects will stretch and contract as long as the gravitational wave is passing by.

Light particles called photons travel at the speed of light. If you bounce them in between two mirrors, this system can act as a light clock. Every round trip – bounce and bounce – is then a clock *tick*.

A light clock is very precise and the ticks, or round-trip time, $t_{rt}$, are separated in time by twice the distance, $L$, between the mirrors divided by the speed of light, $c$, as

$$t_{rt} = \frac{2L}{c}.$$

When a gravitational wave passes by, it stretches and squeezes $L$ ever so slightly. This changes $t_{rt}$, and our clock goes faster or slower from the moment the gravitational wave interacts with the system until the moment the wave has passed by.

We call two mirrors with light bouncing back-and-forth between them a *cavity*. For the maximum effect of stretching and squeezing, the gravitational wave has to travel along a path perpendicular to the cavity, that is, the line between the two mirrors. Now, we just need a system to monitor the change in stretching and squeezing of this cavity.

We choose to use two cavities, make them as long as we practically can, and build them in an L-shape configuration in a quiet place on our planet. We then

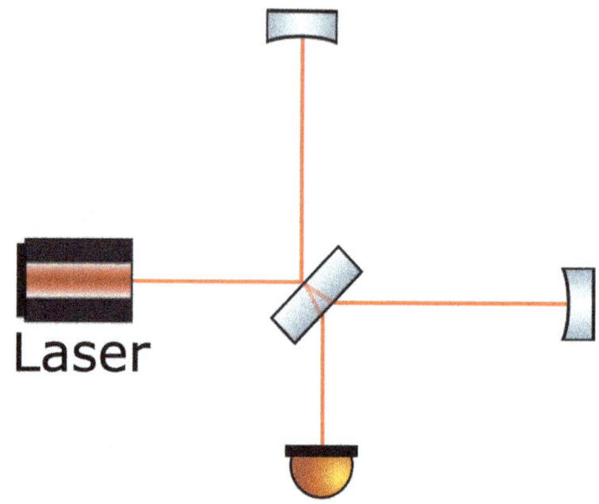


13.3
A simple Michelson interferometer.
(*Source: Copyright by Joris van Heijningen*)

make them part of something scientists call a Michelson interferometer, as shown in Figure 13.3. In Figure 13.4, we see a special kind of interferometer: a *dual recycled Fabry-Perot* Michelson interferometer.

We see a laser beam hitting a beam splitter from the left. A beam splitter can be described as a *bad* mirror. 50% of the light is reflected (in this case upwards), and 50% of the light is transmitted and continues its path into the *horizontal* cavity. We see a laser beam, after passing the input optics and transmitting through the Power Recycling Mirror (PRM), hitting a beam splitter from the left. This light enters the two four-kilometre long cavities that stretch and contract as the wave passes by. There are other cavities to enhance the signal, most notably the power recycling cavity – made by the PRM and the input mirrors, and a signal recycling cavity made by the Signal Recycling Mirror (SRM) and the input mirrors. A low-noise high-power laser and signal detection optics with optical squeezing suppress

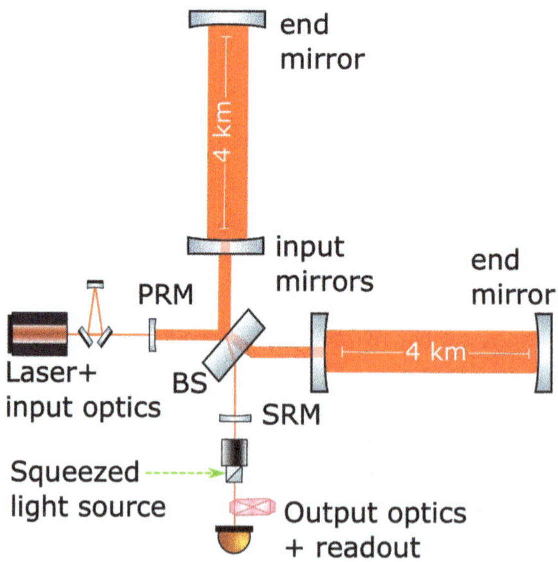


13.4
This sketch provides a simplified overview of a gravitational wave detector, a so-called dual recycled Fabry-Perot Michelson interferometer.
(*Source: Copyright by Joris van Heijningen*)

photon shot noise. When the light comes back from its hundreds of kilometres long travel after many bounces in the cavity, it will meet the light coming back from the other cavity at the beam splitter. Because of the wave nature of light, the photons *interfere* with each other, and this process ultimately decides whether the photons go back left towards the laser or travel downward towards a photodetector. The photodetector essentially counts how many photons arrive there, and this amount per second holds information of the differential cavity length change – if any occurred.

All gravitational wave detectors are set up such that hardly any light reaches the photodetector while we are in observation mode, waiting for a wave to give a signal. When the wave stretches and squeezes the cavities, interference causes *more* or *less* light particles to be counted at this output port. Basically, that is how the gravitational wave is measured. To reduce noise in the measurement process, we actually want only little light at the output port – more on this *shot noise* later.

Any length change of the cavities will result in the photons arriving back to the beam splitter with a different phase or, equivalently, at a different time than would be expected without gravitational wave interaction. A passing gravitational wave has a differential effect on the two cavities, also called arms, of the interferometer. When one arm stretches, the other is squeezed and vice versa. LIGO, which extremely precisely monitors the distance between mirrors four kilometres apart, has to make a distance measurement four thousand times less precise than that of the effect of a gravitational wave on a metre-long stick. Why four thousand times less precise? Because the stretching is a relative phenomenon – large objects get stretched more in an absolute sense, see Figure 13.5 (top) and (bottom). The whole effort is the most precise distance measurement to date.

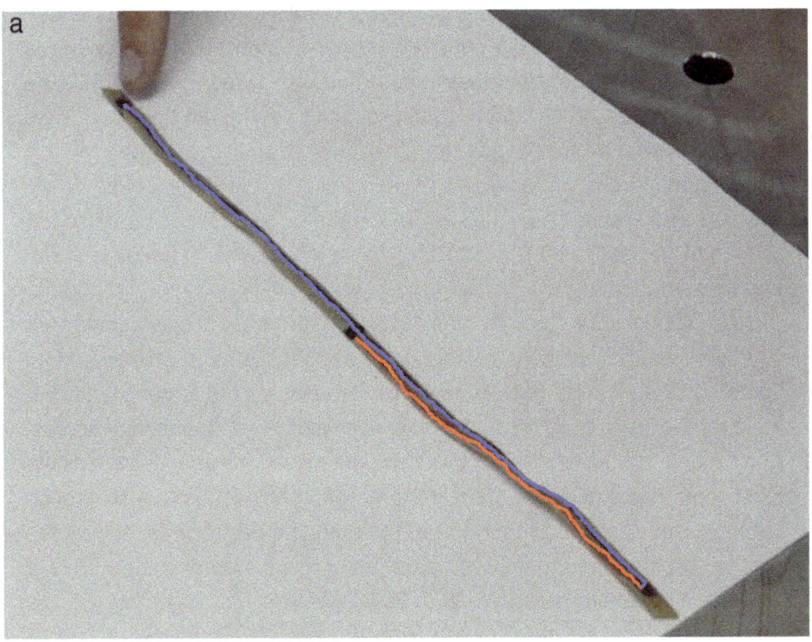


13.5
A simple and instructive way to explain why an interferometric gravitational wave detector arm has to be a few kilometres long. When comparing (top) a red tube and a blue tube that is twice as long, we see (bottom) that a longer tube gives a bigger effect when being stretched. (Video stills from https://www.hetklokhuis.nl/tv-uitzending/4164/Zwaartekrachtsgolven; scene 8:20–9:00.)

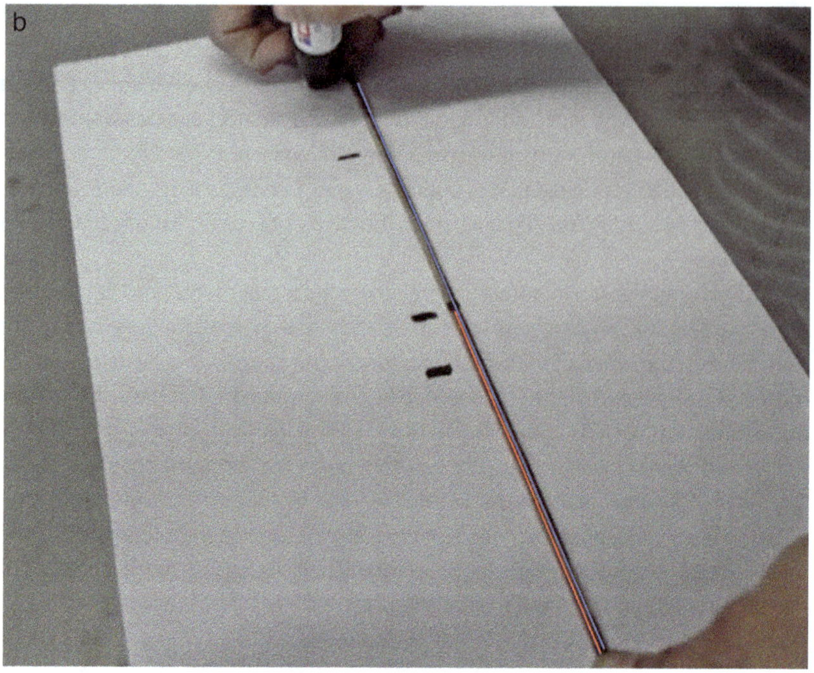


13.5 *(continued)*

**CLASS EXERCISE: HOW MANY PHOTONS DO YOU NEED TO DETECT GRAVITATIONAL WAVES?**

Gravitational wave detectors, opinion polls, and the tossing of coins have one thing in common: a simple mathematical law about uncertainty. Suppose we toss a coin ten times, hoping for heads, what is the most likely outcome? Since a coin has two sides, head or tail, the chance of head is 50-50 or 50% or a probability of 0.5, so the most likely outcome would be five heads (and five tails). But we all know that this is not guaranteed; it would not be surprising to get four heads or six heads, but it would be rare to get no heads or ten heads. Actually, the chance of 10 heads is $0.5^{10} = 1/1024$, so the chance of 10 heads is about one in a thousand.

Now, suppose we toss our coin 100 times, or better still, we could toss 100 coins at once and tally the results. The most likely outcome would be 50 heads (and 50 tails), but we would be rather surprised to get exactly 50. If we repeat the tossing many times, we can plot the distribution of scores. Most scores are between 40 and 60, and the mean, or average, score is 50. Each score has a deviation from the average, which is just the difference between that score and the average. The standard deviation ($\sigma$) shows the spread of score s around the average score or mean. Approximately 68% of scores are between $-\sigma$ and $+\sigma$, with 95% between $-2\sigma$ and $+2\sigma$.

We can also look at the standard error, which shows the uncertainty around the values we obtain. Standard errors are also the standard deviations of values we obtain from repeated sampling or repeated tossing of coins in this case. So here, we could say that the expected score is 50, but the standard error is ±10. Note that 10 is the square root of 100, the number of coins in each toss. This is no coincidence! The standard error is ±10%.

Scientists commonly use $N$ to indicate the sample size. Having tried $N = 10$ and $N = 100$, we will now be more ambitious and toss 10,000 coins, so $N = 10{,}000$. Again, the most likely outcome would be 5,000, but we would be very surprised to get exactly 5,000. If I were to repeat the experiment enough times, I could again plot the distribution of scores. Again, it would be a curve peaked at 5,000, and this time I would find most scores between 4,900 and 5,100. The absolute or mean error has increased from ±10 to ±100, but averaging has helped reduce the relative error to ±1%. The standard error is $\sqrt{N}$, and the relative error, also called the fractional error, is called $N/\sqrt{N}$. We can see with $\sqrt{N}$ that the larger the sample size, $N$, the smaller the standard error will be.

Now, suppose someone suggests that all the coins are unbalanced and there will be a 5% excess of heads. Our task is to test this. How many coins must we toss to be able to detect this small bias? Based on the above results, $N = 100$ tosses has a fractional standard error of 10%, which would not be enough to uncover the bias. However, if we choose $N = 1000$, $\sqrt{N}$ is 33, and the fractional error is about 3%. This sample size would barely detect the bias. It would be better to choose $N = 10{,}000$ because 5% bias would be five times larger than the standard error.

The $\sqrt{N}$ law of statistics describes any random processes. A particular case is opinion polls, where they ask a number of people how they are going to vote. If they chose $N = 100$, the error would be 10%. Seeing the margin for winning an election is usually only a few per cent, $N = 100$ would give no useful information. But it costs money to collect opinions. If you want to improve the precision ten times, you have to pay 100 times as much. For this reason, most opinion polls use $N = 1{,}000$, with uncertainty 3%. This is sufficient to justify a newspaper headline; it is not enough to accurately predict election outcomes.

All the above mathematics is fundamental to laser interferometer gravitational wave observatory detectors. These detectors use a stream of photons to measure tiny motions of their mirrors. The question is: how many photons do you need, and how do they affect the measurement?

In a laser interferometer, a motion of a mirror of half a laser wavelength is sufficient to switch the photon path from the output to the input. If you used a single photon, you could just detect motion equal to one-fourth of a laser wavelength, about 250 nm ($2.5 \times 10^{-7}$ m). The photons are like the coins. With 100 photons, you could improve your precision ten-fold.

So, how many photons do we need to be able to measure $10^{-19}$ m? We need a number equal to the square of the improvement factor: the answer is $(2.5 \times 10^{-7}/10^{-19})^2$, or about $6 \times 10^{24}$ photons. That corresponds to the hundreds of kilowatts of laser power in the actual detectors. There are two other ways that the $\sqrt{N}$ law is visible in the detectors. The detectors measure gravitational waves that are changing the spacing between the mirrors between 10 and 1,000 times a second. If you are to detect at 1,000 Hz, you must be able to measure in a millisecond. The $N$-value at 1,000 Hz is the number of photons in one millisecond, which is 100 times fewer than the number at 10 Hz (0.1 s).

But photons arrive randomly: at one instant there are more, at another instant there are less. The fluctuations in the number of photons follow the $\sqrt{N}$ law, and because $N$ is smaller for high frequencies, the shot noise gets larger. At low-frequency, something else happens: now the problem is due to the absolute errors, not the fractional errors because the absolute fluctuation in the number of photons means there is a fluctuating force acting on the mirrors due to the momentum of the photons. This fluctuation is proportional to $\sqrt{N}$. This radiation pressure noise causes the noise level to rise at low frequency. So, we have seen that the $\sqrt{N}$ law works everywhere we sample random variables whether these be coins, people, or photons.

## WHAT ARE THE CHALLENGES OF BUILDING A WORKING DETECTOR?

Since the 90s, a lot of work has been done to set up a global network of gravitational wave detectors. The two LIGO detectors in Hanford, Washington State and Livingston, Louisiana in the United States of America and the European Virgo detector near Pisa, Italy are already operational. KAGRA, an underground detector in Japan and LIGO India (from 2025 onwards) will strengthen the network in the near future. Future detectors are in an advanced state of design. Their sensitivity will exceed present detectors by at least one order of magnitude across the current frequency range. With more sensitive detectors such as Einstein Telescope or Cosmic Explorer, we can not only crank up the number of detections per year but also look further (back) into the Universe, all the way back to the Big Bang!

The measurement procedure perhaps sounds relatively easy to achieve, but it still took the gravitational wave community more than 50 years. The interferometer technology had been around since 1881, the year in which Albert A. Michelson invented it. However, the laser – Light Amplified by Stimulated Emission of Radiation – was only invented in 1960 when Theodore Maiman built the first laser at Hughes Research Laboratories.

Combining these two ideas, two Russian scientists wrote a paper (Gertsenshtein & Pustovoit, 1962) that subsequently inspired Rainer Weiss and Kip Thorne to think it was possible to detect gravitational waves. Weiss wrote an MIT technical report (Weiss, 1972) which described the challenges ahead. These challenges, and more insights by others – most notably by Ron Drever – would keep several thousand physicists and engineers busy for the next decades. All of this culminated in that first gravitational wave detection in 2015 and continues to be fruitful stimulation for fundamental physics and astronomy. Now we will dive into some of the technical challenges of gravitational wave detection. These challenges consist of different noises that make it harder to detect gravitational wave signals.

**CLASS EXERCISE: GETTING MORE FAMILIAR WITH POWERS OF TEN**

Powers of ten are convenient when dealing with large or small numbers. Rather than using all the ink in the world for writing out zeros, we use, for instance, $10^3$ for a thousand, $10^7$ for ten million, and $10^{-6}$ for a millionth. We saw earlier that the accuracy with which the first detection of gravitational waves was made was comparable to measuring the distance to our nearest star ($4 \times 10^{16}$ m) with the precision of a human hair ($\sim 10^{-5}$ m or a hundredth of a millimetre). Our detectors are unfortunately much smaller ($4 \times 10^3$ m) than the distance to Alpha Centauri, so how much the detectors stretch and squeeze is much smaller, that is, $10^{-5} \times (10^3/10^{16}) = 10^{-18}$ m. How small is that?

If hydrogen atoms ($\sim 10^{-10}$ m) were the size of the Earth ($\sim 10^7$ m), the atom nucleus, a proton, would be the size of a football pitch ($\sim 10^2$ m), and the stretch and squeeze by a gravitational wave would be the size of your phone ($\sim 10^{-1}$ m)!

The energy detected when those mirrors moved $10^{-18}$ m was the result of the energy carried by the gravitational waves. In the last second of the collision, three solar masses ($3 \times 2 \times 10^{30}$ kg) of gravitational wave energy were emitted. How much energy is that? We can use Einstein's famous equation, $E = mc^2$, and find out this is $6 \times 10^{30} \times (3 \times 10^8)^2 = 5 \times 10^{47}$ J. That is, $10^{27}$ or a billion billion billion times as much as the yearly global energy use!

In the gravitational wave detector, the photons we use have a wavelength of $1.064 \times 10^{-6}$ m, and therefore each carries an energy $E = hc/\lambda = 2 \times 10^{-19}$ J. As we shall see later, hundreds of kilowatts ($10^5$ J/s) of photons are whizzing around our detector. This means at any given location in the cavities, $10^5/10^{-19} = 10^{24}$ photons fly by per second. We will see later when studying statistics why we need so many photons.

It is also intuitive to look at sizes of objects we know to get a feeling for the powers of ten. A classic short movie, easily found on the web, was made by Charles and Ray Eames in 1977 named "Powers of Ten". A scrollable, more recent, representation can be found at http://quantumcosmos.ca.

### *Seismic noise*

Seismic noise is the ever-present motion of the Earth and its effect on all components of the detector. We need all main optics of the interferometer –

input/output optics, beam splitter and most importantly, the four mirrors – to be interacting with the small perturbation that is the passing gravitational wave. The interaction causes an apparent change in length smaller than $10^{-18}$ m of the kilometre-scale cavities. If the mirrors were directly attached to the Earth, then they would move with several hundreds of nanometres. This is a factor of 100 billion more than the effect we are trying to measure. Therefore, we need to decouple the motion of the mirrors by suspending them using vibration isolation systems.

To suppress the vibrations, we work with harmonic oscillators. To quickly understand what a harmonic oscillator is in this context, imagine (or go find) an unrolled yo-yo. Hold the yo-yo at the end of the string. Your hand is the vibration you want to suppress – the Earth's vibrations – and the yo-yo is the mirror. If you slowly move your hand back and forth, there is no suppression of the vibrations; the yo-yo moves as much as your hand. There is a frequency of back-and-forth hand motion at which you get lots of yo-yo motion; this is called the resonance frequency, and there is vibration amplification at that frequency. This is the price we have to pay for the behaviour of the system above the resonance frequency. Now, move your hand quickly back and forth, and you see that the yo-yo is not following; above the resonant frequency, there is vibration suppression.

We mitigate the price we have to pay – the amplification at the resonance frequency – with control technology; sensors measure the movement, and a real-time computer tells us when a (small) force needs to be sent to the system (the yo-yo in our previous example) to damp the resonant amplification at the resonance frequency. You can imagine that many of these control systems are running continuously to keep those huge interferometers – in practice, a complex optical arrangement with dozens of suspended mirrors – aligned during *science mode*.

The above describes *passive* vibration isolation. The Virgo and KAGRA detectors use passive vibration isolation in the form of chained harmonic oscillatory systems to isolate the detector elements. LIGO additionally uses *active* isolation for the first few stages of isolation: large platforms supported by springs where sensors measure the vibration that is to be cancelled. This results in a quiet platform from which passive chains of pendula can be hung.

### *Thermal noise*

Thermal noise is the effect temperature has on the detector. Temperature can be described as the motion of atoms and, as the detectors try to measure mirror motions much smaller than an atom, these thermal motions cause noise in our detector signals. Gravitational wave detectors use high-power laser beams to determine the position of the input and end mirror surfaces. A beautiful (not only from a science perspective, see Figure 13.6) coating is applied to the mirrors to make them highly reflective. At room temperature, the coating surface atoms move individually with magnitudes larger than the aimed-for precision. We probe, however, a very large number of atoms to average out the incoherently moving atoms to determine an average mirror motion. Despite this, the thermal noise

13.6
Operations Specialist, Gary Traylor inspecting the test mass of the LIGO interferometer in Livingston, Louisiana. (*Source: Copyright by Matt Heintze/Caltech/MIT/LIGO Lab*)

associated with these coatings is limiting us in the most sensitive frequency interval of the detectors around several hundreds of Hertz.

The suspensions of the mirrors are also made of material operating at room temperature and suffer from similar issues. Its effects can cause thermal noise at low frequency (10–50 Hz) if poorly designed. Operating at very low temperatures, which we call cryogenic operation, can help avoid thermal loss due to the suspension of the mirrors. Currently, KAGRA is the Japanese second effort towards building a kilometre-scale cryogenic interferometer, and all next generation of detectors will thankfully learn from their experience.

### *Quantum noise*

Quantum noise is a collective noun to encompass all noises associated with the fact that the light beams we use are actually made up of photons. Gravitational wave detectors use light to measure gravitational waves. However, the detectors do not count the individual photons; typically, they continuously measure several hundreds of milliwatts at the output port. The noise associated with the counting statistics is a problem for us. This noise is called *shot noise*, and arises out of the uncertainty of the arrival times of the individual photons and the effect this has on the precise intensity measurement at the output port. This is one of the quantum noises we have to deal with, and shot noise typically is dominant in the high-frequency range above 500 Hz. It can be mitigated by using higher power (more photons) in the cavities. Why don't we increase the power to megalomaniac proportions? Apart from the technical difficulties of keeping a kilometre cavity aligned when all these photons bounce around ($10^{24}$ photons per second bounce off any given cavity mirror), there is another quantum noise we should design for: *radiation pressure noise*. Cranking up the power will cause more photons per second to bounce off the mirrors. These mirrors are suspended and feel a photon pressure, which moves these objects around. If the beam of light was a continuous beam, this would not be an issue as the pressure would be constant. However, we are dealing with light particles and uncertainty in their arrival times. Therefore, the mirrors will be moving around because of the varying radiation pressure. This typically causes noise at low frequency; the noise gets less if we decrease the power. So, there you have it: detectors need to optimise the radiation pressure noise at low frequency with the quantum shot noise at high frequency.

***Technical noise***

The above noises go hand in hand with different design choices, and they are well understood within the gravitational wave research community. However, operating a kilometre scale interferometer is not an easy task. We need many feedback loops to keep the detector in check; these feedback loops can also introduce noise – technical noise. The detector designs look very promising on paper, but when they are built, we often encounter subtle couplings that make life hard for gravitational wave physicists. For instance, some parts of the mirrors have imperfections that make operators want to move the light beam slightly. But every time you change the angle of the mirror, the length of the cavity is changed ever so slightly as well.

## CONCLUSION

In this chapter, we have detailed a very recent (2017 Physics Nobel prize!) application and extreme validation of Einstein's general theory of relativity. By showing what effects the dynamical curvature of space and time can have on the way we study the Universe, students can see the purpose of using and validating Einstein's famous theory. Ripples in spacetime allow us to study parts and objects in the Universe that are dark to our telescopes that detect light and other parts of the electromagnetic spectrum. For the first time, we can study exotic star corpses such as black holes and neutron stars. In the process, we discovered where gold comes from and how to explain the abundance of many other heavy elements in our Universe. Studying the numerical values and concepts described in this chapter, students can get acquainted with experiments that span many orders of magnitude – powers of ten – and see the great lengths instrumental physicists go through to test Einstein's theories and open up a new field within astronomy.

The future of gravitational wave detection is as bright as the signals that we see on t-shirts, ties, and tattoos these days. Future designs will go cryogenic to 123 K or 10–20 K temperatures, and upgrades generally involve making everything bigger and better. Suspensions? Longer! Power? Higher! Beam tubes? Longer! This will allow us to learn and study even more exotic phenomena in the (dark) Universe.

## FURTHER READING

Chapter 4: The difficult birth of gravitational wave astronomy
Chapter 11: Gravitational lensing as a focal point for teaching general relativity
Chapter 17: Gold: Einstein's Metal

## REFERENCES

Abbott, B. P. et al. (2016). "Observation of gravitational waves from a binary black hole merger". *Phys. Rev. Lett.*, 116, 061102.

Gertsenshtein, M. E. & Pustovoit, V. I. (1962). "On the detection of low frequency gravitational waves". *Sov. Phys. JETP*, 16, 433.

Weiss, R. (1972). "Electromagnetically coupled broadband gravitational wave antenna". *Quarterly Progress Report of the Research Laboratory of Electronics of the Massachusetts Institute of Technology*, 105(54).

# 14 Using the language of gravity to teach about space, time, and matter in general relativity

*Magdalena Kersting*

**WHAT?**

This chapter explores the links between space, time, and matter in general relativity. Presenting three instructional models and two lesson plans, Magdalena Kersting suggests how the vocabulary of gravity can enrich teaching and learning of general relativity.

**FOR WHOM?**

This chapter is intended for readers who look for instructional models and lesson plans to teach concepts of general relativity at the upper secondary school level, and also, for those who wish to learn more about the pedagogical perspective of talking physics.

## INTRODUCTION

For most people, the force of gravity is a ubiquitous background fact. We are so used to navigate a world governed by the laws of gravity that we take gravitational phenomena for granted. Yet, once we start thinking about these phenomena more carefully, the concept of gravity becomes remarkably elusive. Gravity, as a familiar phenomenon, offers an instructional entryway to Einsteinian concepts of space, time, and matter while providing an example of how different scientific models can describe the same physical phenomena. There is a great need for authentic instructional approaches that introduce students to the current scientific models and that create awareness of the way scientific knowledge is produced, constantly re-examined, and further refined by scientists.

This chapter presents an opportunity to re-examine the ways gravity is taught in schools. Doing so, we explore the links between space, time, and matter and suggest three instructional approaches and two lesson plans that help secondary school students build conceptual understanding of gravity. By unpacking the surprising links between gravity, space, and time, the chapter places particular emphasis on the role of language in learning physics.

The presentation of this chapter follows a three-fold structure: First, we present a pedagogical perspective that emphasises the close link between language and physics. This perspective can help teachers and educators facilitate productive discussions when introducing Einsteinian conceptions of gravity in their classrooms. To illustrate our pedagogical viewpoint, we present the vocabulary of gravity that physicists routinely use to describe gravitational phenomena. Second, we describe three instructional models that allow addressing the Einsteinian concepts of space and time through gravity. Third, we present two lesson plans that use the language of gravity and the instructional models to introduce secondary school students to space, time, and matter in general relativity.

## PEDAGOGICAL PERSPECTIVE: LANGUAGE AND SCIENCE

There is general agreement among science educators that the learning of science is both a scientific and a linguistic act (e.g., Bratkovich, 2018). Psychologist Lev Vygotsky (1962) proposed that the development of scientific concepts and the development of word meanings are the same process. According to this view, scientific thought comes into existence with the acquisition of scientific vocabulary. Indeed, some physics educators have gone so far as to claim that "the mastery of science is mainly a matter of learning how to talk science" (Lemke, 1990, p. 153). This linguistic perspective becomes especially relevant in Einsteinian physics because the advanced mathematical formalism of relativity and quantum physics is often inaccessible at the high-school level. "Talking physics" is one successful instructional approach to address this challenge in the learning domain of Einsteinian physics (Bungum, Bøe, & Henriksen, 2018; Kersting, Henriksen, Bøe, & Angell, 2018).

Just as with learning a foreign language, fluency in physics requires practice at speaking, not just listening. Teachers can capitalise on that insight by inviting students to use written and oral language to promote conceptual understanding. Successful strategies to let students integrate Einsteinian concepts into their existing scientific vocabulary include the following (list adapted from Kersting et al., 2018):

1. Use open discussion tasks that probe conceptual understanding of Einsteinian physics concepts to give students the opportunity to reason with their peers.
2. Let students summarise their understanding of Einsteinian physics concepts in written exercises to let them practise the use of new physics vocabulary.
3. Use plenary discussions guided by the teacher to consolidate understanding and resolve misconceptions.
4. Raise awareness that some confusions and contradictions in physics discussions are not inherent to Einsteinian physics; rather, they are a byproduct of the limited vocabulary that everyday language offers.
5. Emphasise that qualitative understanding of Einsteinian physics can be made rigorous by advanced mathematics.

### *The vocabulary of gravity*

To make the pedagogical perspective of talking physics more tangible, we now take a look at three different vocabularies that physicists routinely employ to talk about gravity. Often, different ways of talking about gravity enable different intuitions of gravitational phenomena (Chandler, 1994). Therefore, fluency in the vocabulary of gravity is beneficial in teaching and learning about space, time, and matter in general relativity.

#### *Gravitational force*

With the formulation of his three laws of mechanics, Newton described the motion of objects using the concepts of inertia and force. The vocabulary of forces is widespread in physics. A force is linked to the intuitive notion of pushing or pulling. The force of gravity is attractive and acts between massive objects, e.g. by pulling an apple down to the centre of the Earth. More precisely, a force is an interaction that leads to objects being accelerated. The gravitational acceleration of an apple (or any other object) in free fall is independent of the object; gravity is, therefore, universal. Mathematically, a force is a vector with a magnitude and direction that describes how much it will change the motion of an object.

Teaching students the vocabulary of forces and inertia has several pedagogical advantages. First, forces play a fundamental role in the way physicists describe the universe. An early introduction to the force concepts helps students become acquainted with classical mechanics and the mathematical formalism of vectors. Second, since students are familiar with the sensations of pushing and pulling, instructors can describe forces by appealing to students' intuitions of motion (Young, Freedman, Sandin, & Lewis Ford, 1999). In particular, the explanation that gravity pulls things downwards is appealing because it can be "fluently formulated in language, and [is] thus easily accessible to conscious elaboration, analysis, and reflection" (Kapon & diSessa, 2012, p. 267). Third, Newtonian mechanics offers a coherent conceptual system that has the potential to impact students' learning similarly as it did impact scientists adopting the system in the first place (Hestenes, Wells, & Swackhamer, 1992).

However, physics education research has shown that the force concept presents students with conceptual challenges. Often, students' intuitions and their commonsense beliefs about motion are incompatible with Newtonian concepts (Hestenes et al., 1992). Many students leave physics instruction with misconceptions of forces. These misconceptions seem to exist independent of the instructor or mode of instruction. Moreover, the force of gravity appears to act, somewhat mysteriously, at a distance. There is no apparent connection between the Earth, the Sun, and the moon and no apparent mediator between these massive objects; yet, gravity holds them together in the solar system. From a pedagogical perspective, the force of gravity, thus, might appear counter-intuitive and for some, even absurd or impossible (Chandler, 1994).

### *Gravitational field*

The introduction of the notion of fields shifts the discussion of gravity to a higher level of abstraction. Fields are powerful conceptual tools in physics. Treating fields as physical entities that possess energy and momentum represents an important paradigm in modern physics (Feynman, 1970). A field is a physical entity that has a value for each point in spacetime, and that value can change with time. Although the most studied fields in physics are those that describe the fundamental forces, it is entirely possible to use the vocabulary of field theory without referring to the notion of forces. Thus, the field of gravity provides an alternative vocabulary to describe gravitational phenomena.

The link between the gravitational field and force is via the gravitational potential energy. A physical object with mass has potential energy in relation to another massive object. It is this potential energy that we associate with the gravitational field, and the force of gravity is the action of the gravitational field at a certain point in spacetime. By multiplying the gravitational field of an object at a specific location with the mass of a given test particle, we obtain the gravitational force that the object exerts on the test particle at that location. In other words, the gravitational force is the gradient of the gravitational potential energy field.

From a pedagogical perspective, the vocabulary of fields seems to be more abstract than that of forces. In contrast to forces, fields cannot easily be associated with pulls and pushes or with any other concrete physical entity. Chandler suggests that "perhaps as science becomes more advanced, its theories become increasingly counter-intuitive, and the more we learn about gravity, the less sense it will make" (Chandler, 1994, p. 165).

Yet, despite the higher level of abstraction, the vocabulary of fields has a broader scope than the notion of the gravitational force. For example, we can describe the fall of an apple off a tree via a potential well. The potential could, of course, represent the force of gravity, but it could also represent gravitational time distortion (Einstein, 1907).

### *Gravity and geometry*

Einstein introduced the language of differential geometry to account for the universality of gravity. Based on the principle of equivalence and the insight that, locally, one cannot distinguish gravity from acceleration, Einstein linked the physics of gravity to the mathematics of geometry. Differential geometry describes spacetime as a mathematical structure that is locally flat, but that can be curved on a global scale – just like the Earth appears flat to us but is a curved sphere in reality.

While the human mind struggles to perceive four dimensions, the vocabulary of differential geometry allows physicists to work with the abstract concept of spacetime and make precise predictions about gravitational phenomena. Moreover, replacing Newton's mysterious force of gravity with a geometric mechanism, Einstein's vocabulary offers an explanation rather than a mere description of gravity.

However, the enormous explanatory power of differential geometry comes with one great disadvantage, namely, its inaccessibility at the high-school level. The advanced mathematical formalism renders Einstein's vocabulary of gravity incomprehensible to many (Levrini, 2002). Even though many instructional models use students' geometric intuitions to present key ideas of general relativity, there remains the challenge of separating the mathematical formalism from the conceptual foundations of Einstein's theory of gravity.

## INSTRUCTIONAL MODELS THAT LINK GRAVITY, SPACE, AND TIME

In this section, we present three instructional models of gravity in line with an Einsteinian perspective. Each model capitalises on the close link between gravity, space, and time in a different way. Using the linguistic explorations of the previous sections as our backdrop, we unpack instructional opportunities and limitations of each model before presenting a comprehensive lesson plan in the next section.

### *Gravity and space – The rubber sheet model*

The popular rubber sheet model visualises the dynamic interplay between spacetime and matter (Figure 14.1). The model compares spacetime to a stretched fabric: marbles or balls represent massive objects that curve the material which, in turn, changes the marbles' trajectories on the sheet. Historically, the rubber sheet model has become a widely used instructional tool to make sense of four-dimensional curved spacetime. The use of malleable fabrics such as rubber and cloth to illustrate features of relativity reaches back to the very beginning of popular accounts of the theory of relativity (Russell, 1925).

Linguistically, the rubber sheet model is appealing because it illustrates the catchy phrase "spacetime tells matter how to move, matter tells spacetime how to curve" (Wheeler, 1998). The model thereby relates Einstein's vocabulary of geometry to the vocabulary of forces and movement with which students are familiar. Moreover, its main elements "confirm so much to our expectations from everyday experience that the model can be presented as a narrative" (Pössel, 2018, p. 13). In comparing spacetime to a rubber sheet, the model establishes an analogical link between the rubber sheet (an everyday object students know well) and the abstract concept of spacetime. Based on the assumption that conceptual understanding of general relativity is linked to an active level of word knowledge, the rubber sheet model can serve both as a productive analogy and as a feature of relativistic thought that students develop to reason about gravity (Kersting & Steier, 2018).

#### *Educational opportunities and limitations of the rubber sheet model*

An instructional model works when it satisfies its purpose, that is, conveying a key idea of an abstract scientific concept. Educators and physicists hold divided opinions on the educational value of the rubber sheet model (Kersting & Steier, 2018).

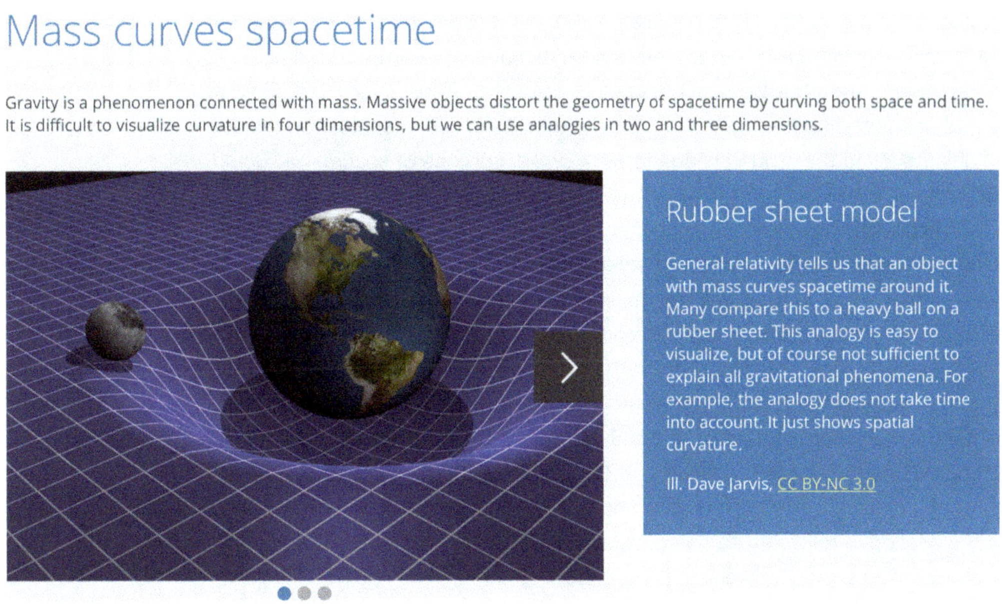


14.1 The rubber sheet model of gravity as illustrated in the digital learning environment *General Relativity* at www.viten.no/relativity. (*Source: Copyright by Project ReleQuant*)

Strengths of the rubber sheet model include its visual power, simplicity, and intuitive explanation of gravity. The deformed sheet presents a mechanism of how gravity arises, and this analogy has great explanatory power: it is suitable to show orbital motions, curved space, and photon trajectories (Kersting & Steier, 2018). Besides, the rubber sheet is a dynamic continuum and is, therefore, a "welcome antidote to people whose mental picture of gravity includes a boundary for, say, the Earth's gravity" (Pössel, 2018). Thus, the rubber sheet model can counteract the development of misconceptions related to the idea that there is no gravity on the moon or in outer space.

Yet, despite these educational opportunities, no instructional model comes without limitations (Kersting & Steier, 2018). Since the rubber sheet model depicts four-dimensional spacetime as a two-dimensional spatial sheet and suppresses the time dimension, the model emphasises the role of space over the role of time. This emphasis on spatial curvature is misleading, since we experience gravity on the Earth mostly because of warped time and not curved space (Kersting, 2019). Besides, the rubber sheet model obscures the intrinsic nature of spacetime curvature. The universe does not curve into an "unseen dimension" like the two-dimensional rubber sheet that bends into the third dimension of space. Additionally, the rubber sheet model makes use of gravity to explain gravity and, thus, obscures the physical mechanism of gravitational phenomena in general relativity.

### *Gravity and time – The ageing Earth model*

By interpreting gravity as a manifestation of spacetime curvature, general relativity has a greater explanatory scope than Newtonian gravity. Still, on the Earth, we mostly experience the laws of physics in line with Newton's theory. Physicists call gravitational conditions that reduce to this classical framework the

"Newtonian limit". In this limit, particles move slowly compared to the speed of light, and the gravitational field is weak and not changing in time. Remarkably, in the Newtonian limit, gravity is an effect of warped time. Thus, in contrast to what the rubber sheet model suggests, the gravity that makes our planets orbit the Sun and objects fall to the ground is due to *temporal* distortions of spacetime (Gould, 2016; Pössel, 2018).

Time dilation is one measurable effect of how gravity affects time. Depending on the gravitational potential and its gradient, time might pass more slowly or more quickly (Figure 14.2). In everyday life, the effects of time dilation are not noticeable. Still, there are ways to relate students' daily experience to this relativistic phenomenon. One such way goes back to Richard Feynman who observed that the centre of the Earth is younger than its surface due to gravitational time dilation. A hypothetical measurement of the age of the Sun or the Earth will yield different results depending on whether the measurement is performed at the surface or near the centre of the galactic object. This difference in age is "a fascinating demonstration of time dilation in relativity, and as such a very illustrative example for use in the classroom" (Uggerhøj, Mikkelsen, & Faye, 2016, p. 8).

When introducing the ageing Earth as an instructional model, teachers can use the vocabulary of the gravitational field to help students build understanding

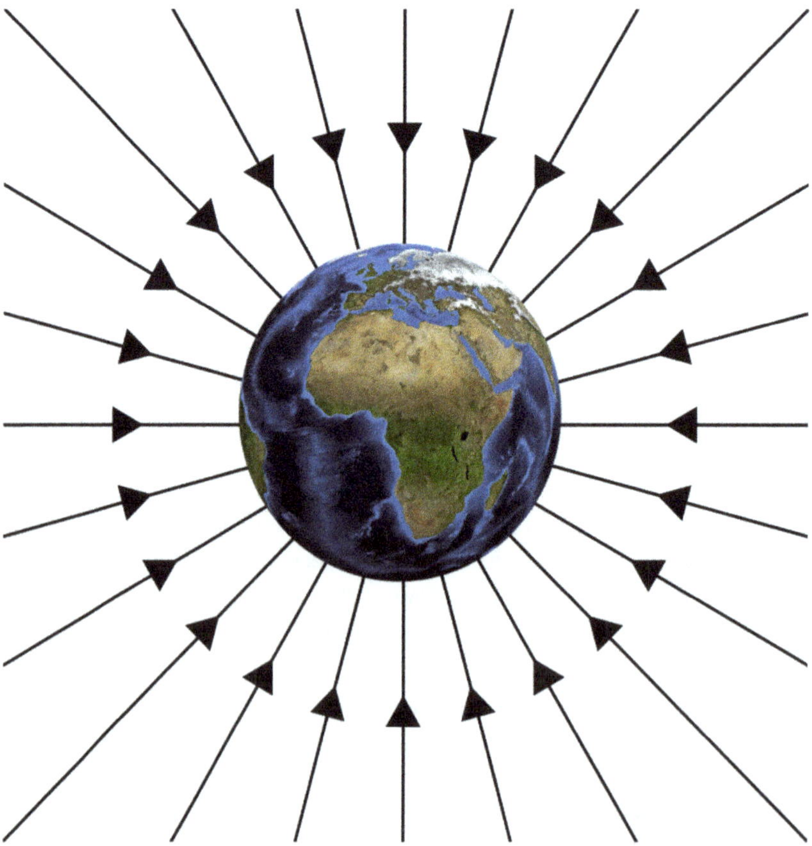

14.2
The gravitational field of the Earth is not homogenous and gets stronger towards the centre of the Earth. The difference in the strength in the gravitational field leads to the phenomenon of time dilation. This phenomenon is further explained in the digital learning environment *General Relativity* at www.viten.no/relativity. (*Source: Copyright by Project ReleQuant*)

of how gravity and time are closely interlinked. Gravitational time dilation might seem abstract and far away from students' lived experiences. The ageing Earth model provides a concrete example of how the difference in the gravitational potential leads to time passing differently at the surface and the centre of the Earth. Moreover, the example of the ageing Earth allows a quantitative treatment of gravitational time dilation. Here, teachers can invite students to use mathematics as well as the vocabulary of fields to understand the link between time and gravity in general relativity (Uggerhøj et al., 2016).

*Educational opportunities and limitations of the ageing Earth model*
Gravitational time dilation is a minuscule effect when considered in everyday life. Thus, it can be challenging for students to relate to time dilation because the phenomenon seems counter-intuitive. While the ageing Earth model provides an example that presents a measurable time difference of several years, the challenge of relating to the scale of the effect remains. Students often struggle to understand long time scales such as the age of the Earth or cosmic distance scales such as the size of the Earth and other planets. Nonetheless, the ageing Earth model is pedagogically valuable because it presents a complete solution to the relativity of ageing due to differences in the gravitational field. Students can, therefore, practice the use of the field vocabulary in addition to becoming familiar with quantitative treatments of time dilation.

***Gravity and time - The warped-time model***
Arguably, it is challenging to visualise temporal distortions. Thus, many instructional models of Einsteinian gravity do not take the time dimension into account. In the previous section, we have seen how we can link time and gravity by drawing on the phenomenon of time dilation. Here, we present an interactive warped-time model that illustrates a different approach to connecting time and gravity.

Physics students routinely use height-time diagrams to depict movement through space as a function of time. An interactive height-time diagram lies at the heart of the warped-time model that is freely accessible at the Norwegian learning platform Viten (Figure 14.3): www.viten.no/relativtiy.

The warped-time model illustrates that we experience gravity on the Earth because of warped time (Kersting, 2019). By drawing and discussing the trajectory of an object in free fall, students can explore the physics of gravity according to Newton and Einstein. In a traditional height-time diagram, the trajectory of a freely falling object corresponds to a parabola. Newton explains this trajectory as the result of the force of gravity accelerating an object down to the ground. To move from a Newtonian to an Einsteinian perspective, students warp the time-axis of the height-time diagram upwards by moving a slider. In this curved diagram, the parabola trajectory of a freely falling object gets stretched into a straight line. Instead of interpreting gravity as a force, Einstein understands gravity as a geometric phenomenon. An object in free fall follows the straightest possible path in a curved space. This

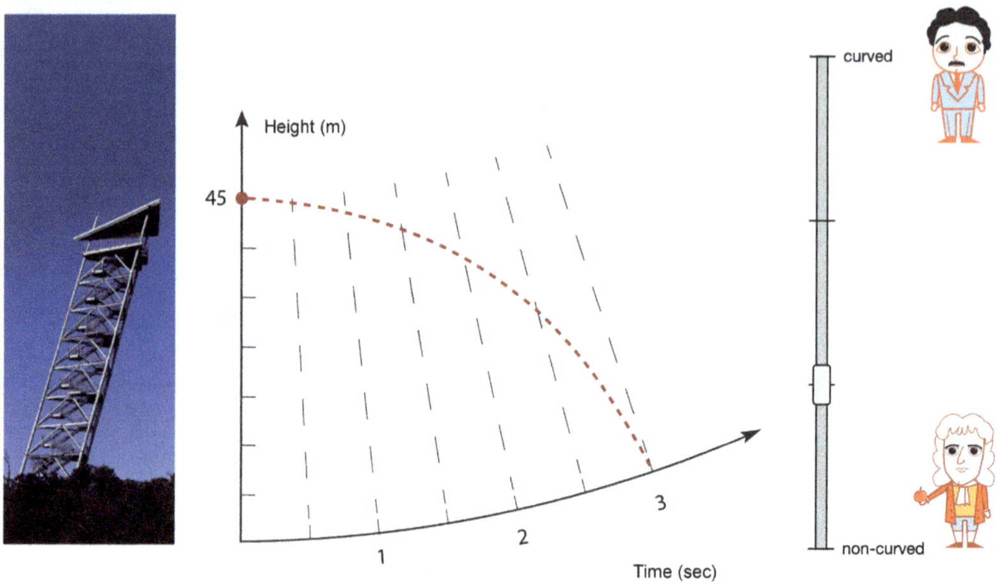


14.3
The warped-time model allows comparing Einstein and Newton's models of gravity by drawing on the physics of free fall. The warped-time model is available at www.viten.no/relativity. (*Source: Copyright by Project ReleQuant*)

observation generalises Newton's first law to the realms of four-dimensional curved spacetime.

*Educational opportunities and limitations of the warped-time model*

The warped-time model compares Einstein and Newton's theories of gravity by using a representational tool that high-school students are familiar with. Height-time diagrams allow students linking their previous knowledge of movement in a gravitational field to a relativistic model of gravity. Moreover, thought experiments are powerful tools to communicate relativistic concepts to high-school students (Kersting et al., 2018; Velentzas & Halkia, 2013). Taking the physics of free fall as its starting point, the warped-time model uses one of Einstein's most famous thought experiments that students can easily relate to.

However, just like the rubber sheet model, the warped-time model has inherent limitations because the model translates the physics of movement in four-dimensional spacetime to a two-dimensional visualisation. One limitation is the exaggerated warping of the time-axis: the mass of the Earth is not big enough to warp time to an extent as presented in the simulation. Besides not being true to scale, the curved visualisation of the warped-time diagram is not a correct coordinate depiction either: a straight line in the diagram does not correspond to a straight line in spacetime. Rather than being geometrically precise, the model serves the pedagogical purpose of promoting a qualitative understanding of Einstein's theory (Kersting, Toellner, Blair, & Burman, 2020).

## LESSON PLAN: USING THE VOCABULARY OF GRAVITY TO TEACH ABOUT SPACE, TIME, AND MATTER IN GENERAL RELATIVITY

In this section, we present the lesson plan of one double lesson (90 minutes) that introduces secondary school students to the concepts of space, time, and matter in general relativity. A second lesson plan can be downloaded in the e-resources. Both lessons invite students to "talk physics" and to practice using the vocabulary of gravity in discussions and writing tasks. Lesson 1 is an introductory lesson that encourages students to explore the links between gravity, curved space, and warped time through qualitative reasoning and discussions with peers based on the rubber sheet and the warped-time model. Lesson 2 is a more advanced lesson that builds on concepts developed in lesson 1 to guide students to a deeper understanding of gravitational time dilation using the ageing Earth model and elementary calculations. Table 14.1 gives a brief overview of how the content of the two lessons links to common topics of physics curricula.

### *Lesson 1 – Linking gravity to the geometry of curved space and warped time*

In this introductory lesson of 90 minutes, you engage your students in qualitative thinking processes that lead to an exploration of geometry as the cause of gravitational phenomena. The lesson aims to foster qualitative understanding of the four-dimensional nature of spacetime by inviting students to talk physics. Working in pairs or small groups, students follow an instructional sequence from the digital learning environment *General Relativity* that is freely accessible at www.viten.no/relativity.

#### *Learning goals*

After completing this lesson, students will be able to

- use the vocabulary of forces to describe Newton's model of gravity
- use the vocabulary of curved space and warped time to describe Einstein's model of gravity
  - describe general relativity as a theory relating space, time, and gravity
  - describe the universe as having three spatial dimensions and one time dimension
  - explain how gravity can be interpreted as a geometric phenomenon
  - describe how mass curves spacetime and how the curvature of spacetime influences the movement of mass
- compare Newton and Einstein's model of gravity and conclude that general relativity has a greater explanatory scope

#### *Instructional models and approaches*

In this lesson, students will work with the **rubber sheet** and the **warped-time model** in the digital learning environment *General Relativity*. Open discussion tasks, writing tasks, and plenary discussions challenge students to practice the **vocabulary of the gravitational force and geometry** in the context of general relativity.

***Table 14.1* Overview of how the lesson plans link to common topics in physics curricula**

| *Curriculum topic* | *Connection to lesson plans* |
| --- | --- |
| Gravity and weight | In Newton's theory, gravity is an attractive force between massive objects that gives weight to physical objects. In Einstein's theory, gravity is not a force, but the consequence of the curved geometry of space and time. Weight is the magnitude of the force that deviates objects from their straight path through curved spacetime. Objects in free fall are weightless. |
| Force and acceleration | A force is an interaction between objects that leads to acceleration of these objects. In Newton's theory, gravity is an attractive force between massive objects that causes the acceleration of falling bodies. In Einstein's theory, gravity is not a force, and objects in free fall do not accelerate. |
| Time dilation | Time is not absolute, but influenced by gravity. Time goes more slowly further down in a gravitational field than higher up. Observers can test this by moving a clock into the vicinity of a massive object, leaving it there for a while, and moving it away again for a direct comparison with a clock that has stayed farther away. |
| Nature of science | Science is an attempt to explain natural phenomena. Scientists constantly refine and re-examine scientific knowledge which is always subject to change. Einstein replaced Newton's force model of gravity with his spacetime model of gravity. Einstein's theory is our current best understanding of gravity because it fits the experimental evidence better than Newton's theory. |
| Scientific inquiry and scientific models | Scientific inquiry is the process of discovery in science that involves asking questions, making predictions, testing hypotheses, and studying the natural world. As part of that process, scientists build models to explain natural phenomena. There can be different models for the same physical phenomenon. Both Newton's force model and Einstein's spacetime model provide explanations of gravity. Einstein's model, though, has a greater explanatory scope than Newton's model. Only Einstein's model predicts that time is influenced by gravity. |

In pairs or small groups, students work through parts of modules 1 (the principle of equivalence) and 3 (curved spacetime) at their own pace. Alternatively, you can pace the flow of the lesson by guiding your students through each part of the module, making sure to allow enough time for group and plenary discussions.

Optional additions to the lesson include a hands-on activity with a physical rubber sheet model or a discussion of gravitational bending of light (as outlined in the second part of the module "Curved Spacetime" at www.viten.no/relativity).

*Instructional sequence (90 minutes)*

The instructional sequence follows parts of the presentation in modules 1 (the principle of equivalence) and 3 (curved spacetime) of the digital learning environment *General Relativity*. In Table 14.2, we provide an overview of the instructional steps.

***Table 14.2*** **Overview of the instructional steps of lesson 1**

| *Activity* | *Module* | *Time* | *Comment* |
|---|---|---|---|
| Writing task: What do you know about gravity? | 1 | 5 min | This writing task is meant as a warm-up for the students to recollect their knowledge of Newton's model of gravity. |
| Video: General relativity in a nutshell | 1 | 5 min | Watch this short introduction video which presents the key ideas of general relativity that students will explore in module 3. |
| Plenary discussion: why a new theory? | 1 | 5 min | Facilitate a plenary discussion on the need for a new model of gravity to raise awareness of the nature of science and scientific models. |
| Move to module 3 | 3 | 5 min | From here onwards, let students form pairs/small groups and let them work through the learning resources at their own pace. |
| Group work from panel "gravity is not a force" to panel "summary warped time" | 3 | 60 min | To practise the vocabulary of gravity, students will encounter<br>• an interactive discussion task "geometry in curved spaces"<br>• a discussion task on the rubber sheet model followed by a writing task to summarise the discussion<br>• an interactive discussion task "curved time" that introduces the warped-time model |

*(Continued)*

*Table 14.2 (Continued)*

| *Activity* | *Module* | *Time* | *Comment* |
|---|---|---|---|
| Plenary discussion: summing up together – curved spacetime | 3 | 10 min | Wrap up the lesson with a plenary discussion to give students the opportunity to ask questions and consolidate their knowledge of general relativity. You can use these prompts to get the discussion started:<br>Both Newton and Einstein have their own theory of gravity. Summarise Newton's and Einstein's explanations.<br>In which situations should you use Newton's theory, in which should you use Einstein's? |

## CONCLUSION

The familiar phenomenon of gravity offers different instructional entryways to Einsteinian concepts of space, time, and matter. In this chapter, we have used the vocabulary of gravity to present three instructional models that link gravitational phenomena to curved space and warped time. While the rubber sheet model illustrates gravity as a manifestation of curved space, the ageing Earth model and the warped-time model invite students to explore the links between gravity and time. By offering teachers several instructional models, all of which are complimentary and capitalise on different vocabularies of gravity, this chapter promotes a contemporary way of teaching gravity. First, the pedagogical emphasis on language fosters conceptual understanding and qualitative reasoning in a learning domain that often has a reputation of being inaccessible to younger learners. Second, the two lesson plans provide examples of how different scientific models can describe the same physical phenomena. The lesson plans, therefore, offer opportunities to introduce students to the nature of science and raise awareness of the way scientists continuously produce and re-examine scientific knowledge. Despite its ubiquitous presence, gravity has eluded human understanding for centuries and has challenged some of the keenest minds in the history of science. It is by emphasising this historical struggle and by illustrating the links between space, time, and matter that students can obtain a comprehensive picture of gravity in line with our current best understanding of the universe.

## FURTHER READING

E-resources: Lesson 2 – Exploring how the gravitational potential influences the rate at which time passes

Chapter 10: Models and analogies in teaching general relativity

Chapter 19: Designing learning resources and investigating student learning in general relativity and quantum physics in Norway

## REFERENCES


Bratkovich, M. O. (2018). Shining light on language for, in, and as science content. *Science and Education, 27*(7–8), 769–782.

Bungum, B., Bøe, M. V., & Henriksen, E. K. (2018). Quantum talk: How small-group discussions may enhance students' understanding in quantum physics. *Science Education, 102*(4), 856–877.

Chandler, M. (1994). Philosophy of gravity : Intuitions of four-dimensional curved spacetime. *Science & Education, 3*(2), 155–176.

Einstein, A. (1907). Relativitätsprinzip und die aus demselben gezogenen Folgerungen [On the Relativity Principle and the Conclusions Drawn from It]. *Jahrbuch Der Radioaktivität, 4*, 411–462.

Feynman, R. (1970). *The Feynman lectures on physics vol. I*. Addison-Wesley Longman.

Gould, R. R. (2016). Why does a ball fall?: A new visualization for Einstein's model of gravity. *American Journal of Physics, 84*(5), 396–402.

Hestenes, D., Wells, M., & Swackhamer, G. (1992). Force concept inventory. *The Physics Teacher, 30*(3), 141–158.

Kapon, S., & diSessa, A. A. (2012). Reasoning through instructional analogies. *Cognition and Instruction, 30*(3), 261–310.

Kersting, M. (2019). Free fall in curved spacetime – How to visualise gravity in general relativity. *Physics Education, 54*(3), 035008.

Kersting, M., Henriksen, E. K., Bøe, M. V., & Angell, C. (2018). General relativity in upper secondary school: Design and evaluation of an online learning environment using the model of educational reconstruction. *Physical Review Physics Education Research, 14*(1), 010130-1–010130–18.

Kersting, M., & Steier, R. (2018). Understanding curved spacetime —The role of the rubber sheet analogy in learning general relativity. *Science & Education, 27*(7), 593–623.

Kersting, M., Toellner, R., Blair, D., & Burman, R. (2020). Gravity and warped time – Clarifying conceptual confusions in general relativity. *Physics Education, 55*, 015023.

Lemke, J. L. (1990). *Talking science: Language, learning, and values*. Norwood, New Jersey: Ablex Publishing Corporation.

Levrini, O. (2002). Reconstructing the basic concepts of general relativity from an educational and cultural point of view. *Science & Education, 11*(3), 263–278.

Pössel, M. (2018). "Relatively complicated? Using models to teach general relativity at different levels." arxiv.org/abs/1812.11589.

Russell, B. (1925). *ABC of relativity*. London: Allen & Unwin.

Uggerhøj, U. I., Mikkelsen, R. E., & Faye, J. (2016). "The young centre of the Earth." *European Journal of Physics*, 37(3).

Velentzas, A., & Halkia, K. (2013). "The use of thought experiments in teaching physics to upper secondary-level students: Two examples from the theory of relativity. *International Journal of Science Education, 35*(18), 3026–3049.

Vygotsky, L. (1962). *Thought and language*. Cambridge: M.I.T. Press.

Wheeler, J. A. (1998). *Geons, black holes, and quantum foam: A life in physics*. New York: W.W. Norton & Company.

Young, H. D., Freedman, R. A., Sandin, T. R., & Lewis Ford, A. (1999). *Sears and Zemansky's university physics* (10th ed.). Addison-Wesley.

# Quantum physics

## 15 Introducing quantum physics with toy photons

*Tejinder Kaur and David Blair*

**WHAT?**

This chapter centres on hands-on activities which were developed to teach quantum physics concepts at an early age. The physical models and analogies allow students to understand key concepts of the quantum world and the limitations of analogical models.

**FOR WHOM?**

This chapter is intended for readers who are interested in a fun approach for teaching the fundamentals of quantum physics at any level from primary to high school.

### INTRODUCTION

The recognition that light comes in the little packets we now call photons, by Max Planck and Albert Einstein, did away with the idea that light is a simple wave. This chapter aims to present the fundamental concepts of quantum physics in a form which is intuitive and easily digested (Box 15.1). We first introduce the particle nature of light and the statistical nature of quantum phenomena. Simple thought experiments illustrate the surprising weirdness of the quantum world.

**BOX 15.1: KEY CONCEPTS OF QUANTUM PHYSICS**

- Photons: light particles
- Heisenberg's uncertainty principle
- Photoelectric effect
- Diffraction
- Interference

Quantum physics describes how the universe works on scales from small macroscopic objects down to objects much smaller than atoms. It has enormous applications in our daily life. Indeed, it is estimated that more than 50% of current economic activity is based on quantum technologies (e.g., computers, smartphones, lasers, medical imaging, drug developments, and new materials). All of our modern gadgets, and especially the smartphones we all depend on, have Einsteinian physics inside them. Everyone should understand key physics concepts behind these gadgets. Otherwise, our devices may as well be magical!

Quantum physics was traditionally excluded from the school curriculum because it was considered too difficult. Teachers, who often had inadequate training in the subject are not well-equipped to teach quantum concepts at the school level. This itself might have arisen because at the tertiary level, quantum physics is often taught in an abstract mathematical formalism that fails to give an intuitive understanding of the core concepts.

Our goal is to introduce important key concepts of quantum physics with no abstract mathematics. Richard Feynman was already aware that school physics had outdated concepts in 1985. In his book, *QED: The Strange Theory of Light and Matter* (Feynman, 1985), he emphasised the importance of introducing the concepts of quantum physics:

"I want to emphasise that light comes in this form – particles. It is very important to know that light behaves like particles, especially for those of you who have gone to school, where you were probably told something about light behaving like waves".

The goal of the Einstein-First project in Western Australia is to introduce the concepts and the language of quantum physics at an early age so all students can have a basic understanding of the world they experience. Learning a language at a young age is easy. The language of quantum physics, including words like *photons*, *electrons*, *quantum uncertainty*, and *probability*, is relevant to the devices we use every day.

We have used the approach of teaching through physical models and analogies, partly as a way of changing the emphasis on abstract qualities to that of physical activities that both teachers and students can appreciate readily. Rather than introducing a concept, and using an activity to illustrate it, we prefer to introduce the activity and use it to define the concept. The set of activities are more memorable than the set of concepts because they harness the full breadth of human memory – context-dependent memory, social memory, tactile memory, etc.

Teaching through models and analogies is a well-known method among researchers, scientists, and teachers. Many studies have shown that learning through models and analogies has a lasting impact on students' understanding. Harrison and Treagust (1996) suggest that models and analogies can be effective in teaching science concepts. Analogies are considered the subset of models which involve the comparison between two objects that are similar in a few aspects. The Einstein-First project has tested models and analogies at the primary and secondary level with great success. Research results show that children respond very positively to our programme and accept the basic concepts of quantum physics taught to them through activities (Choudhary et al., 2018, Kaur et al., 2018). A surprise was that girls' improvement factors were greater than boys' in several separate trials (Kaur et al., 2020).

Following Feynman's advice, we introduce light by first emphasising its particle nature. However, from the youngest age groups onwards, we keep the central idea that light has both particle-like and wave-like properties. We introduce it in the broad context of De Broglie's law by using two words, *bulletiness* and *waviness* and the aphorism:

*Everything has bulletiness; Everything has waviness.*
*Bulletiness follow the maths of number; Waves follow the maths of arrows.*

In this aphorism, bulletiness refers to the photon's momentum, and the ability of high-energy photons to be destructive. Waviness refers to the fact that photons have a wavelength and an associated frequency. Because frequency and wavelength are not independent, we need to introduce both quantities. We emphasise the bulletiness by showing videos of single photons arriving (https://www.youtube.com/watch?v=e30BuTnhJng) that also highlight the intrinsic randomness of photon arrival. We use the word "everything" to emphasise that these properties are common to all things, whether normally thought of as radiation (like light) or as particles (like atoms or cricket balls). The allusion to mathematics is important. The bulletiness of photons allows them to be counted: we can discuss the number of photons per second and calculate their numbers. However, waviness is best described by vectors (the mathematics of arrows) in which two vectors can add to zero (called destructive interference) or to a length equal to the sum of the arrows (called constructive interference). In this way, the above aphorism condenses quantum physics and its mathematics into a memorable form.

De Broglie's law states that waviness and bulletiness are a universal property: the greater the momentum, the smaller the wavelength (Box 15.2). When the momentum is large like that of a tennis ball or an ultra-high energy photon, the waviness (wavelength) is tiny, and its momentum dominates the system. When the momentum is extremely small like that of a radio photon, its waviness dominates the system because the wavelength is large. The destructiveness of objects depends on momentum. For this reason, under most circumstances, radio waves are harmless, but ultraviolet photons can damage our DNA.

## PARTICLE NATURE OF LIGHT

Analogue photons in the form of Nerf gun bullets are used to understand the bulletiness of light. Nerf guns illustrate various concepts such as imaging, photon

### BOX 15.2: BASIC FORMULAE OF QUANTUM PHYSICS: ENERGY AND MOMENTUM OF PHOTONS, CONNECTION BETWEEN FREQUENCY AND WAVELENGTH, AND DE BROGLIE'S LAW

Energy of a photon, $E = h \times f$

Planck's discovery required light energy quanta with energy $E = h \times f$, where $h$ is a tiny number called Planck's constant = $6.63 \times 10^{-34}$. For visible light, the frequency, $f$ is $-10^{15}$ hertz.

Momentum of a photon, $\boldsymbol{p} = \frac{\boldsymbol{E}}{\boldsymbol{c}}$

Connection between frequency, $f$ and wavelength, $\lambda$: $\boldsymbol{f} = \frac{\boldsymbol{c}}{\lambda}$

where, $c$ is the speed of light = 300,000 kilometres per second.

De Broglie's law: $\lambda = \frac{\boldsymbol{h}}{\boldsymbol{p}}$

For objects of mass $m$, moving at low velocity, $p = mv$, where $v$ is the velocity. In this case, the wave is called a *matter wave*.

momentum, the photoelectric effect, and radiation pressure effects. In this context, students learn that the energy and momentum of a photon depend on its frequency, with energy given by Planck's formula, and momentum given by $E/c$ (Box 15.2). Depending on the age of the students, we go into more or less detail.

#### *Nerf gun photons*

Nerf gun bullets come with either blunt caps or with suction caps that can stick to surfaces such as whiteboards, allowing their positions to be registered. A stream of Nerf bullets represents a beam of light. First, we investigate this stream of "photons" to get a clear understanding that photons are generally widely spaced and behave independently. This becomes important when we explore interference. Once we have explored the stream of photons, we use them to explore photography, the uncertainty principle, and the photoelectric effect.

#### *The spacing between real photons: A thought experiment*

Before exploring the surprising properties of the quantum world, it is very useful to appreciate that photons are sparse and widely separated. This statement in itself is questionable because quantum mechanics violates local reality. This means that while we always detect light as photons, we cannot really say anything about photons except when we measure them. One could argue the philosophical point that photons do not exist until they are detected by some measuring device. However, the time-spacing relation between photon pulses is used frequently in science and technology. An interesting example is

**BOX 15.3: STUDENT ACTIVITY WITH NERF GUN BULLET STREAM**

Consider a Nerf gun that fires 10 bullets per second, at a speed of 10 metres per second. How far apart will these bullets be?

**In 1 second, the distance travelled by bullets = 10 metres**

**In 1 second, the number of bullets = 10**

$$\textbf{Spacing between the bullets} = \frac{\textbf{Distance travelled in 1 second}}{\textbf{Number of bullets in 1 second}}$$

$$= \frac{10}{10} = 1 \textbf{ metre}$$

Hence, the bullets are 1 metre apart. Using a multiple shot Nerf gun to send a stream of toy photons, students can make a slow-motion video and estimate the real numbers for bullets per second, bullet speed, and bullet spacing.

a laser ranging to the moon, which is used to measure the distance between the Earth and the moon that increases by about ten millimetres per year (Murphy, 2013).

In this activity, we use observations of the motion of Nerf gun bullets to investigate the distances between real photons. Using simple mathematics, students can calculate the spacing between successive photons by applying the familiar formula, distance = speed × time. The calculation of the spacing between bullets is given in Box 15.3, which also describes an activity in which students video the bullet stream and make numerical estimates.

Having introduced students to the observations that light comes as photons and that the spacing between moving objects can be easily calculated when their speed and count rate (number of objects moving past a given point per second) are known, we can investigate the spacing between real photons by extending the above activity into a thought experiment.

For this thought experiment, we start with the traditional scale for the brightness of stars: zeroth magnitude is the brightest (e.g., the star Sirius), down to sixth magnitude (just visible with the naked eye). A sixth magnitude star is approximately 250 times dimmer than a zeroth magnitude star. For the zeroth magnitude star, the light flux is roughly 1 million photons per second per square centimetre. We are interested in knowing how many photons per second enter the human eye. The calculation is given in Box 15.4. We discover that the photon stream from a just-visible star is about one photon per thousand kilometres.

***Photography with toy photons***

Photography with Nerf gun photons (with suction caps) can emphasise the particle nature of light. Our "photographs" are silhouette images created by Nerf

**BOX 15.4: DISTANCE BETWEEN PHOTONS**

The aperture of the human eye, the pupil, is roughly 3 millimetres in diameter. If you look at a sixth magnitude star, approximately 300 photons per second enter the human eye. Applying the logic we used for Nerf gun bullets (and knowing the speed of light), students can calculate the distance between successive photons

**Distance between successive photons =**

$$\frac{\mathbf{300,000,000}}{\mathbf{300}} = \mathbf{1\ million\ metres\ or\ 1,000\ kilometres}$$

Thus, when a photon from a sixth magnitude star enters our eye, the next one is 1000 kilometres away! This is, of course, a remarkable result as we perceive starlight as relatively constant. Even for the brightest, zeroth magnitude star, the photons are 4 kilometres apart.

gun bullets that stick to a whiteboard, but not to clothing. In a real camera, a Charge-Coupled Device (CCD) registers photons. In our activity, the whiteboard registers "photons" (Figure 15.1).

We ask students to "photograph" each other using Nerf gun bullets. One student stands against a glossy wall or whiteboard, while others "illuminate" the subject with bullets. Bullets stick only to the glossy surface and create a silhouette "photograph" of the student. We explain to students that photons have properties analogous to those of bullets, including energy and momentum.

### *Heisenberg's uncertainty principle with analogue photons*

In 1927, Heisenberg realised that uncertainty was an intrinsic part of quantum physics. According to this principle, it is impossible to measure the position and the momentum of a particle at the same time. If we try to measure the position of a particle, it will be uncertain because the act of observing it influences its position as photons transfer momentum to the object. Quantum uncertainty extends beyond measurement uncertainty, as discussed in Chapter 3, but here we restrict ourselves to measurement uncertainty.

Since the Heisenberg uncertainty principle arises from photon momentum, it can be visualised using Nerf guns and an object of comparatively low mass such as a balloon. As shown in Box 15.5, students hang balloons containing different masses of water. When students try to "photograph" the lowest mass balloon suspended by a string, the momentum of the bullets creates a large position displacement of the balloon causing intrinsic fuzziness in the "image" of the balloon. On the other hand, when bullets hit the heaviest balloon, its displacement is much smaller because momentum is mass × velocity. (Strictly, momentum is mass × velocity when the velocity is low compared with the speed of light.)

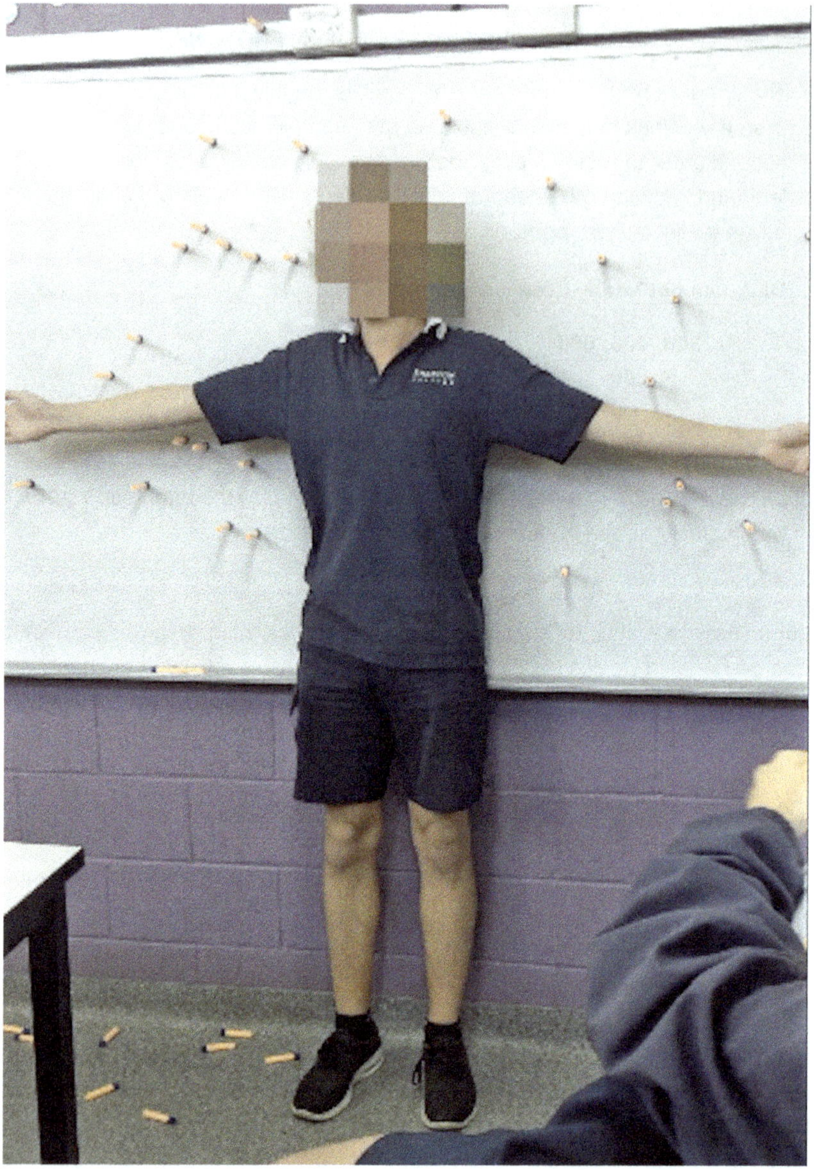

15.1
A student being "photographed" by other students using Nerf guns. The activity is highly memorable and illustrates the idea that cameras register photons. (*Source: Copyright by Jyoti Kaur/ Einstein First*)

Real measurements with light disturb both position and momentum. The effect is of central importance in gravitational wave detectors that use laser light to make precise measurements of the position of mirrors. The momentum from randomly arriving photons transfers to the suspended mirrors in the detector that leads to unavoidable uncertainty in the measurements. We have adapted this phenomenon into an engaging activity called the Nerf gun challenge, which is described in Box 15.5.

**BOX 15.5: NERF GUN CHALLENGE**

The Nerf gun challenge consists of a set of balloons and an accompanying poster that outlines the physics involved. (The poster can be downloaded in the e-resources.) The balloons are filled with varying masses of water. When a Nerf bullet is fired at a balloon, its position oscillates as momentum is transferred from the bullet to the balloon. The extent of the oscillation is inversely proportional to the mass of the balloon; thus, the greatest oscillation is observed for the lightest balloon, and the heaviest experiences negligible movement. This "uncertainty" in the position of the balloon as a function of its size resembles the uncertainty in the position of an atom when "observed" by a photon. The challenge can be set up such that students use their observations to identify the masses of each balloon according to a set of possible answers.

### *The Nerf gun photoelectric effect*

The photoelectric effect occurs when electrons are ejected from the surface of a metal when irradiated by light of high enough frequency. The electrons of metal are bound with a characteristic strength, which is known as its *work function*. The higher the work function of a metal, the stronger the bond between the metal and its electrons, and the more energy is required to sever this bond. Since the energy and momentum of photons are proportional to frequency, the light must exceed a certain frequency to induce electron emissions. The minimum frequency of light required to eject an electron from a given metal is known as the threshold frequency. Beyond this threshold, increasing frequency results in electrons being expelled at a higher speed. We model this phenomenon in a simple interactive activity using Nerf guns.

Nerf gun bullets again provide our stream of toy "photons", but this time we use ping pong balls to represent electrons. They are confined inside bowls. The height of the bowl represents the work function for a particular metal. Bowls with higher rims represent metals with higher work function. They need more momentum to be transferred to the "electrons" for them to escape.

To conduct the activity, we place ping pong balls in a set of bowls, as shown in Figure 15.2. Students stand a few metres away from the array and prepare to fire their Nerf guns. Students can be asked to predict the consequences of firing Nerf gun bullets at the array.

Since all Nerf gun bullets originate from the same source, we can assume they each have the same energy. In terms of our analogy, this corresponds to photons of the same frequency (and thus energy). Upon striking the ping pong ball, the bullet transfers some of its energy, causing the ball to oscillate in its bowl. In a shallow bowl (that is, a metal with a low work function), the transfer of energy should be sufficient to eject the ball from the bowl (that is, the

15.2
Students firing Nerf gun bullets at ping pong balls resting in bowls of varying depths. Ping pong balls in shallow bowls are ejected, while those in larger bowls are not. This represents the photoelectric effect where electrons in metals with low work functions are ejected when irradiated by light exceeding the threshold frequency of the metal. (*Source: Copyright by Jyoti Kaur/ Einstein First*)

threshold frequency of the metal has been exceeded). Ejected balls correspond to photoelectrons. For deeper bowls (that is, metals with higher work functions), the threshold frequency will not be exceeded, and consequently, the energy will be insufficient to eject photoelectrons.

Using the same apparatus, we can assemble an alternative activity that emphasises the threshold frequency rather than the work function. Arrange the bowls in a line of ascending depth, and place a single ping pong ball in each. By firing Nerf gun bullets at bowls in order of ascending depth, students can find the bowl depth that corresponds to the threshold frequency for their Nerf gun photon energy.

Having explored the bulletiness of photons, we will now go on to explore the phenomenon of interference that was once used to "prove" that light is a wave. With classical waves, interference is the cancellation or reinforcement of two waves when they overlap. However, for widely spaced photons, we have to find a different explanation for the same phenomena.

## INTERFERENCE

In this section, we want to emphasise diffraction and interference phenomena in quantum contexts. The *interference of light occurs whenever photons can take two alternative paths.* To do this, we primarily use laser pointers to produce observable interference patterns, but interpret the observable phenomena as being photons. For example, the activities emphasise the characteristic light and dark bands of interference, which correspond to regions where the probability of photons arriving is higher or lower, respectively.

We consider four activities to demonstrate the quantum nature of interference phenomena of light:

- Toy model interferometers illustrate two alternative paths for travelling to a screen in a *Michelson interferometer.*
- Single-photon interference and simple vector addition illustrate interference patterns created by the arrival of one photon at a time. We use these patterns and the mathematics of arrows to introduce the concept of zero and highest probability of photon arrival.
- Hair diffraction illustrates light taking paths on either side of a human hair to create diffraction patterns on a projecting wall.
- Soap bubble interference illustrates two ways that light can reflect off a soap film, resulting in interesting interference phenomena.

In every case, when light can take alternative paths, we observe interference patterns of light and dark. A difference in path length of one path compared with the other path of just half a wavelength is sufficient to change from constructive interference (bright light, waves adding up) to destructive interference (complete darkness, waves cancelling). Because interference changes require only half a wavelength change (for green light, this is 250 nanometres, smaller than a virus!), interferometers are exquisitely sensitive.

### *Toy model interferometer*

We begin with a toy Michelson interferometer that allows students to experience the workings of a *beam splitter* that creates the alternative paths in the context of gravitational wave detectors. The Michelson interferometer is chosen because of its historical importance. In 1887, Michelson and Morley used such a device to show that the speed of light is constant. In 2015, physicists used Michelson interferometer to detect gravitational waves for the first time.

In a Michelson interferometer, the light from a light source splits into two parts through a 50%-50% beam splitter. The beam splitter is like a half-silvered

15.3
(a) A diagram of a Michelson interferometer in which light splits into two parts to get an interference pattern of dark and bright patterns; (b) a beam from a green laser light is split by a CD case into two light beams directed towards the two suspended mirrors, which are perpendicular to each other. This toy interferometer allows students to experience the requirement of sending light beams to two separate mirrors; and (c) a photo of a table-top Michelson interferometer used to demonstrate how a gravitational wave interferometer works. (*Source: Copyright by Jyoti Kaur/ Einstein First*)

mirror set at 45 degrees so that half the photons turn left, and the rest travel straight. This gives the light two alternative paths 90 degrees apart, as shown in Figure 15.3a. Figure 15.3b is a photograph of a toy interferometer consisting of a beam splitter made out of a clear plastic CD case and two suspended toy "mirrors". The purpose of the toy interferometer is firstly to allow students to experience the requirement of sending light beams to two separate mirrors. Students quickly learn the critical aspects of reflected and transmitted beams that enable them to orient a laser pointer and beam splitter appropriately so that they illuminate both mirrors. In a real interferometer, both beams have to reflect and recombine at the screen. Also, students can fire Nerf gun bullets at the mirrors and measure the perturbation, thereby recognising that interference experiments are still susceptible to quantum uncertainty due to photon momentum.

Referring to Figure 15.3a, we now examine the return paths. Both end mirrors return their beams to the beam splitter where they have a choice of returning to the laser or the screen. When the alignment is imperfect, the screen will show bands of bright and dark. Where the screen is dark, the photons return to the laser. If light was a wave, we would understand this process as waves adding or cancelling. The same happens for photons, even if they arrive one at a time as we discuss further below. Finally, Figure 15.3c shows a Michelson interferometer designed to demonstrate the measurement of the alternate stretching and shrinking of its arms as gravitational waves pass. In real detectors and toy interferometer, the mirrors hang freely so that they can respond to space itself.

### *Radiation pressure activity with suspended mirrors*

A toy interferometer with Nerf gun photons and suspended toy mirrors can illustrate the effect of photon momentum on mirrors in gravitational wave

15.4
A photo of two suspended toy mirrors of different masses. Students measure the mirror recoil when Nerf gun photons bounce off them. (*Source: Copyright by Jyoti Kaur/ Einstein First*)

detectors (Figure 15.4). A set of "mirrors" of different masses, each suspended by a pair of long wires, has Nerf gun photons reflected off the mirrors. Students use rulers and videos to estimate the mirror recoil and mirror fluctuation (centimetres for low mass mirrors). If the mirror has a reflective surface, students can measure the motion of a reflected laser-pointer beam. This activity demonstrates how the Heisenberg uncertainly principle affects gravitational wave detectors. The random arrival of photons causes *radiation pressure noise*, and for this reason, very massive mirrors are used to reduce this effect.

***Single-photon interference and simple vector addition***

We observe interference patterns of bright and dark fringes even if the photons are very sparse. While some physicists explain this phenomenon in terms of wave functions, we avoid such interpretation and simply observe that the probability of the arrival of photons is low at some places and high at others. Single-photon interference videos show photons arriving one at a time at different places, gradually building up an interference pattern of dark fringes, where probability is zero, and bright fringes where the probability is 1. Individual photons have created the pattern in Figure 15.5. This pattern is similar to the patterns that were historically used to "prove" that light is a wave.

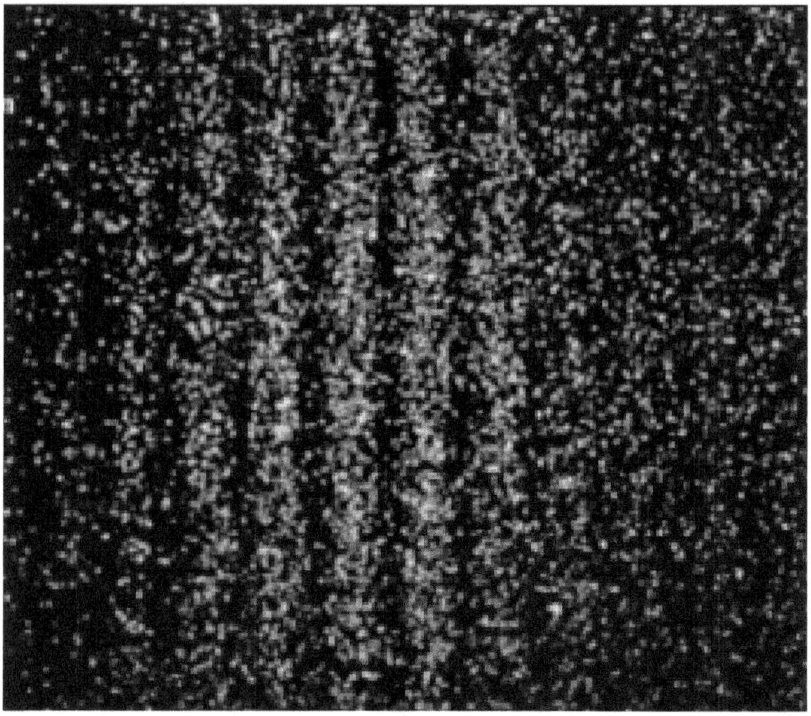

15.5
Image of an interference pattern created by the arrival of one photon at a time. (Credit: Image still from https://www.youtube.com/watch?v=5xOlqdPRX8k, Creative Commons Attribution license (reuse allowed).) (*Source: Copyright by Quantum Interference Patterns by Quantum Laser Pointer, https://www.youtube.com/watch?v=5xOlqdPRX8k, Creative Commons Attribution license (reuse allowed)*)

We use single-photon interference to introduce primary and middle school students to the mathematics of arrows mentioned in the above aphorism. Students are provided with arrows and asked to add them in two ways: (a) make the arrows add up to zero (corresponding to zero probability of photon arrival), or (b) make the arrows add up to two (corresponding to the highest probability of photon arrival). Students rapidly grasp the idea of adding arrows by placing them head to tail to obtain varying resultants (Figure 15.6). This minimal introduction to vector mathematics is designed to pave the way towards the vector mathematics needed for more advanced quantum physics. Having investigated the concepts of interference of photons, we now offer student activities with green laser pointers. In these activities, students are able to make exciting observations when light has two alternative paths.

***Diffraction of laser light through a human hair***

Before commencing this activity, we first illustrate classical diffraction by showing students photographs of ocean waves diffracting around islands. We explain that waves from either side of the islands, redirected by the landmass, interfere with each other and create interference patterns as the waves add and subtract. Using this illustration as an analogy for the upcoming activity, we substitute a laser beam as the wave, a single human hair as the island, and a screen to display the wave intensity.

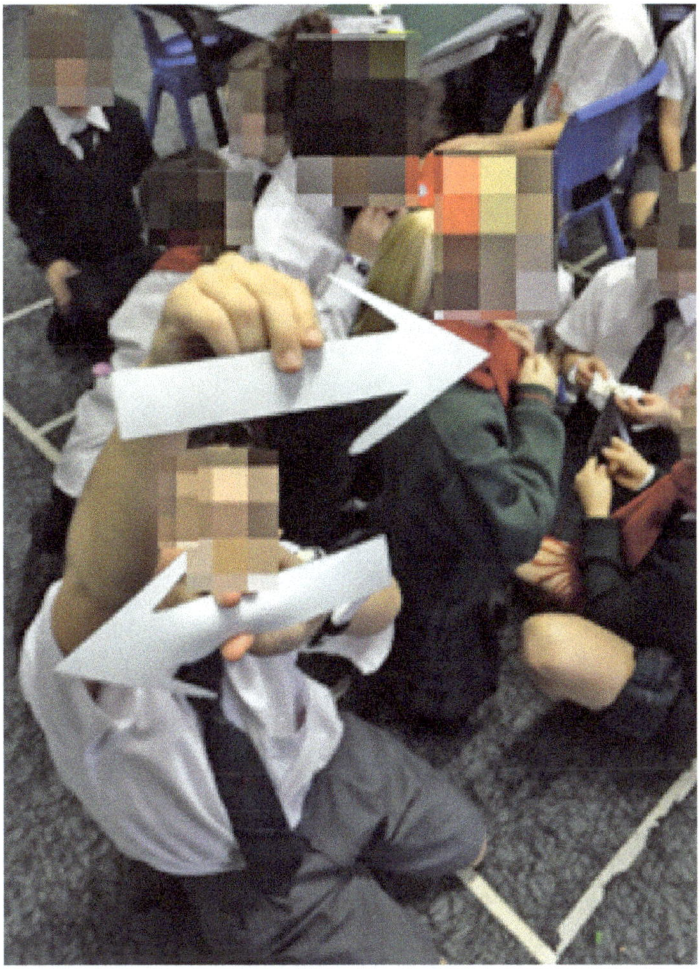

15.6
An eight-year-old student showing how 1 + 1 can be equal to zero with two arrows. Students of this age quickly pick up the ideas of 1 + 1 = 2 or 1 + 1 = 0 with the help of arrows. (*Source: Copyright by Jyoti Kaur/ Einstein First*)

Students provide hair strands which are stuck over a gap in a cardboard holder, and placed in the beam of a green laser pointer, and directed to a screen 2–5 metres away. Students observe an intensity pattern with bands of light and dark, as illustrated in Figure 15.7. It is important that students appreciate that the photons create a brightness distribution similar to that of a wave. Light bands are equivalent to places in the ocean where the wave amplitude is large, and dark bands are equivalent to those places where the waves are small.

With the understanding that photons are spaced far apart, it is not easy to believe that the photons themselves are adding and subtracting like the ocean waves. We avoid the question of what is interfering. However, we may point out that some physicists use the name *wave function* to describe the mathematical quantity that interferes and defines the *probability* that a photon will be detected at a certain point.

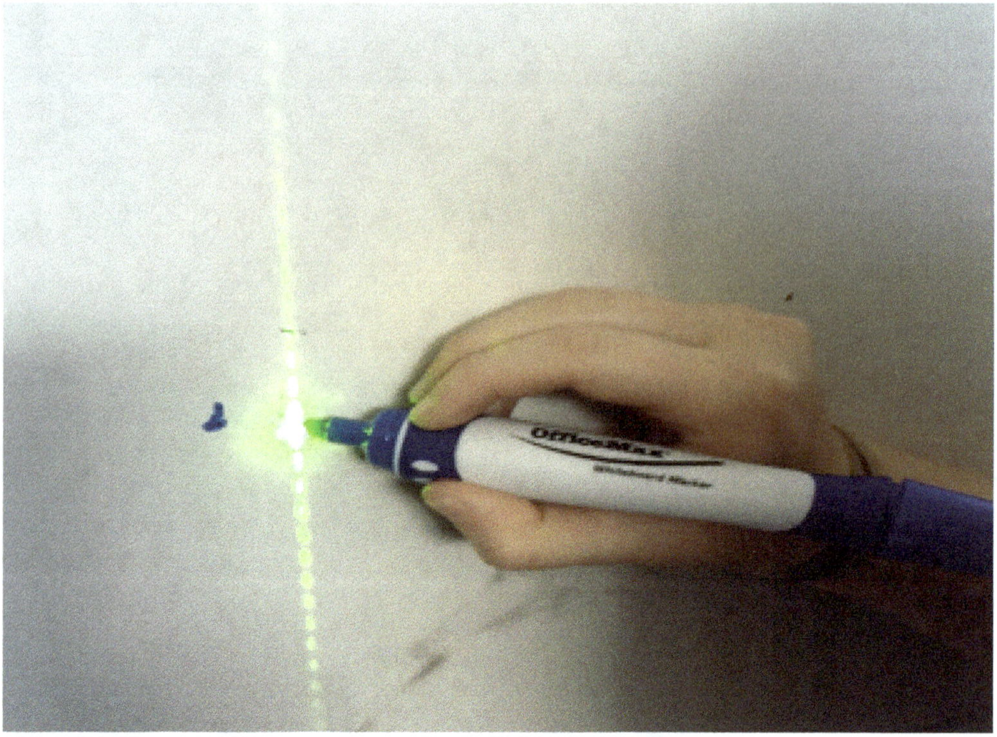

15.7
Vertical interference pattern created by the diffraction of green light through a horizontal human hair. The bright and dark band of light represent regions of constructive and destructive inference, respectively. (*Source: Copyright by Jyoti Kaur/ Einstein First*)

Just as the interference pattern on the island depends on both the size of the island and the wavelength of the waves, the laser interference pattern depends on both the diameter of the hair and the wavelength of the laser light. By measuring the spacing between successive bands in the interference pattern (Figure 15.7), students can calculate the width of their own hair, using a straightforward formula in which known quantities are given numerically (Box 15.6). Students can compare the thickness of their hair with their classmates' hair.

### *Soap film interference*

Another vivid and beautiful illustration of quantum interference uses reflective interference of light in soap bubble films (Box 15.7). Light reflected from the front and rear surfaces of a soap film produce high contrast interference patterns that can be projected onto a screen (Figures 15.8 and 15.9). This interference pattern arises from the phase difference between light reflected from either side of the soap film wall.

It is useful to present both the classical concept and a quantum version that emphasises the sparseness of photons. Figure 15.9a shows what you would expect if light were a continuous wave. Depending on the soap film thickness, the reflected waves may be in phase, so they add up and create a high intensity or out of phase, to create darkness.

**BOX 15.6: ACTIVITY: MEASURE THE DIAMETER OF YOUR HAIR**

The basic formula is

$$d = \frac{\lambda L}{x}$$

where, $d$ is the diameter of human hair, $\lambda$ is the wavelength of laser light (for green laser light, $\lambda = 532$ nanometres), $L$ is the distance from hair to the screen (say 2 metres), and $x$ is the distance from the centre of the pattern to the centre of first dark area which students measure in millimetres.

Hair diameters, like the fibres in textile fabric, are usually measured in microns, $10^{-6}$ metres. Then, the hair diameter in microns resolves to:

**Hair diameter *d* (in microns) is given by** $d = \frac{532 \times 10^{-9} \times 2}{x}$

$$= \frac{1.06}{x} \text{ micrometre}$$

By putting the value of $x$, students calculate the diameter of their own hair.

This can be elaborated to plot the distribution of hair sizes in a class, and to compare hair with other fibres.

15.8
Interference pattern created by green light through a soap bubble. Light reflected from either side of the soap film wall interferes, creating characteristic regions of constructive and destructive interference. Despite the solidity and permanence of the interference pattern, only a single photon is present in the soap bubble at any time. (*Source: Copyright by Jyoti Kaur/ Einstein First*)

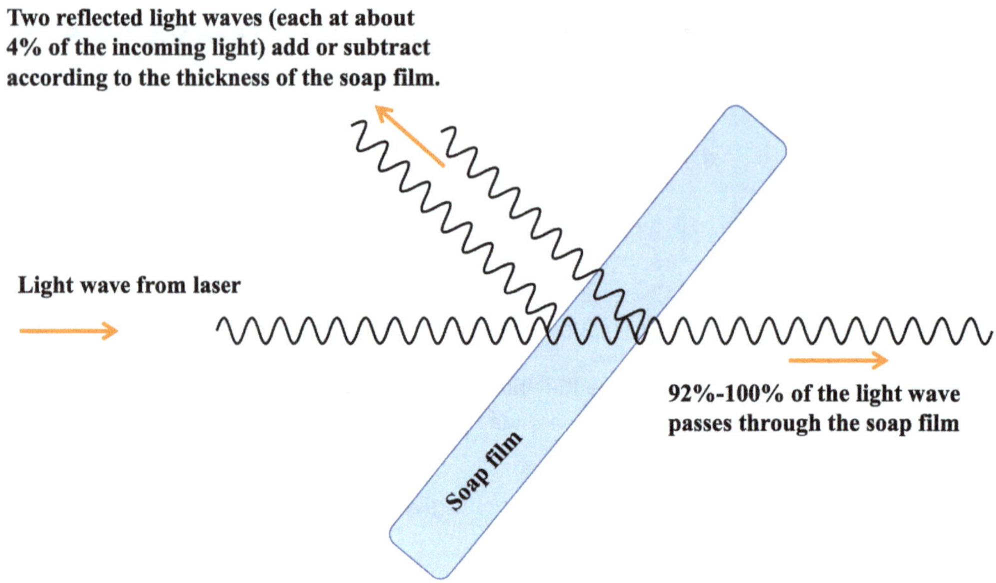


15.9 (a)
Light treated as a continuous wave reflected from a soap film. While the majority (92–100%) of the wave will pass through the soap film unperturbed, a small fraction of the wave will be reflected from the front and the rear wall of the soap film, producing dark and bright patterns. (*Source: Copyright by Jyoti Kaur/Einstein First*)

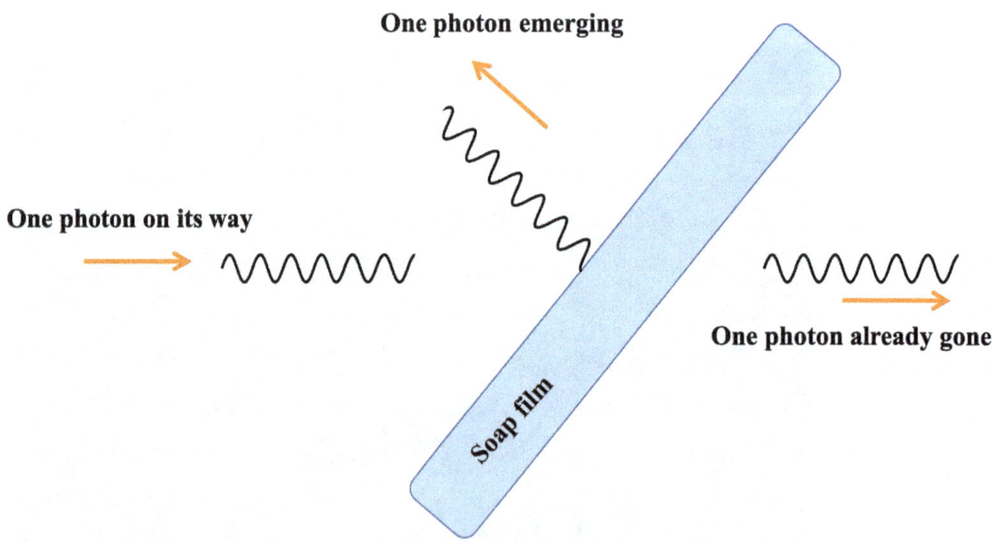


15.9 (b)
Light treated as photons reflected from a soap film. Unlike a continuous wave that can be split with reduced intensity, photons cannot be split. Yet, characteristic wave interference patterns still emerge. (*Source: Copyright by Jyoti Kaur/Einstein First*)

**BOX 15.7: SOAP FILM INTERFERENCE ACTIVITY**

Mount a laser pointer on a stand to project vertically downwards through an irregular centimetre-sized wire loop set at an angle of roughly 45 degrees. Lift a small cup of soapy water to immerse the loop and look for the reflected spot on a nearby wall. The film should have some convex surface so that the reflected beam is expanded. Adjust the beam position on the film. Students will observe swirling interference patterns that change as the film evaporates (Figure 15.8).

Figure 15.9b shows how we must imagine it actually occurring. The fact that interference occurs, even with a very dim light source, when the spacing between photons greatly exceeds the thickness of the soap film, demonstrates the weirdness of the quantum world.

The source of the interference is the two "walls" of the soap film, a structure of only a few micrometres thickness. Since we have already calculated that individual photons are separated by distances often measured in kilometres, we are presented with a puzzling scenario. At any instant, there can never be more that one photon present inside the soap film, yet the interference pattern that emerges appears permanent. If photons do not cause the interference pattern, what is the cause? We can invoke the idea of a wave function or a *probability wave*, which determines the probability that a photon will arrive at a particular location. The probability follows the same mathematics that the intensity would follow for a real continuous wave. It may not be clear what actually interferes, but it always happens. It is important to show the connection with single-particle interference images because the only difference between them is the number of photons contributing to the image.

## LIMITATIONS OF THIS QUANTUM MODEL

All the models outlined in this section, while being useful tools for appreciating the abstract concepts of Einsteinian physics, have limitations that must be emphasised. There are limitations in the sense that the correspondence between model and reality breaks down. Identifying precisely where and how the analogy breaks down, in a thorough discussion with students, is vital for obtaining a deep understanding of the relevant concepts.

While using toy bullets to represent photons is useful for illustrating various properties of photons, as well as individual phenomena such as the photoelectric effect and the uncertainty principle, there are numerous limitations. Nerf gun bullets have non-zero rest mass and cannot propagate at the speed of light, nor can they produce interference patterns. Their energy is dependent on their velocity, and their trajectories are strongly distorted by gravity. Students can identify these limitations in a group discussion.

The discussion about the spacing between photons assumes that photons are really particle-like. However, the reality is that physicists do not have a clear, simple understanding of what a photon is like. Since photons are true quantum objects that possess both waviness and bulliteness, we cannot easily imagine the nature of a photon. When we detect a photon, it is detected like a particle, but before the detection, we cannot say what it is like.

Students need to know that quantum mechanics is a mathematical theory, based on vector algebra. It has been tested to enormous precision and shown to be correct. However, to this day, people argue about the meaning of quantum reality. Richard Feynman said that if you meet a person who says they understand quantum mechanics, they are either stupid or lying! Yet because the theory works, we have been able to use it to design the components of computers, lasers, and mobile phones, and without the theory, we would never have been able to create such marvellous devices.

## CONCLUSION

Throughout this chapter, we have shown that many quantum phenomena can be understood by using the popular children's toy called Nerf guns. Students enjoy the hands-on activities which help them develop an Einsteinian language at an early age. The activities also help foster students' interest in physics, which is often considered boring because the physics curriculum fails to address modern topics.

Previous research has shown that physical models and analogies provide students with an insight into complex topics. The activities described in this chapter are low cost and easily created. The activities combined with thought experiments, web resources, and laser pointer experiments help students understand the particle nature of light and the weirdness of quantum interference. A discussion of the limitations of the models contributes to their learning. High-school students can extend their knowledge by completing simple related mathematical calculations. The Einstein-First project has tested these activities with both primary and secondary school students. Positive research results obtained from various programmes show that these models and analogies enhance students' understanding, enjoyment, and their attitudes towards learning modern physics (Kaur et al., 2018;).

## FURTHER READING

E-resources: Soap film interference, hair interference, hear gravitational waves
Chapter 3: The difficult birth of quantum physics
Chapter 16: Teaching quantum physics to middle and high-school students using phasor wheels

## REFERENCES

Feynman, R. (1985). *QED: The strange theory of light and matter*. Princeton NJ: Princeton University Press.

Harrison, A. G. & Treagust, D. F. (1996). "Secondary students' mental models of atoms and molecules: implications for teaching chemistry." *Science Education*, 80(5), 509–534.

Kaur, T. et al. (2018). "Determining the intelligibility of Einsteinian concepts with middle school students." *Research in Science Education*, 1–28.

Kaur, T. et al. (2020). "Gender response to Einsteinian physics interventions in school." *Physics Education*, 55, 035029.

Choudhary, R. et al. (2018). "Can a short intervention focused on gravitational waves and quantum physics improves students' understanding and attitude?" *Physics Education*, 53, 065020.

Murphy, T. W. (2013). "Lunar laser ranging: the millimeter challenge." *Reports on Progress in Physics*, 76(7).

# 16 Teaching quantum physics to middle and high-school students using phasor wheels

*Rahul Choudhary and David Blair*

**WHAT?**

This chapter builds on and extends Chapter 15 to introduce the powerful Feynman path integral approach to quantum mechanics. Rahul Choudhary and David Blair use graphical methods to provide insights into the rules of the quantum world to show how everyday phenomena have a quantum origin.

**FOR WHOM?**

This chapter is intended for readers who are interested in a non-mathematical approach to quantum physics at the middle and high-school level.

## INTRODUCTION

Einstein's discovery of light quanta played a major role in the birth of quantum mechanics. Quantum mechanics is probabilistic, and the lack of causality bothered him for the rest of his life. By 1938, Bohr, Schrödinger, Heisenberg, and others had extensively developed the theory of quantum mechanics. Despite its success, Einstein remained deeply concerned about the co-existence of wave and particle aspects.

Forty years later, Richard Feynman, who inspired this chapter, echoed Einstein's dilemma, "*We cannot expect light to behave as particles on Tuesdays and waves on Wednesdays*" (Feynman, 1979). In 1948, Feynman had developed a compelling interpretation that became the foundation of quantum electrodynamics. His approach, formally known as quantum path integrals, is encapsulated by the phrase "*all possible paths*". The power of his approach is its ability to explain *why* light takes the path of least time when it reflects, refracts, or merely travels in a straight line. It also exposes the profound weirdness of the quantum world when we look at it with classical eyes. We explore these ideas in this chapter, and present classroom approaches suitable for high schools.

We have developed our use of Feynman's interpretation from a set of lectures at the University of Auckland (Feynman, 1979), and his closely related book *QED: The Strange Theory of Light and Matter* (Feynman, 1985). Rather than imposing conventional macroscopic thinking on the world, Feynman took an inventive approach that makes us familiar with the rules of the quantum world. His approach emphasises the fact that we always detect light as "particles". These discrete packets of energy may be a tiny burst of current as a photoelectron is emitted in a camera sensor, or an ion current in a nerve cell of the eye when a photon breaks a chemical bond. Modern quantum physicists joke that a photon is merely a click in a detector!

Feynman's path integral formulation uses *phasors* to calculate the photon arrival probability. The method closely parallels techniques discovered in the 19th century for adding classical waves such as water waves. The name *path integral* refers to a summation of all possible paths. The word *possible* allows us to consider paths that classical physics would dismiss as impossible such as violating the law of reflection.

The path integral approach connects closely to the principle of least action. We invoke least action when we walk around an obstacle or take a short cut at the corner of a garden path. In physics, action is a quantity with dimensions of energy × time or momentum × length. The parabolic path of a stone thrown into the air is one that minimises the action. For a photon with constant energy (Energy = Planck constant × Frequency, $E = hf$), the action is proportional to the travel time. In this case, the principle of least action says that photons should always take the fastest possible path. The question is, how does a single photon *know* to take the path of least action?

In this chapter, we evaluate path integrals graphically with the aid of phasor wheels, starting from two possible paths. We apply this to multiple possible paths, and finally, we *delete* paths. The phenomena of deletion lead to surprising *actual* paths, which can easily be observed in simple activities suitable for classrooms.

To make the concept accessible and tangible for students, *we use Nerf gun photons* and *phasor wheels* to emphasise the two characteristic aspects of photons: *bulletiness* and *waviness* (see Chapter 15). The phasor wheels connect the waviness to the phasor vectors. We complement the models with laser experiments and videos of single-quantum interference.

## MODERN UNDERSTANDING OF LIGHT

Einstein made the bold prediction that light comes as a stream of particles called "light quanta" (Einstein, 1905). He showed that his simple but revolutionary prediction explained the photoelectric effect, and concurred with Max Planck's predictions regarding black body radiation in 1900. Einstein completed the particle-like picture of light quanta by proposing that they must also possess momentum (Einstein, 1917). The name "photon", proposed in 1926 (Lewis, 1926) contributed to the rapid general acceptance of the concept by 1928. However, Einstein remained very uncomfortable with the implications.

Our much more confident interpretation of the quantum nature of light today is a result of many beautiful experiments. These experiments use sensitive photon detectors that allow physicists to confirm the quantum theory of light in detail. Scientists have shown that human eyes can sense single photons (Tinsley et al., 2016), and videos of single-photon interference[1] show that photons arrive randomly, but with enough of them, they create interference patterns. The observation of interference patterns was once used as the conclusive "proof" that light is a wave. Single-photon interference forces us to attribute the wave characteristics to the photon arrival probability rather than the interference of actual waves.

### *The bulletiness and waviness of photons*

Photons, like bullets, carry momentum that applies forces to mechanical objects. If a large number of photons impact on a surface, they create *radiation pressure*. Although the mechanical forces are very tiny in the context of daily life, radiation pressure is important in sensitive instruments like gravitational wave detectors. Wavelength and frequency describe the waviness of photons, and it is essential to emphasise that wavelength and frequency are not independent (that is, frequency × wavelength = speed of light).

In general, for photons across the electromagnetic spectrum, large values of wavelength imply low momentum. This relationship means that radio photons have very low momentum: they are dominated by their wave-like behaviour, and their bulletiness is negligible. Thus, we use the term radio *waves* even though they, too, are photons. In contrast, bulletiness dominates X-rays, and these photons can damage living cells. For visible light, both properties are essential.

While the waviness of photons relates to a probability wave that describes the chance of photons arriving at some point, the mathematics of probability waves is the same as that of classical waves. One represents the *probability amplitude*, and the other describes the *wave amplitude*. The intensity of a wave depends on the square of its amplitude. Likewise, the number of photons per second arriving at a particular location depends on the square of the probability amplitude.

### *The adding of waves using phasors*

We start with a thought experiment that relates to everyday experience. Let us consider a single frequency radio wave broadcast from a transmitter to a mobile phone, as illustrated in Figure 16.1. Signals arrive at a mobile phone via reflections from six different buildings. However, the seventh building blocks the direct path. Each path has a different length. The receiver detects the sum of the electric fields from all six waves, each with the same frequency, but with different phases and different amplitudes. The question is, how strong is the phone signal?

Up until 1893, the solution was to write an equation for each received signal and then add them up using algebra and trigonometry, which was very

1 https://www.youtube.com/watch?v=MbLzh1Y9POQ.

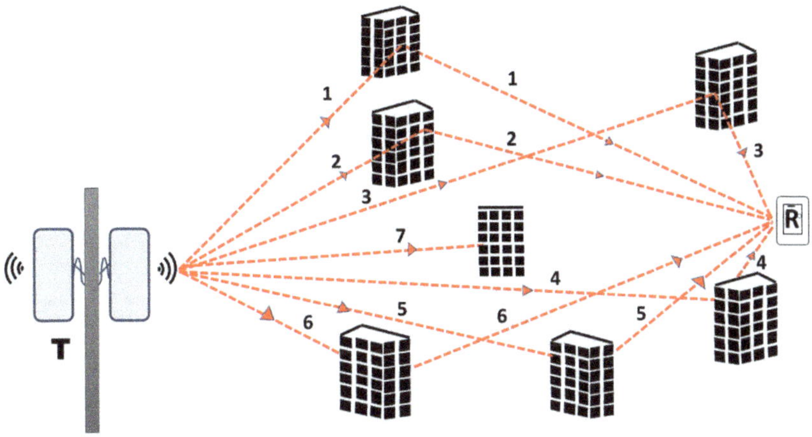


16.1
Radio waves, when reflected off a set of six buildings and blocked by the seventh, will have different amplitudes and phases. The reflection angle and the path length of the photons determine the amplitude. (*Source: Copyright by Rahul Choudhary/Einstein First*)

complicated. It is shown as follows where $A_1 \ldots A_6$ are different amplitudes, $\phi_1 \ldots \phi_6$ different phases, $f$ the frequency, and $t$ the time:

$$\text{Total signal, } E = A_1 \sin(2\pi ft + \phi_1) + A_2 \sin(2\pi ft + \phi_2) + A_3 \sin(2\pi ft + \phi_3) + A_4 \sin(2\pi ft + \phi_4) + A_5 \sin(2\pi ft + \phi_5) + A_6 \sin(2\pi ft + \phi_6)$$

Oliver Heaviside and Charles Steinmetz independently discovered the method of phasors for solving this type of problem (Heaviside, 1893; Steinmetz, 1893). Their discovery dramatically simplified the addition of multiple waves, by turning the problem into a simple one of adding vectors. Phasors are vectors of length equal to the amplitude of the wave, and with a phase angle (that is, arrow direction) that rotates 360 degrees ($2\pi$ radians) once every cycle, as illustrated in Figure 16.2. In general, we only care about the *relative phase* between waves of the same frequency at a particular point because the absolute value of phase has no meaning. The absolute phase depends on where we start measuring the wave.

The sinusoidal function is the *y*-axis projection of a point on the wheel. The corresponding phasor at each point in the wave is shown. It is important to emphasise that waves progress in both space and time. At a fixed point in space, we observe an oscillation in time: the phasors rotate in time. If time is frozen, (like in a photograph) the oscillation is in space: the phasors change direction along the distance axis. Here, we consider only relative phases at fixed points in space and an instant in time so that we do not have to worry about the phasors rotating.

Heaviside and Steinmetz realised that we could add up multiple waves by simply taking the vector sum of all their phasors. The sum is obtained by placing all the phasors head to tail. The *resultant* phasor (the arrow from the tail of the first phasor to the head of the last) describes the amplitude and phase of the summed waves.

The phase of the resultant wave is not important. We care about the intensity of waves like the intensity of light or the number of signal bars on a mobile

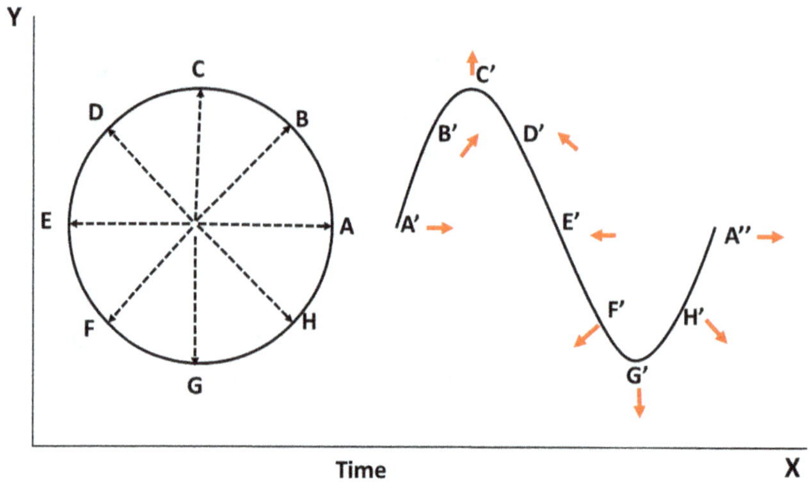


16.2
One complete rotation of the wheel is equivalent to one full cycle (wavelength) of the wave. The phasors are shown in red at different locations on the wave. A is a point in the circle whose corresponding phase is marked as A′ on the wave. After one complete rotation, the phasor has returned to its direction at A, indicated as A″, and so on. The connection is shown better in this link: https://thumbs.gfycat.com/DistantGenuineCurassow-size_restricted.gif. (*Source: Copyright by Rahul Choudhary/Einstein First*)

phone. This itself depends on the *square* of the amplitude. For this reason, we use the resultant phasor to inscribe a circle, which has an area proportional to the intensity of the resultant wave (see Figure 16.4c).

Chapter 15 introduces the simplest case of phasor addition for primary school students. This is the case of constructive and destructive interference of two waves, where the phasors are either co-aligned or anti-aligned. We introduce this simplest case, by simply saying that the addition of light follows the mathematics of arrows. Co-aligned arrows represent the case 1 + 1 = 2, (bright light, large waves), while anti-aligned arrows represent 1 + 1 = 0 (darkness, wave cancellation). With this preparation, students are ready for the next cycle of learning presented here.

We now introduce phasor wheels, which will allow us to apply the method of phasors to the six-path problem for the mobile phone signal.

### *Phasor wheels*

To make the concept accessible to middle school students, we developed phasor wheels illustrated in Figure 16.3. The wheel creates a tangible, visible connection between rotation, wave motion, and phasor direction in a simple, low-cost gadget. It causes a vertical rod to move in an approximate sinusoidal manner. For a small radius and a long rod, the approximation is adequate to make a clear connection between rotation and wave motion.

The phasor itself is indicated as an arrow on the wheel. The length of the phasor is proportional to the wave amplitude. Note that rolling is just a convenient means of creating rotation and measuring the phase for various paths. To give clarity on the rotating wheel and prevent confusion, teachers can ask students to imagine wheels spinning at high speed (high-frequency waves) as they walk across the classroom.

For classroom demonstration, we built a phasor wheel (Figure 16.3a and b) of diameter 25 cm (approximately) with a rod attached to its centre and an arrow marked on it. Smaller models use cup lids (Figure 16.3c) or small wheels to allow interference experiments to be performed on paper. The wheel is connected to a handle that can be used to roll it. The phasor arrow completes one turn for every rotation, and the attached vertical rod undergoes one up-down motion, tracing a roughly sinusoidal pattern (if projected on a whiteboard or screen) as it moves forward. It shows that one wavelength is equal to one cycle of rotation. To prevent mistakes, we impose the convention typically used in physics that phase increases anti-clockwise. It is easy to flip the wheel and get incorrect results by measuring a path with negative (clockwise) phase.

16.3 (a)
A wooden phasor wheel used to show the connection between waves and phasors. The wood is cut into a circle of 25 centimetre diameter. It is connected to a handle and a stick with bolt and screw. The handle is used to roll the wheel while the stick goes up and down. (*Source: Copyright by Rahul Choudhary/Einstein First*)

16.3 (b) Close-up view of the phasor wheel. (*Source: Copyright by Rahul Choudhary/Einstein First*)

With the help of phasor wheels, we are ready to tackle the six-path problem in the context of both classical and quantum waves. For the six paths in Figure 16.1, we can imagine rolling a phasor wheel along each path, to determine the phasor direction at the receiver. We attribute six phases and for simplicity, equal amplitudes to each of them, leading to the phasors (labelled 1 to 6) shown in Figure 16.4a. The vectors are summed by placing them head to tail, as shown in Figure 16.4b. The resultant is shown dotted. Figure 16.4c shows that the same resultant is obtained when the phasors are added in a different order. Both figures show the circles that represent the intensity of the resultant wave. Figure 16.4d shows what happens if the path length for path 4 is decreased by a half-wavelength causing phasor 4 to flip 180 degrees. Suddenly, the resultant phasor has reduced. This represents a weak spot in the phone reception.

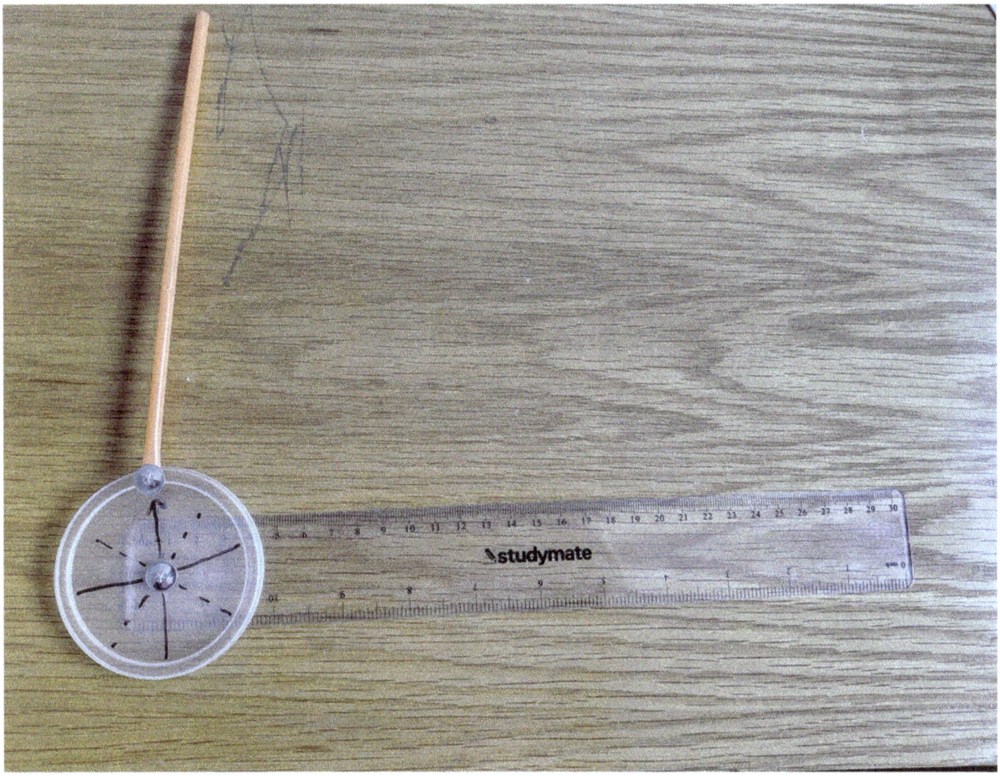


16.3 (c)
A smaller phasor wheel used for photon path integrals on paper. A small cup lid is attached to a drinking straw and a ruler using drawing pins. (*Source: Copyright by Rahul Choudhary/Einstein First*)

We have seen how all possible paths the signal can take to influence the intensity of the mobile phone signal strongly. A small change in one path can have a strong effect on the total signal. A strong signal might mean one of the signals is dominant, or that all the reflected paths have aligned phasors. Since the phasors rotate once per wavelength, and mobile phone signals have a wavelength of less than 1 metre, small movements of a phone can sometimes significantly alter the signal strength.

We do not need to use phasor wheels to solve the six-paths problem. We just need to measure the six distances in units of wavelength, ignore the integers, and express the fractional wavelengths as fractions of a rotation, measured in degrees or radians. We also need an estimate of the amplitude of each wave,

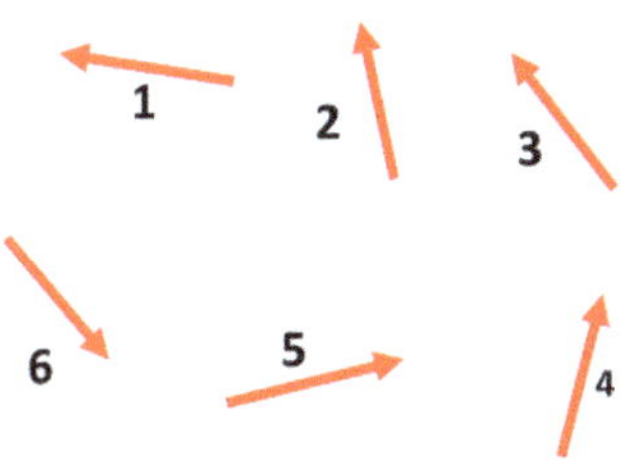


16.4 (a)
Arbitrarily chosen six individual phasors of equal amplitude. (*Source: Copyright by Rahul Choudhary/Einstein First*)

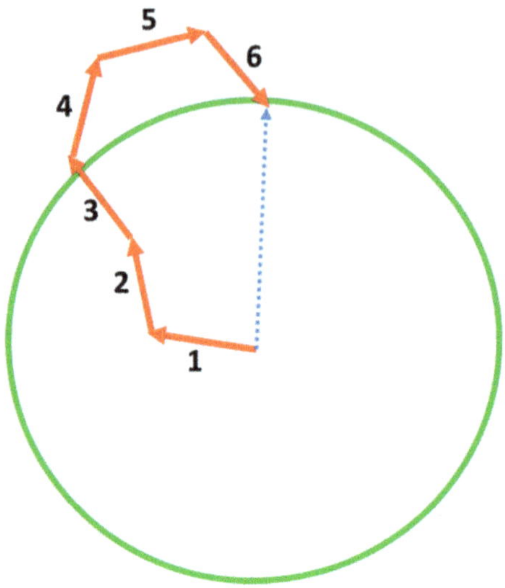


16.4 (b)
The vector addition of the phasors produces a resultant about three times larger than the individual phasors. The circular area (proportional to probability amplitude) represents the intensity of the resultant wave or the photon arrival probability. (*Source: Copyright by Rahul Choudhary/Einstein First*)

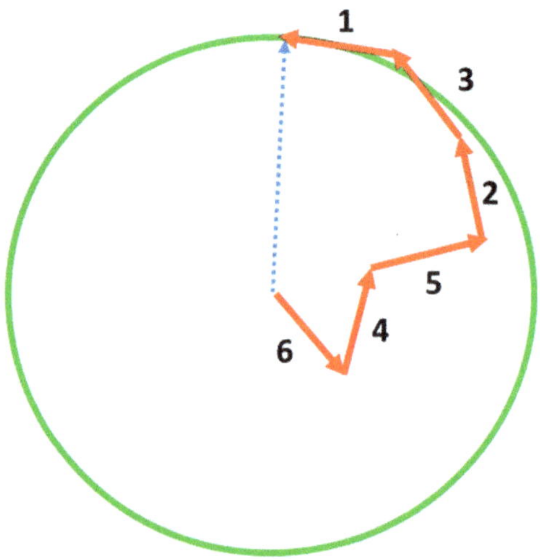


16.4 (c)
Changing the order of addition gives the same resultant and the same intensity circle. (*Source: Copyright by Rahul Choudhary/Einstein First*)

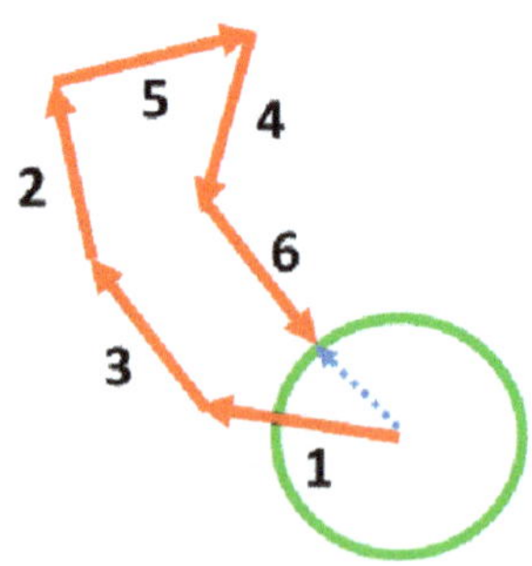


16.4 (d)
On flipping phasor 4 by 180 degrees, the resultant drops almost to one-fourth of the original resultant. (*Source: Copyright by Rahul Choudhary/ Einstein First*)

which depends on how well radio waves reflect off each building. In Figure 16.4, we chose arbitrary phasors, but engineers who build mobile phone networks must consider these interference effects very carefully.

The method of phasors that we applied to radio waves works equally for visible photons and even for particles such as electrons and molecules. We use phasor wheels to study the conventional double-slit interferometer that was originally used to show the wave nature of light. Today, in the knowledge that light comes as photons, it illustrates the application of the method as well as illuminates the weirdness of quantum mechanics.

Figure 16.5a shows the double-slit experiment in a classical setup. Figure 16.5b shows how we experiment using a laser beam. If projected onto a screen, the laser light creates a pattern of bright and dark bands. We will show how we can construct the interference patterns in Figure 16.5c with phasor wheels.

A plane wave impinges on two slits $S_1$ and $S_2$: at these points, the waves are identical. The wave from $S_1$ has to travel a different distance than the wave

16.5 (a) We arbitrarily choose two different points at P for wave 1 and wave 2. The corresponding phasors at point P are shown and added adjacent to it, and a circle indicates the intensity. (*Source: Copyright by Rahul Choudhary/Einstein First*)

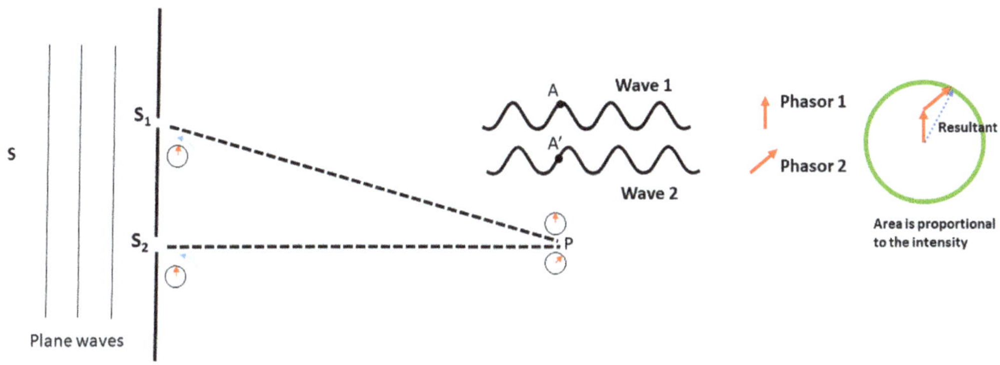


16.5 (b) Experimental setup consisting of a laser, slits, screen, and a camera charge-coupled device (CCD) sensor. A concave lens expands the beam that passes through the two slits. The slits and screen are 1 metre apart. The camera is the screen that records photons arriving at different locations. (*Source: Copyright by Rahul Choudhary/Einstein First*)

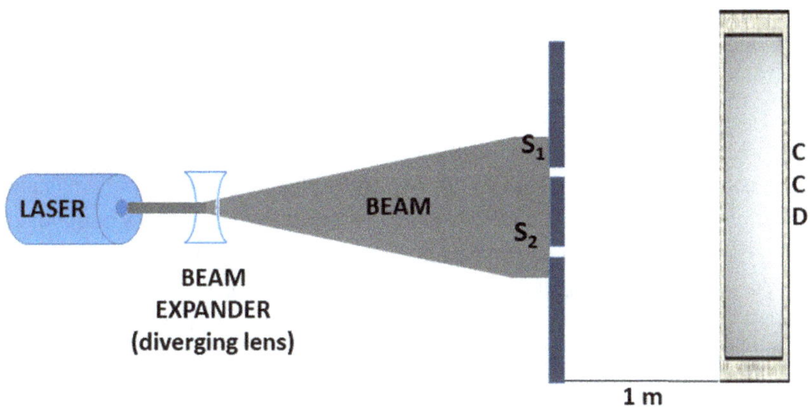

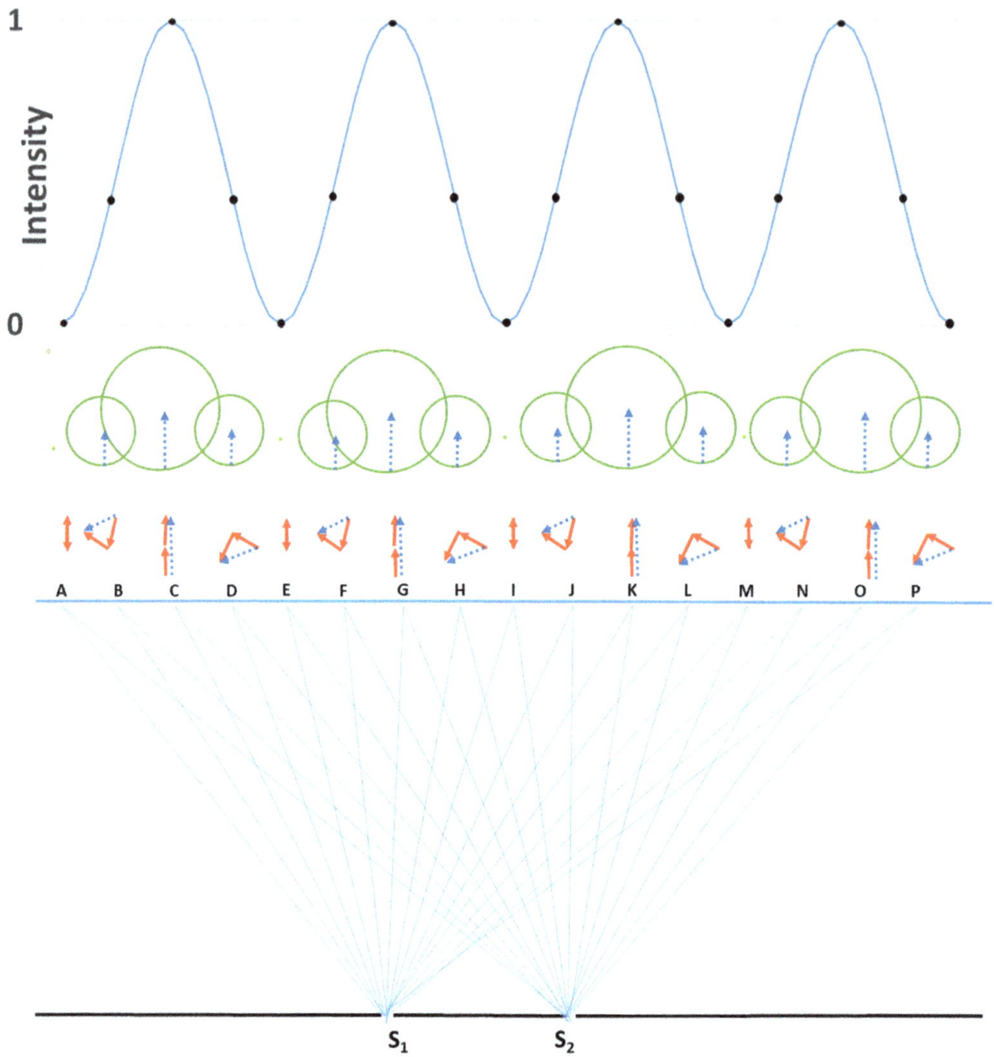


16.5 (c)
After graphically adding the phasors, we obtain the probability amplitude of photons travelling from $S_1$ and $S_2$ represented by dotted arrows at points A, B, ..., P. The circles represent the intensity at each point, which varies in a sine-squared function as shown. (*Source: Copyright by Rahul Choudhary/Einstein First*)

from $S_2$. To determine the amplitude, we compare the phases of both waves at P. This could be done by stretching out two wave trains along each path indicated as phasor 1 and phasor 2 adjacent to it. The length of the phasors depends on the amplitude of the wave. By the rule of vector addition, we obtain the resultant, which is the probability amplitude. The area of the circle represents the light intensity when there are large numbers of photons, and the probability of a single photon arriving. Note that the numbers are always relative because the brightness of the light source has not been specified.

***Double-slit interference with phasor wheels***

In this exercise, students determine the probability of photons reaching the screen at points A, B, C, …, P from the slits $S_1$ and $S_2$ (see Figure 16.5c). They must carefully roll the phasor wheels along each of the paths from the slits to the screen positions. Printed sheets showing the trajectories can facilitate this activity. Students will note that the wheels may take multiple turns from slit to screen. We are only interested in the rotation modulo 360 degrees at the point of detection (screen). Suppose it was 273.1 rotations along the first path, the final phasor angle will be 0.1 rotations or 36 degrees. Similarly, if the phasor rotation along the second path to the same point of interest is 275.5 rotations, the phasor angle for this path will be 0.5 rotations or 180 degrees.

In Figure 16.5c, we obtain the phasor angles for each pair of paths $S_1A$, $S_2A$, then $S_1B$, $S_2B$, and so on. In each case, we determine the resultant phasors. The resultants for each screen point (represented by dotted arrows) are shown. Each resultant gives us the probability amplitude, the square of which is proportional to the number of photons arriving per second at a point (intensity). The probability amplitude for all the points varies in a sinusoidal pattern so that the intensity follows a sine-squared function. Students can be assigned to find multiple peaks and troughs in the interferogram. They can also find the probability amplitude at different distances between slits and screen, and compare the difference.

We have now seen how the method of phasors allows us to calculate intensity pattern created when light impinges on two slits, such as those in Figure 16.5b.

We can make a connection to the real world by viewing videos of single-molecule (https://youtu.be/wDx8tu-iX8U) and single-photon interference (https://youtu.be/MbLzh1Y9POQ). The videos show that the intensity variation is sinusoidal. They consolidate the idea that photons and molecules arrive randomly in time, yet they display the properties of waves. The bright and dark bands of the interference pattern are places where two phasors are aligned and counter-aligned, respectively.

Figure 16.5b is a typical apparatus for interference with the screen about a metre from the slits. Now, imagine what would happen if we attenuated the laser light until there were only about 1000 photons per second (one photon per millisecond) entering the slits. Photons travel at $3 \times 10^8$ metres per second or 3 nanoseconds per metre. A familiar example of this thinking is that 60 kilometre per hour is 1 minute per kilometre. Since the apparatus is one metre in size, each photon is inside the apparatus for only 3 nanoseconds. If photons are arriving once per millisecond, 99.9997% of the time, there is no photon present at all. Videos show that interference patterns are still created, just as the phasor method predicts.

A more easily imagined equivalent situation is a pedestrian crossing that takes ten seconds to cross, with pedestrians arriving at the rate of one person per month. Most of the time, the crossing would be deserted. The chance of two people finding themselves on the passage together is negligible.

Since the phasor method still works for single photons, we ask what the phasors actually represent. It should be something wavy, something with a wavelength. Here, we call it a probability wave, but some physicists prefer *wave functions* or *virtual photons*. The bottom line is this: *if light can take two or more alternative paths, then a phenomenon called interference occurs, in which photons arrive at different places with a different probability.* The method of phasors allows us to calculate the exact probability of photons arriving at each point.

### *The method of all possible paths using phasors*

An even bigger surprise awaits us. Feynman path integrals apply exactly the same method to light paths that we might consider impossible. In just a few steps, we can discover an exceptional aspect of the quantum world, which could be stated as "*paths apparently untrodden actually contribute to observations*".

We start with another analogy. Imagine two villages separated by a mountain, and connected by several paths, some easier than others. When a landslide blocks one path, suddenly the number of arriving travellers increases. Then, we build a bridge that makes that path both easier and shorter. Suddenly, the number of travellers arriving drops to zero. Very puzzled, we decide to find out what is happening to the people. We set up counting stations on each path. The moment we start determining how many people are on each path, people start arriving at the second village. The moment we stop checking, the people stop arriving.

If we replace the travellers by phone signals and look back at Figure 16.4, we can see that if the blocking eliminated one phasor, it could have caused *less* cancellation. The bridge could have lengthened one phasor such that it caused the resultant to drop to zero.

This village paths analogy also shows something new: if we determine which path a photon travels, the interference disappears. Interference only happens when all paths are *possible*, but they must not be *determined*. When paths are determined, interference vanishes! This quantum behaviour is counterintuitive, but it is the reality of the quantum world. Even Feynman jokingly remarked, "*If you don't like it, go to another universe*".

### *The principle of least action*

We discussed the topic of least action in the introduction. The profundity of this principle is that it applies to any system, including quantum mechanics. To understand this principle, let us consider the following situation. Imagine a drowning swimmer at the beach. A lifeguard, as shown in Figure 16.6a, has to reach the swimmer as quickly as possible. What path does the lifeguard choose? The figure shows twelve possible paths. One lifeguard might say, "A straight line is the shortest path". Another might say, "I can run much faster than I can swim so that I will minimise my swimming distance". A stupid lifeguard will take a short path (trajectory 12) to the water and then swim in a straight line. A clever lifeguard works out that a compromise path (the bold line in Figure 16.6a) that takes into account the ratio of running speed to swimming speed is the fastest path. The clever lifeguard *calculates* the optimum path. Photons do the same thing: the beach is air, photons travel slower in water or glass, and without calculation, they take the shortest path. We call it *refraction*. The path of the light is bent. Both cases are examples of the principle of least action.

Feynman's method of all possible paths assumes a large number of possible paths. We saw that the phasor direction depends only on the fractions of a wavelength difference between different paths, so we need to plot the phasor rotation for each path. If we assume say hundreds of paths slightly offset from each other (instead of the 12 shown for the swimmer), we can draw a graph of travel time versus path number as shown in Figure 16.6b. For paths far from the optimum, the travel time decreases in equal steps from one path to the next on the left side of the diagram, but increases in equal steps on the other side. However, near the optimum, each adjoining path will be only slightly different.

16.6 (a)
A set of 12 possible paths taken by a lifeguard to save a drowning person. The smart lifeguard will estimate the fastest path to reach the swimmer, which is the path of least action, as discussed in the text. (*Source: Copyright by Rahul Choudhary/Einstein First*)

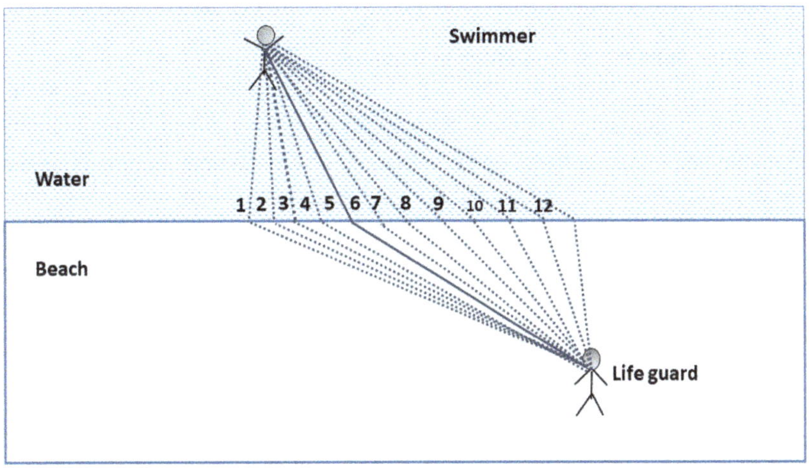


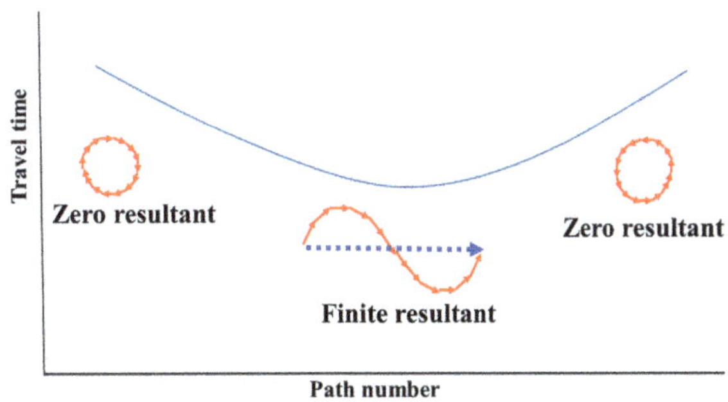


16.6 (b)
Phasor wheel rotations for possible photon paths. In general, when the path length as measured by phasor rotation is a linear function of trajectory position, phasors make a uniform increment in angle to create complete circles that have a zero resultant. Only at turning points, where phasor increments reverse in sign is there a place where the phasor resultant has a finite size. (*Source: Copyright by Rahul Choudhary/Einstein First*)

We can draw the phasors for nearby paths, such as 20 paths on the left, and the same on the right. The equal steps in travel time mean that when we add up the phasors, the paths on the left create circles in the clockwise direction, while those on the right side create circles in the anti-clockwise direction. Both close on each other, giving a resultant of zero. Only in one part of the curve do the phasors not add to zero. This is the centre where a circle in one direction turns into a circle in the other direction. The phasor resultant tells us that the probability of photons arriving is only high for the optimum path.

Students can plot the trajectory in this thought experiment with phasor wheels. However, in this case, we must take into account the fact that the wavelength is reduced in a medium where light travels slower. This means that we have to switch between a larger and smaller diameter phasor wheel at the interface. The ratio of diameters should match the speed ratio between air ($3 \times 10^8$ metre per second), and water (about $2 \times 10^8$ metre per second) while making sure that the phasor directions match from the big to the small wheel.

***Why does light travel in a straight line?***

Now, it becomes apparent why light travels in a straight line. The diagram for alternative paths could look like Figure 16.7a, or it could involve more elaborate deviations from straightness. Whatever we choose, the path length versus trajectory graph has two parts – a roughly linear part for large deviations, and a turning point where the path length passes through a minimum. The characteristic phasor patterns are circles (with a zero resultant) for large deviations and an s-shape near the centre. The s-shape always has a large resultant. This means that photons taking the straight trajectory have a high arrival probability, while other possible paths have zero probability.

In Figure 16.7a, we consider photons travelling with 15 possible paths between A and B. In Figure 16.7b, the resultant gives the probability amplitude of the photon travelling from A to B. It is evident that the phasors, which are near

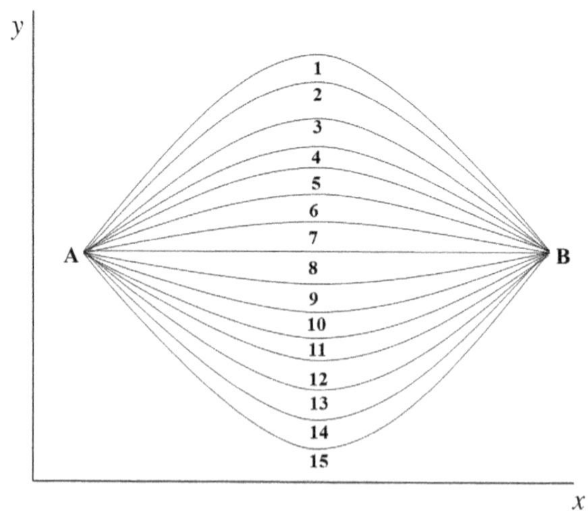


16.7 (a)
A photon travelling from point A to B (in one single medium) showing 15 possible paths. The *y*-axis is exaggerated since the total phase rotation (shown below) is less than one wavelength. (*Source: Copyright by Rahul Choudhary/Einstein First*)

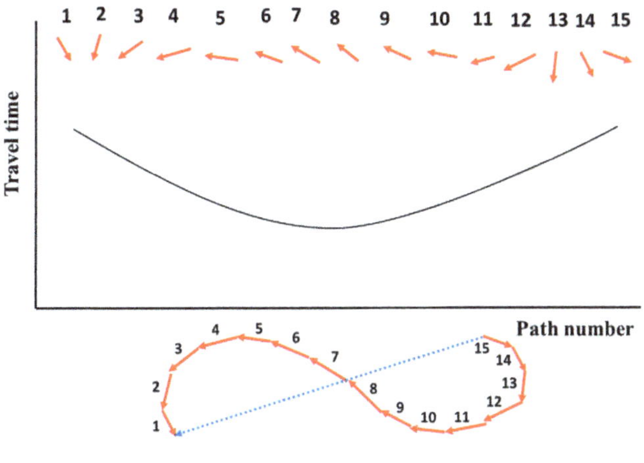


16.7 (b)
It is showing the origin of the s-shaped phasor diagram. Paths 5–11 contribute to a significant resultant because the phasor angle is nearly constant at the turning point. The longer paths form circles, which add to zero. (*Source: Copyright by Rahul Choudhary/Einstein First*)

the centre (5–11), contribute a significant part of the resultant. Photons will have a larger probability amplitude for paths which are in the centre because they form a larger resultant altogether. Phasors for longer paths will form a near-closed ring to create small resultant. In this case, the paths that we would typically consider untrodden by photons are the longer paths. Therefore, the photons will have a larger probability to take the shortest path, which is the straight line.

### *Can we trick the photons?*

Of course, it is not a trick! The easiest way to illustrate this "trick" is to consider light from a source S reflecting off a mirror, as illustrated in Figure 16.8a and 16.8b. Again, we find paths that we might consider impossible. All the preceding arguments follow, and we observe phasor patterns as illustrated. The path length is minimum, and here, the phasors create the usual s-pattern with a large resultant. Large deviations create full circles with zero resultant. The minimum path occurs for the angle of incidence equal to the angle of reflection.

Now, we can change the situation by adding a succession of black stripes that absorb any incoming photons as shown in Figure 16.8c and 16.8d. Phasors for any path that passes a black stripe will have zero length. Suddenly, we have

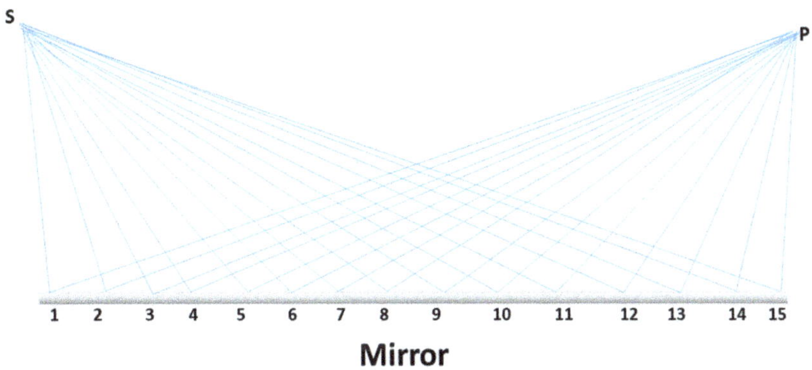


16.8 (a)
A photon travelling from S to P will have all the possible paths (15 shown here for simplicity) before it reaches P. These paths in the mirror are just samples of a large number of paths that are not shown here. (*Source: Copyright by Rahul Choudhary/Einstein First*)

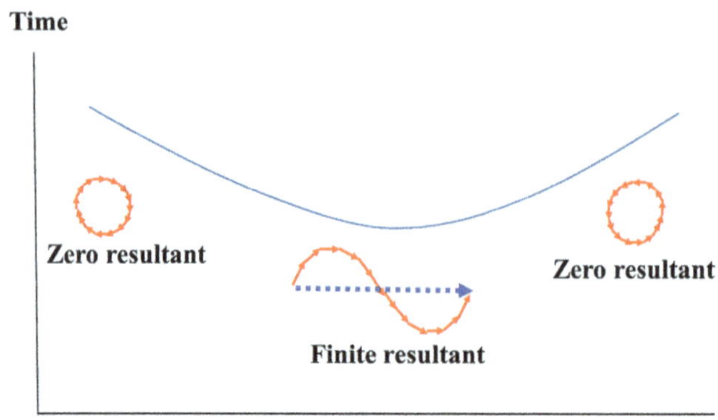


16.8 (b)
The *y*-axis is the time axis that represents the path length of the photons. (*Source: Copyright by Rahul Choudhary/Einstein First*)

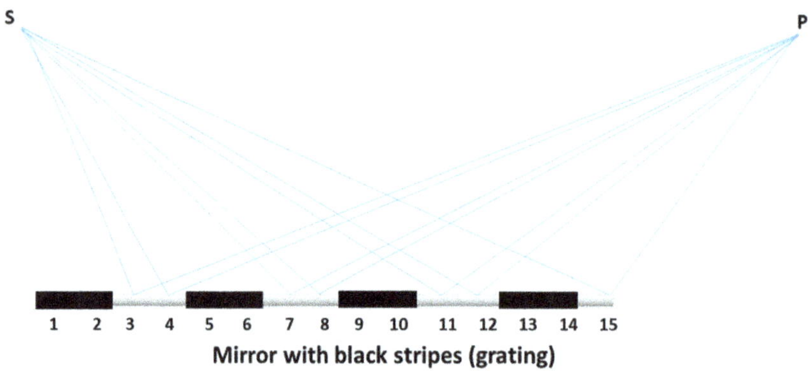


16.8 (c)
Here, we make some paths impossible by adding black stripes. (*Source: Copyright by Rahul Choudhary/Einstein First*)

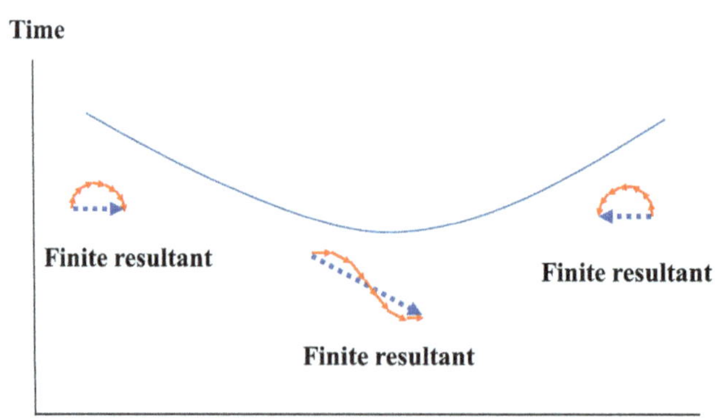


16.8 (d)
If the black bands correspond to half a turn of the wheel (half a wavelength of path difference), then we will only have half-circles of phasors. This will create a finite amplitude, with corresponding intensity given by the inscribed circles. Now, three different reflection conditions are possible, as shown. (*Source: Copyright by Rahul Choudhary/Einstein First*)

reduced the amplitude for half of the possible paths. If the blackening scale is comparable to half a wavelength, the circles will become half circles. There is a finite amplitude for angles for which the angle of incidence does not equal the angle of reflection. Similar to our mountain paths analogy, we block some paths untrodden, and more difficult paths are trodden instead.

### CLASSROOM ACTIVITY: LASER EXPERIMENTS WITH CD AND RULER

The situation explored in Figure 16.8 can be easily observed by shining a laser beam at a highly oblique angle off the millimetre scale lines of a simple standard plastic ruler. First, we need to reflect the laser off the unmarked region of the ruler to see a simple reflection on a wall. Then, we move the beam sideways onto the ruler scale. Suddenly new paths appear. The same experiment can also be performed with a compact disc and by reflecting laser light off the wing of a butterfly. Both are micro-structured like the striped mirror.

#### *Limitations of this model*

Neither phasor wheels nor Nerf gun bullets accurately represent photons. The striped mirror experiment tells us that photons are not localised; they do take all possible paths. The non-locality allows them to interact in ways that challenge our intuition. Photons do not have a fixed location in space until measured. We need to emphasise how frequency and wavelength connect in phasor wheels explicitly. If a photon interacts with an absorber along the path, the probability amplitude of the photon reduces. Waves are not always sine waves, and white light has many wavelengths. The phasor wheels with fixed diameter apply to one wavelength only. Another limitation is that a single phasor wheel cannot account for photons travelling from one medium to another, where the wavelength changes without changing the wheel diameter.

## CONCLUSION

Single-photon counting experiments and laser technology allow us to see deeply into the quantum world. This quantum world describes a reality in which all possible paths contribute to everything we observe. Phasor wheels allow the introduction of Feynman's path approach to quantum physics for middle and high-school students. We aim to provide a sense of the counterintuitive nature of the universe while making students familiar with essential concepts such as vectors and probability.

The quantum concepts in this chapter may seem bizarre and counterintuitive to adults, but from our experience, they appear to be less odd to children who are ready to accept new ideas. We trialled the content of this chapter with 14-year-old students. Students relate to the phasor wheel approach because it serves as a tactile and memorable tool for cementing the core concepts. The key to teaching quantum concepts is our observation-based approach supported by thought experiments, activities, videos, and simple laser experiments. The videos of single-particle interference consolidate the idea of photons and their intrinsic randomness. By emphasising the quantum nature of interference patterns, we hope to avoid much of the confusion of wave-particle duality.

## FURTHER READING

Chapter 3: The difficult birth of quantum physics
Chapter 15: Introducing quantum physics with toy photons
Chapter 21: Einstein-First: Modernising the school physics curriculum in Western Australia and beyond

## REFERENCES

Einstein, A. (1905). Ueber einen die Erzeugung und Verwandlung des Lichts betreffenden heuristischen Gesichtspunkt. *Annalen Der Physik*, *17*, 132–148.

Einstein, A. (1917). The quantum theory of radiation. *Physikalische Zeitschrift*, *18*, 121.

Feynman, R. (1979). Science videos, The Douglas Robb memorial lectures, http://www.vega.org.uk/video/subseries/8.

Feynman, R. (1985). *QED: The strange theory of light and matter (Alix G. Mautner memorial lectures)*. Princeton, NJ: Princeton University Press.

Heaviside, O. (1893). "On operators in physical mathematics." *Proceedings of the Royal Society of London*, *52*, 504–529.

Lewis G. N. (1926). The conservation of photons. *Nature*, *118*(2981), 874–875.

Steinmetz, C. P. (1893). "Complex quantities and their use in electrical engineering." *Proceedings of the International Electrical Congress, Conference of the AIEE: American Institute of Electrical Engineers Proceedings, Chicago (33–74)*, Chicago: AIEE.

Tinsley, J. N. et al. (2016). Direct detection of a single photon by humans. *Nature Communications*, *7*.

# 17 Gold, Einstein's Metal

*David Blair*

**WHAT?**

This chapter uses gold as a vehicle for showing the relevance of Einsteinian physics. Special relativity combined with quantum mechanics explains the unique colour and properties of gold. General relativity and gravitational waves are the key to understanding how gold was created and spread throughout our galaxy, finally to be found on our planet, the Earth.

**FOR WHOM?**

This chapter is intended for teachers who look for material for a middle-to-upper senior school module on Einsteinian physics that combines relativity, nuclear physics, atomic physics and astrophysics.

## INTRODUCTION

Everyone loves gold. Gold is heavy, brilliant yellow, resistant to corrosion, and very rare. It has been collected and treasured since the birth of civilisation. It has caused wars, genocide, and innumerable crimes. Governments collect gold in huge vaults, and in the past, based their financial systems on it. Alchemists over the ages laboured vainly to make it. People wear it and are buried with it. What gives gold that allure that surpasses all other metals? Why is it yellow? Where is it made in the universe?

All the answers have come from Einsteinian physics. Without special relativity, general relativity, and quantum mechanics, gold would still be a mystery. This chapter summarises the discoveries that finally gave satisfactory answers to all questions. Gold is no longer a mystery. Instead, it is a wonderful case study for the importance of Einsteinian physics to understand the universe.

### *Overview – On the trail of gold*

This chapter is like a murder mystery, but in this case, the body in question is gold. We begin with an autopsy, in which we discover the structure of gold. Then, we ask what aspect of its structure gives gold its unique qualities. Our lead began in the 1930s, but was only resolved in the 1970s when Einstein's theory of relativity explained the colour of gold and its chemistry.

Having examined the body, we go on to investigate the ancestry of gold. What was the sequence of events that brought us to gold? First, we need to understand the origin of common atoms like carbon or iron. We start our investigation from the Big Bang, and then examine the interior of stars. This way, we discover how the common elements arose, but then the trail of evidence becomes confused.

Searching the stars, we find a new clue: supernova explosions, the explosions of whole stars. We follow this lead and discover how relativity and quantum physics work together to allow new stable states of matter to exist in stars. But then, we discover how relativity makes these stars collapse. An informer suggests that the collapse drives a vast explosion and also the creation of gold.

Following this lead, we find no sign of gold. From this seeming dead end, a witness suggests that double neutron stars can coalesce due to powerful gravitational wave emission. We find that these events explain not one, but two mysteries – the mystery of enormously powerful gamma-ray bursts that hit the Earth every day, and the mystery of the making of gold. But we find that all the evidences are circumstantial. We need to catch the culprit. For 28 years, no proof could be found.

The breakthrough in the case took place on 17 August 2017. Astronomers found irrefutable proof that gold was created by Einsteinian physics and gravitational waves (Box 17.1).

## BOX 17.1: HOW TO MAKE GOLD: THE TIMELINE OF DISCOVERY

Centuries ago alchemists thought they could make gold by cooking up chemicals. It became clear that this was impossible once we understood that matter was made up of atoms.

Scientists began to understand the structure of atoms with their small central nucleus and a cloud of electrons in 1913. The missing neutral component was noticed in 1919 and named neutrons in 1921. They were not discovered until 1934.

The idea of neutron stars took shape from 1934 to 1939, and coalescing binary neutron stars were predicted 23 years later. Five years after this, neutron stars were discovered, and in 1976, the first *binary* neutron star system was discovered. In 1989, it was predicted that coalescing neutron stars would create gold and other heavy elements, as well as give out powerful bursts of gamma-rays.

In 2017, gravitational wave detectors detected the coalescence of a pair of neutron stars, followed by a burst of gamma-rays and an optical outburst in line with the 1989 prediction. The light contained the clear signature of heavy elements. The mystery of gold was solved.

## AUTOPSY OF GOLD

In our autopsy, we dissect gold down to the nucleus, then investigate its electrons, and use this information to understand how special relativity and quantum physics explain the unique colour of gold. We begin with an identity check.

### *The identity of gold*

The elements of the periodic table are the outcome of all the possible ways that protons, electrons, and neutrons can be combined to form stable structures called atoms. Gold is a very heavy atom. It has 79 protons squeezed into its tiny nucleus along with 118 neutrons. Because protons have a positive charge, they attract 79 electrons, which arrange themselves into a relatively enormous fuzzy cloud about 50,000 times bigger than the nucleus. It is this cloud of 79 electrons that gives gold its golden hue!

We name atoms by their outside appearance. Matter made of 79-electron atoms is a shiny yellow metal and we call it gold. Add an extra proton and electron and it is a silver metal that is liquid at room temperature – we call it mercury. Change the number of neutrons and the atom might become unstable. We call this instability radioactivity, but the atom will still be gold or mercury.

Isotopes are atoms that have the same number of protons, but different numbers of neutrons. If you make gold with only 116 neutrons, the gold nucleus is nearly stable. Each atom will survive for an average time of 186 days before it blows itself apart. The survival time is called the half-life, and this radioactive isotope is called gold-195 (79 + 116 = 195). If you add further two neutrons to increase the total number of neutrons to 118, you will have gold-197. This isotope is completely stable. It is the only stable isotope of gold.

If you add further one or two neutrons, you get gold-198 or gold-199. Both of these are even more radioactive – they have half-lives of two to three days. Artificially created gold-198, with 119 neutrons, is used for treating some types of cancer. When it decays, it emits an electron and a gamma-ray photon. The electron is emitted from the nucleus when a neutron turns into a proton. Because the number of protons defines atomic elements, this decay turns gold into element number 80, which is mercury. A neutron has become a proton, and the new atom, mercury-198, is also stable. In just two to three days, half of a lump of gold-198 will have turned into the shiny liquid metal mercury.

### *The electrons of gold*

The 79 protons in the gold nucleus create a strong positive electric field that attracts the negatively charged electrons very strongly. Once the number of electrons has reached 79, the charge is neutralised. From outside, there is no longer an electric field to attract further electrons. Because all matter, including electrons, exhibit wave-like behaviour, the electrons swarm around the nucleus in beautiful patterns of probability waves, also called orbitals (Figure 17.1 and Box 17.2).

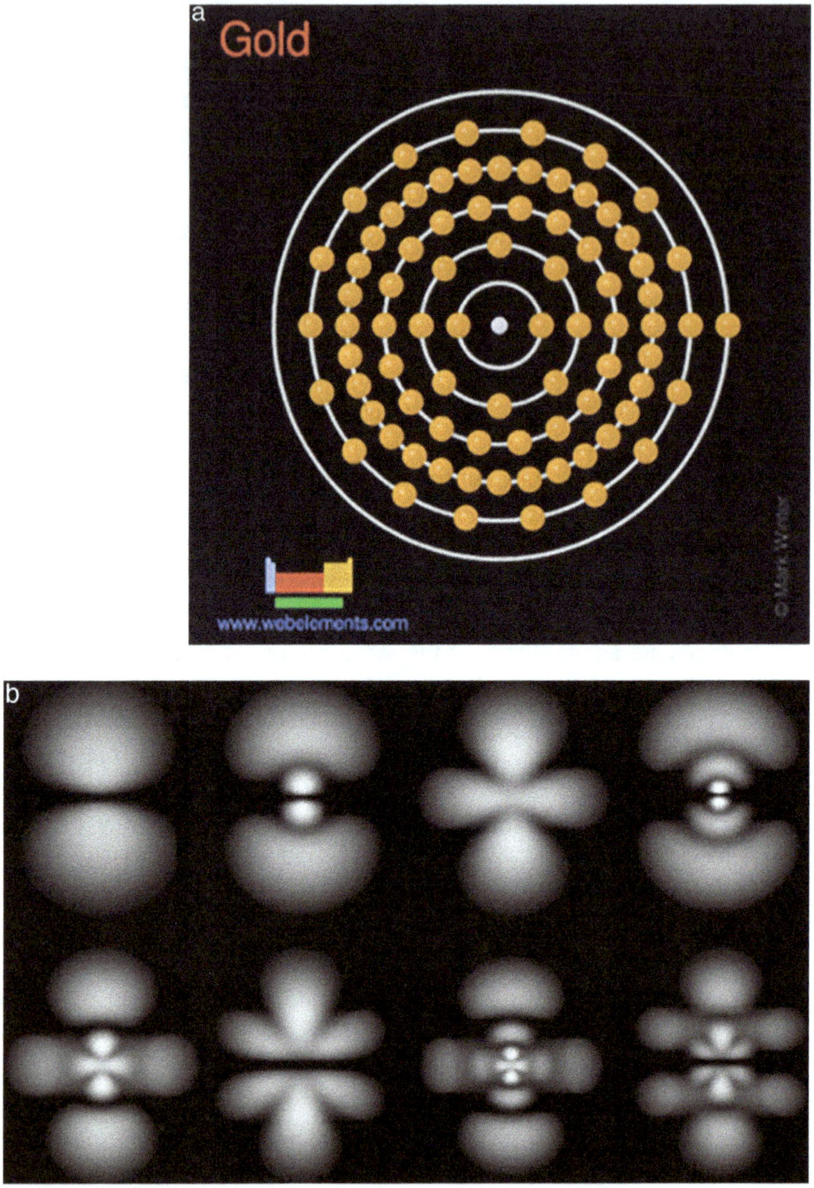


17.1
Electrons in gold are structured in orderly clouds of probability waves. They are not like the orderly pattern of "planets" shown in (a) because the waviness of electrons means that their locations are only defined by probability. Each electron is represented by a cloud of probability density where brightness represents increasing probability, as shown in (b). They are complicated three-dimensional structures, here seen in cross section. For gold, 79 such patterns are superimposed. Figure (a) is useful for categorising the electrons in terms of energy, but not their location. The energy of the last unpaired electron compared to that of the electrons below it determines the colour of gold. (*Source: Figure 17.1a: Figure from webelements.com courtesy Mark Winter Figure 17.1b: courtesy The Physicist at www.askamathematician.com /2012.*)

The mass of gold atoms is determined by the protons and neutrons because these particles carry 99.9% of the total mass of the atom. However, the size of the atom is determined by the size of the swarm of electrons. For gold, the size is 166 picometres. One picometre is a millionth of a millionth of a metre. In all atoms, the electric repulsion between electrons prevents other atoms from getting too close.

Gold is a heavy metal because of so many protons and neutrons, and because the electron swarm is held tight by the attraction of the 79 protons in the nucleus. It is interesting that most types of atoms have very similar sizes because more protons hold the electrons more tightly. For example, compared with gold at 166 picometres, hydrogen atoms with only one proton are 120 picometres in diameter, and the size of aluminium atoms with 13 protons is 184 picometres. So, gold atoms are heavier but not much bigger, and as a result, gold is very dense. It has a density of 19.3 grammes per cubic centimetre or 19.3 tonnes per cubic metre.

### *The colour of gold*

To understand the colour of gold, we have to better understand its cloud of 79 electrons. Because their behaviour is dominated by waviness, we need to consider the wave pattern of the probability waves. Since the electrons are confined, the probability wave of each electron must close on itself in the form of a *standing wave* around the nucleus. Like standing waves in wire loops, as illustrated in Figure 17.2, the standing waves of electrons must be continuous. This means that the standing waves are restricted to having certain numbers of wavelengths such as the four and five wavelength cases illustrated in Figure 17.2. Simple geometry links the radius to the wavelength.

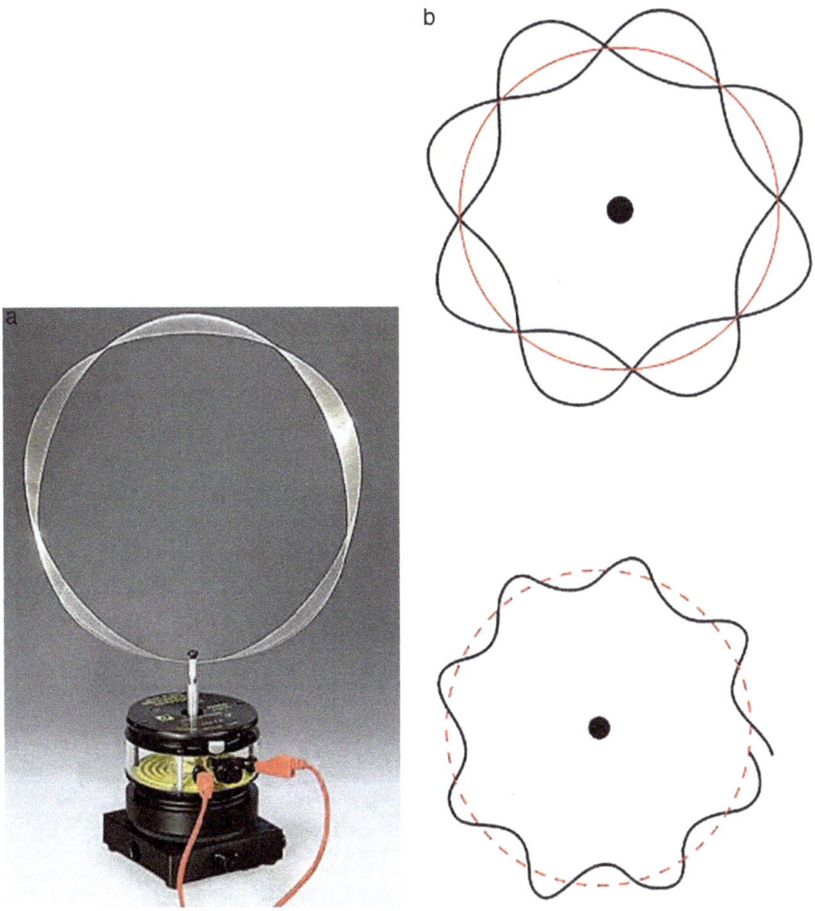


17.2
(a) A classroom experiment with a vibrating wire loop made by the laboratory equipment company PASCO and (b) The top line drawing illustrates an analogous probability wave of an electron circling the nucleus of an atom. The wave must be continuous in its ring around the nucleus. The bottom line drawing shows a disallowed wave form: it has a discontinuous break in the vibration pattern and is not allowed in any physical system. This disallowance of breaks in standing wave loops is the reason that energy levels in atoms are quantised. (*Source: 17.2a: Photo credit: Doug Bail www.pasco.com SF9405 17.2b: Copyright by the author*)

### BOX 17.2: SPECIAL RELATIVITY AND QUANTUM MECHANICS GIVE GOLD ITS GOLDEN COLOUR

Four separate pieces of physics act together to give gold its colour. First, special relativity tells us that mass increases as speeds approach light speed, as shown in Figure 17.3.

Second, the quantisation of energy levels in atoms arises because standing probability waves must be continuous, and therefore, have integer numbers of wave crests.

Third, the electron momentum, which depends on mass, determines the wavelength of the probability wave by *de Broglie's law*, which says that wavelength $\lambda$ depends inversely on momentum *p*, scaled by Planck's constant *h*:

$$\lambda = \frac{h}{p}$$

Finally, photons can only be absorbed if their energy exceeds the energy difference from a filled energy level to a vacant energy level. The momentum increase due to the relativistic mass increase makes the last energy level of gold low enough for blue light to be absorbed while the lower energies, red, yellow, and green, are still reflected.

Electrons have an important property called quantum spin, discussed more in the chapter on atoms. Usually, pairs of electrons with opposing spin directions, share the same wave pattern. Since the gold atom has an odd number of electrons, there remains one unpaired electron in the swarm of electrons. It is the energy level of this very outer unpaired electron that gives gold its unique chemical and physical properties.

The speed of an electron can be thought of as the speed needed for the centripetal force to balance the electric force pulling it towards the nucleus. In

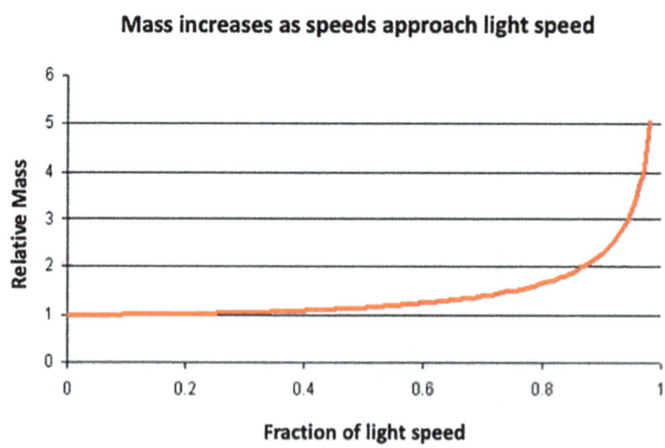


17.3 Special relativity tells us that mass increases as speeds approach light speed. (*Source: authors*)

gold, the very strong electrical attraction from the nucleus causes the outermost electron to move at about 58% of the speed of light. At this speed, you have to consider effects of special relativity.

In 1928, physicist Paul Dirac managed to combine Einstein's theory of relativity with the quantum theory of electrons. By 1930, Bertha Swirles had used Dirac's new theory to work out how to calculate the effects of relativity in atoms. But gold with all its electrons was just too complicated for her to solve without computers, even though the physics is straightforward.

Relativity tells us that mass increases as speeds approach light speed. An electron moving at 58% of light speed turns out to have 22% higher mass, and hence, higher momentum. But de Broglie's law says that wavelength reduces as momentum increases (see Box 17.2). If the wavelength of the standing wave is reduced by 22% (see Figure 17.2), the radius of the standing wave must also reduce by 22%. This means that the energy of the outer energy level reduces.

Metals are made up of atoms that share their outer electrons. The shared electrons make a sea of electrons that can flow freely from atom to atom. An incoming photon makes a disturbance in this sea, a bit like a person jumping into a swimming pool. The electron sea recoils and rebounds, causing a photon to bounce off the surface of the electron sea. That is why metals are shiny.

A photon can only be absorbed by an atom if an electron can accept its energy by jumping to a vacant higher energy level. The bigger the spacing in energy level, the higher the photon energy you need for the photon to be absorbed.

Most metals look silver-white because none of the photons in visible light have enough energy to span the energy gap from the second last energy level to the last one. Thus, they reflect photons of all colours equally. But in gold, the shrinking of the outer electron's energy levels due to relativity is sufficient that the highest energy photons of visible light, blue photons, have sufficient energy to make electrons from the second highest energy level jump into the highest energy level. In this way, blue light is absorbed by gold atoms. With blue removed, gold looks gold!

So, the explanation for the colour of gold is that the relativistic increase in mass of the electron shrinks the energy level, so that blue photons are absorbed instead of being reflected. This was clarified in the 1970s when people used Bertha Swirles' theory with computers to calculate the colour of gold. Just two other metals, copper and cesium share this property.

## THE ANCESTRY OF GOLD

We have completed our autopsy of gold. Now, we need to understand the ancestry of gold. How was it made and how did it arrived on our planet?

### *How to build atoms*

Heavier atoms are built from lighter atoms. Starting from the simplest, the universe builds up atomic nuclei by adding protons and neutrons. The process is called *nucleosynthesis*, the making of atomic nuclei. Nucleosynthesis is not easy because positively charged protons repel each other. They can only stick together once they are close enough together that the strong nuclear force can grab them.

The difficulties arise in making nuclei because of the different nature of the two main forces involved: the *electric* force and the *strong* force (Box 17.3). The electric forces of attraction and repulsion act over long distances, following the same mathematical *inverse square law* as gravity. Like gravity, electric forces act over long distances, but unlike gravity, there can be attraction or repulsion according to whether you have *like charges* (two positives or two negatives) or *unlike charges* (positive and negative).

While the electric force reduces to one-fourth every time you double the distance, it is still strong enough to hold the clouds of electrons in place 100 picometres from the nucleus, just like the Sun holds Pluto in its orbit six billion kilometres from the Sun. The strong nuclear force is quite different. It is strongly attractive at a distance of about 1 femtometre (that is, $10^{-15}$ metres), but falls to zero by about 2.5 femtometres. A proton that is approaching a nucleus has to force itself through a relatively huge electric repulsion barrier before it gets close enough to be grabbed by the strong nuclear force. Thus, it is quite difficult to build atoms by adding to the nucleus one proton at a time.

Once you have a nucleus, the electric charge of the protons attempts to accumulate an equal number of electrons to make it electrically neutral. Because

### BOX 17.3: THE FORCES THAT MAKE MATTER

To understand the universe, we have to know about the four fundamental forces: the strong, the weak, the electric, and the gravitational force.

The *strong nuclear force* holds protons and neutrons very tightly together – it provides very strong attraction at the tiny distances needed to make atomic nuclei.

The *weak nuclear force* is a force that can be attractive or repulsive. It causes neutrons to break up into protons and electrons: a process we call radioactivity, and specifically beta decay.

The *electromagnetic force* acts over long distances and can attract and repel. The electric attraction between electrons and protons holds the cloud of electrons in place in atoms, but the electric repulsion between electrons makes atoms bounce off each other.

The *force of gravity* also acts over long distances. General relativity tells us that the force of gravity arises from the curvature of spacetime. Gravity always attracts so it can accumulate indefinitely. On a big enough scale, gravity can overcome all other forces, like in neutron stars where gravity overcomes the electromagnetic and weak forces to squeeze electrons and protons back together to make neutrons.

there are usually plenty of electrons around, this part of the process would be easy if it wasn't for high-energy photons that act like bullets to blast the electrons away from the atoms. If atoms are in a hot environment (such as our Sun or a star), there are plenty of high-energy photons. Only when the atoms are in a cool environment, like planet Earth, do they become neutral atoms with the proper number of electrons.

### *The first atoms*

The first atoms were created very early and only shortly after the Big Bang that marked the start of the universe as we know it today. By studying clouds of gas not-yet-turned-into-stars, astronomers have been able to determine the conditions when the universe was just a few seconds old. We pick up the story ten seconds after the universe began. The universe was extremely dense and extremely hot. Space itself was a billion times smaller than it is today, but it was expanding extremely fast and carrying with it matter and photons that were cooling rapidly.

Ten seconds after it began, the universe was a "soup" of photons, protons, neutrons, and electrons at a temperature of about one billion degrees. The soup was expanding fast and cooling fast. The billion-degree heat meant that the photons had enough energy to blast apart any protons and neutrons that stuck together. But as the soup cooled, the photons had less and less energy to blast the protons and neutrons apart. This meant they could remain stuck together. Within three minutes, one quarter of the mass of protons and neutrons had turned into helium nuclei consisting of two protons and two neutrons (Figure 17.4).

The universe continued its expansion and cooling. It was so hot that any electrons that were attracted to the nuclei were instantly blasted free again by the photons. This meant that the universe was like a thick white-hot fog. Photons could not travel because they kept colliding with electrons, and then the electrons collided with nuclei and made new photons.

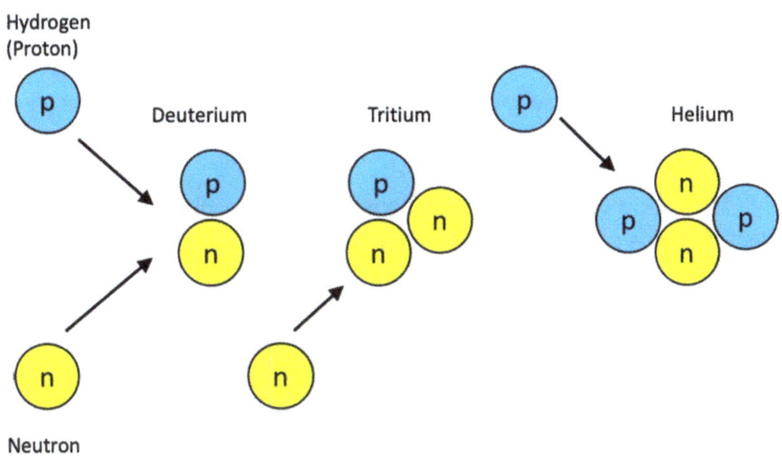


17.4
A sequence of fusion reactions in which protons and neutrons combine to make helium. First, a proton and neutron combine to make the heavy isotope of hydrogen called deuterium. Then, a second neutron is absorbed to make the radioactive form of hydrogen called tritium, and finally, another proton combines to make helium, which is a very stable nucleus. (*Source: Wikimedia Commons*)

Expansion and cooling carried on for 380,000 years. At this time, the entire universe had cooled to a few thousand degrees, cooler than the Sun. It was cool enough that at last, the photons no longer had enough energy to dislodge electrons from their probability wave clouds around the nuclei. Proper hydrogen and helium atoms could now form. There were no more free electrons, and the gas suddenly became transparent in the same way that air is transparent. The photon energy was too low to be able to make an electron change its energy. Note that this is the same quantum physics that makes gold yellow – photon absorption only happens when the photon energy is high enough compared with the energy levels.

This special time in the universe marked the first time the universe had any proper atoms: about 75% by mass hydrogen and 25% helium. Suddenly, the universe was transparent, so the photons could travel on without ever being absorbed. We see these photons today, 13.8 billion years later. The image we see is called the cosmic microwave background, easy to find in an image search. It is the oldest light in the universe. Because it has travelled for 13.8 billion years to reach us, it is an image of the universe when it was young.

But we have digressed: we have only found the first two members of the atom family and we have 77 to go to reach gold!

### *Atoms get made in stars*

Stars and planets are places where gravity creates a dense accumulation of matter. In a star, gas collapses into a dense ball. Collapsing and compressing makes heat, like the heat we feel in a bike pump when we compress gas to force it into a bicycle tire. The heat and the pressure take the hydrogen and helium atoms back to the conditions they experienced a few minutes after the Big Bang. But in stars, the heat and the pressure last for millions of years instead of minutes. This extra time makes all the difference.

We have seen how radioactive atoms have half-lives – the average time to break apart in *fission* reactions. Similarly, *fusion* reactions, in which atoms can grow by accepting new protons or by two nuclei joining together, also have a sort of half-life, an average time for the reaction to occur. However, this fusion half-life is not a pure property of the nuclei. It depends strongly on the temperature and the pressure, and these vary enormously from the hot dense central core of a star to its much cooler surface. Stars get hot through collapse, but then they generate heat by the fusion of nuclei.

The fusion of two nuclei is opposed by electrical repulsion of positive electric charge of the protons. As a result, the half-life of an average hydrogen nucleus (consisting of one proton) in the Sun before it undertakes fusion is about 4 billion years. The protons are only able to fuse at all because of a remarkable feature of quantum physics called *quantum tunnelling*. The quantum uncertainty principle gives protons a finite but tiny probability of being close enough together that the strong nuclear force is able to overcome the electric repulsion barrier. Once close enough, the nuclear force pulls the nuclei together. In the process

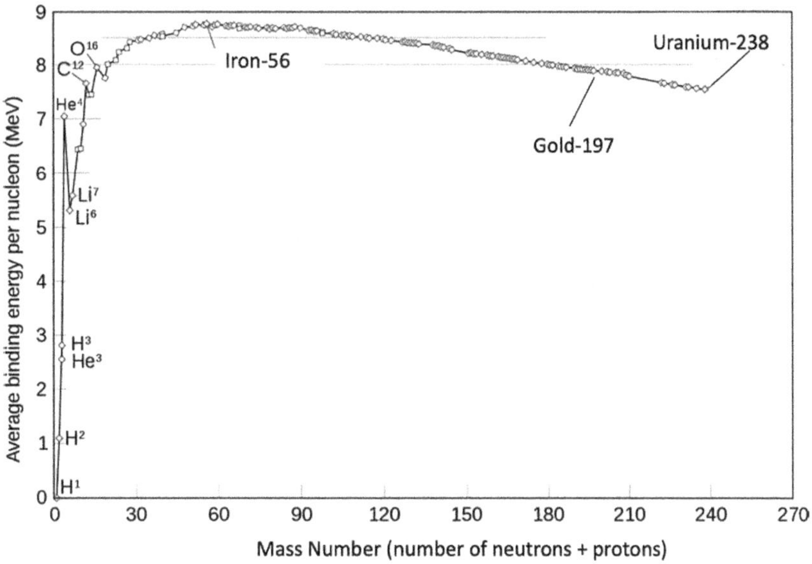


17.5
This graph tells us the energy benefits and costs from building atoms. On the left side of iron, energy is given out every time you add protons or neutrons to make atoms heavier. On the right of iron, you have to provide extra energy to build bigger atoms. Gold-197 releases energy if it is split apart. The energy unit, MeV, is millions of electron volts. (*Source: authors*)

of snapping together, excess energy, called *binding energy*, is released. Most of this energy is released as high-energy photons. The chance of this happening to a single proton in the Sun is very small, like once in 4 billion years. Even though the nuclear fusion reactions are very unlikely for individual protons, they are still happening at an enormous rate in the Sun because of the vast number of protons in the Sun. The rate is just right to keep us warm on our planet, and keep the Sun burning for billions of years.

The fusion reactions that make heavier elements in stars are called *nuclear burning*. This process happens in many steps, sometimes by absorbing protons, or proton-neutron pairs, or helium nuclei. The important feature of nuclear burning is that each reaction *releases* energy. Figure 17.5 shows how the binding energy per nucleon (that is, proton or neutron) depends on the size of the nucleus. The graph peaks at iron. If you want to get the most energy out of nuclei, turn them into iron, either by fusing lighter nuclei, or by breaking up heavier nuclei. The iron peak shows that you have to *provide* energy to make heavier nuclei.

For example, imagine trying to fuse together two nuclei of mass 118. This would make a nucleus of mass 236 that could be uranium. On the graph in Figure 17.5, you can see there is an energy deficit from mass 118 to mass 236. The big nucleus would have less binding energy, so you would have to provide that energy. In particle accelerators, you can make heavy nuclei by smashing one type of nucleus into another. The accelerator provides the extra energy in the form of the kinetic energy of the high-speed nucleus when it smashes into the target nucleus. It is possible to make miniscule amounts of gold in this way.

Because extra energy is needed to make heavy nuclei, stars cannot make them by normal fusion. Iron and nickel are the final product of nuclear burning. At iron we have reached 26 protons and we need to reach 79 to make gold. So, where does gold get made? Astronomers thought they had the answer:

supernova explosions, the cataclysmic explosion of massive stars. We will follow this clue.

***The supernova clue***

Stars maintain their size by pressure. In normal hot stars like the Sun, pressure is created from the momentum of very hot electrons, nuclei and photons pushing outwards. The momentum which depends on the mass and speed of the particles in the star comes from the energy released in the fusion reactions. But fusion ceases once the star's constituents have reached iron. Then, there is no more heating, and the pressure reduces. This makes the core of the star shrink, and gravity squeezes it still further. Could anything stop it from collapsing?

In 1930, a 19-year-old Indian student named Subrahmanyan Chandrasekhar was on a boat trip from Calcutta to England to study astrophysics at Cambridge. On the long voyage, he applied the new theory of quantum mechanics to the mysterious stars called white dwarfs. These stars were tiny – 100 times smaller than most stars – (about the size of the Earth), yet as massive as the Sun – a million times more massive than the Earth. It was thought that all stars must end up like this.

Chandrasekhar realised that if you squeezed matter a million-fold, the electrons would be squeezed out of their orderly clouds, and wander randomly throughout the star. But would it be random? We saw in gold that the 79 electrons arrange themselves in energy levels. They cannot share the same standing wave pattern, and hence cannot have the same energy and spin. This non-sharing is a property common to all the particles that have half a quantum unit of spin (see Chapter 18). They are called fermions, which include protons and neutrons. All fermions have to arrange themselves in a sort of *energy staircase*, starting at the bottom and filling the energy levels upwards.

In an outrageous extrapolation, Chandrasekhar applied the concept of the energy staircase in atoms to the vast scale of a dying star. He showed that near the top of the staircase, the electrons would be approaching light speed, and the bigger the star, the more massive the electrons would be as they got closer and closer to light speed, according to Figure 17.3. In gold, the increasing mass of the high-speed electrons changed its colour. In white dwarf stars, something rather similar happens. The increasing electron mass near light speed makes the star shrink as you add more matter. Add still more and the star collapses. Chandrasekhar predicted that no white dwarf could be heavier than 1.44 times the mass of the Sun. Up to the critical mass, the star behaves like a single vast atom, about the size of the Earth. Above that mass, white dwarfs must collapse. Despite huge scepticism, Chandrasekhar was proved correct and was awarded the Nobel prize fifty years later, in 1983.

A year after Chandrasekhar's work, neutrons were discovered, and almost immediately Swiss astronomer, Fritz Zwicky suggested that there could be stars made of pure neutrons and they might get made in supernova explosions. This extraordinary guess was prophetic! In 1939, Robert Oppenheimer, who later led the Manhattan Project, realised that a star made of neutrons must also obey the non-sharing rule of fermions and have an energy staircase with one step for each neutron. Because neutrons are nearly 2,000 times heavier than electrons, they have 2,000 times more momentum. This lets them exert 2,000 times more pressure. So roughly, a neutron star could be 2,000 times smaller than a white dwarf, but just as massive. The density would be trillions of tonnes per litre, the size about 20 kilometres! But like a white dwarf, once the fastest particles reached light speed, the star would collapse.

Many astronomers disliked the idea of collapsing stars. Chandrasekhar's supervisor, Arthur Eddington suppressed the work of his student stating, "I think there should be a law of Nature to prevent a star from behaving in this absurd way!" (Chandrasekhar, 1990).

Despite this opposition, and long before neutron stars were actually discovered by Jocelyn Bell in 1967, it became the common view that when stars run out of energy, their core would collapse, the pressure would force electrons and protons to fuse together to make neutrons, and you would have a neutron star. The rest of the star would be blasted off into space, and somehow this would contain gold and uranium and all the other heavy elements.

### *Supernova 1987A*

In January 1987, a star exploded in our backyard, a small galaxy just 150,000 light years away (Figure 17.6). Its light was brilliant, its form quite magical, and almost every telescope on the Earth examined its light. Rings of gas were brightly

17.6 Supernova 1987A in the Large Magellanic Cloud. This was the first supernova visible to the naked eye since Kepler's supernova of 1604, and the discovery of the telescope. The rings were created by blast waves expanding into space. (*Source: NASA Goddard Space Flight Center from Greenbelt, MD, USA / Public domain*)

illuminated as ejected material smashed into them, and the abundant light allowed careful analysis. There was plenty of evidence for radioactive iron and nickel, and even barium and strontium, but no sign of the really heavy elements like gold. The trail was cold! Supernovae were innocent bystanders in our mystery.

## THE LAST LINE OF ENQUIRY

One last line of enquiry was left. It still involved neutron stars. This line of evidence takes us back to 1962, when neutron stars were still hypothetical. Freeman Dyson was thinking about how alien civilisations might send powerful signals to each other. Most stars in the universe have companions, and so too could neutron stars. He realised that vast bursts of energy could be transmitted by gravitational waves created by pairs of neutron stars. He wrote, "*the loss of energy by gravitational radiation will bring the two stars closer with ever increasing speed, until in the last second of their lives they plunge together and release a gravitational flash at a frequency of about 200Hz and of unimaginable intensity ... So the death cry of a binary neutron star could be heard on Earth*" (Dyson 1963).

Dyson was another prophet! Neutron stars would not be discovered for five more years. Based only on quantum physics, relativity, and atomic physics, Oppenheimer had imagined neutron stars in 1939. Dyson had taken Oppenheimer's idea, added a second star, and then used Einstein's theory of gravitational waves to make an astonishing prediction of vast bursts of gravitational energy.

In 1976, a pair of neutron stars was discovered, spiralling slowly closer together, just as Dyson had imagined (Lorimer, 2008).The neutron stars of the pair were orbiting each other every 8 hours, and careful measurements over several years showed that the system was spiralling closer together by about a millimetre every year. In 300 million years, the stars will get so close together that they will be extremely distorted by each other's gravity. The cores will merge together, but in the moments before this happens, the balancing of gravity between the stars will depressurise the matter at their surfaces, while flinging material outwards by centrifugal forces. The stars will distort to become tadpole-shaped, long tails of material flying into space while the heads will merge as illustrated in Figure 17.7.

The matter of the neutron stars is as dense as an atomic nucleus, but at facing surfaces of the two stars, the neutrons are suddenly freed of the immense gravitational pressure that created them. Now fission begins. The depressurised neutron matter breaks up into smaller and smaller fragments. Neutrons convert to protons by emitting electrons. The huge density of neutrons plus newly formed protons make extremely unstable oversize nuclei. These nuclei decay almost instantly. Within one to two seconds, the nearly pure neutron matter has converted into heavy nuclei. Most are unstable and continue to decay in the same way that we saw that gold isotopes decay.

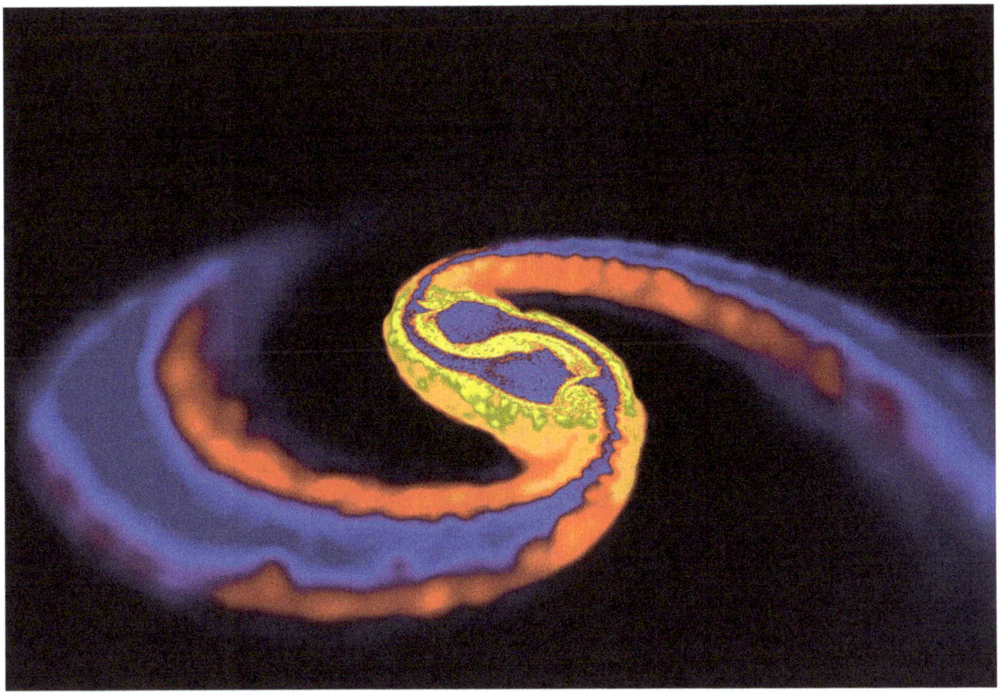

17.7 Computer simulation of two neutron stars coalescing. The cores merge, but long tails of material are thrown into space. (Computer modelling and image by Daniel Price and Stephan Rosswog.) (*Source: Daniel Price (U/ Exeter) and Stephan Rosswog (Int. U/ Bremen) http://users.monash.edu.au/~dprice/research/nsmag/*)

The released neutron matter explodes like a gigantic atom bomb, creating a cloud of highly radioactive heavy elements, which all decay according to their particular half-lives, adding extra energy to the exploding mass of material. A flash of gamma-rays, and a bright explosion called a kilonova could be expected. Much of this was predicted in 1989 (Eichler et al., 1989).

Atom bombs make tiny amounts of gold when a few kilogrammes of uranium or plutonium spontaneously explode. The kilonova explosion should involve up to 100,000 Earth masses of neutron matter. Most of the matter turns into heavy elements, and almost all of this explodes out into space. This includes hundreds of Earth masses worth of gold, and the same for platinum and other heavy elements.

The Earth is 4.6 billion years old, but our galaxy is at least 10 billion years old. Long before the Earth was born, binary neutron stars were coalescing and throwing heavy elements into space. This mixed into huge gas clouds that collapsed to make new stars and planets. This way gold was mixed into the fabric of planet Earth, very rare but very beautiful. Case solved? Not quite!

The theory was in place, but where was the evidence? It all came on 17 August 2017, when the LIGO and Virgo gravitational wave detectors detected the unmistakable siren sound of two neutron stars spiralling together. A few moments later, a burst of gamma-rays was detected by the Fermi gamma ray space telescope. A few hours later, astronomers found the visible light of a kilonova explosion. The light showed unmistakable evidence for heavy elements. Now, the case was solved (Abbott et al., 2017)!

The culprit in our mystery was the gravitational waves of general relativity. These gravitational waves radiated a fraction of the mass-energy of the two neutron stars, allowing them to fertilise the space between the stars with heavy elements, including gold, whose alluring properties are explained by Einstein's special theory of relativity. Truly, gold is Einstein's metal!

## CONCLUSION

Thanks to Einsteinian physics we know how the universe makes gold. The Einstein-First project has used the mystery story in this chapter to motivate classroom instruction from year 3 to year 9. The e-resources contain examples of relevant classroom activities. At the year 3 level, we turned the story into a role-play featuring the scientists mentioned here, as well as gravitational wave physicists. It culminates with two children enacting the coalescence with balloons containing gold-wrapped sweets, which burst and spread their treasures. See www.einsteinianphysics.com.

At the year 9 level, we structured classes around activities to investigate all the aspects of Einsteinian physics encompassed in this chapter. The discovery of the origin of gold provides a unifying link and connects all the topics to our world of experience.

## FURTHER READING

E-resources: Examples of classroom activities
Chapter 4: The difficult birth of gravitational wave astronomy
Chapter 18: Patterns and atoms: The structure of atomic matter

## REFERENCES

Abbott, B. P., Abbott, R., Abbott, T. D., Acernese, F., Ackley, K., Adams, C., Zweizig, J. (2017). GW170817: Observation of gravitational waves from a binary neutron star inspiral. *Physical Review Letters, 119*(16), 30–33.

Chandrasekhar, S. (1990). *Truth and beauty: Aesthetics and motivations in science.* Chicago: University of Chicago Press.

Eichler, D., Livio, M., Piran, T. *et al.* Nucleosynthesis, neutrino bursts and γ-rays from coalescing neutron stars. *Nature* **340**, 126–128 (1989). https://doi.org/10.1038/340126a0

Lorimer, D. R. (2008). Binary and millisecond pulsars. *Living Reviews in Relativity, 11*(8).

# 18 Patterns and atoms

## *The structure of atomic matter*

*David Blair*

**WHAT?**

This chapter connects the quantum physics of atoms to easily observable large-scale wave patterns, and shows how on all scales from atoms to metals to the Sun, these patterns follow simple rules of standing waves that lead to energy quantisation. The presentation builds on earlier chapters in this book that introduce basic quantum concepts such as photons and matter waves.

**FOR WHOM?**

This chapter is intended for teachers with some familiarity with quantum concepts that look for an introduction to the Einsteinian physics of atoms and the materials of modern technology. The material is designed to support both middle and high-school teachers and students.

### INTRODUCTION

> "We are accustomed to speak of heat, of cold, of colour. In reality, there are atoms and space".
>
> Democritus (ca. 460–360 BC)

This astonishing insight of Democritus almost 2500 years ago was not confirmed until the 19th century. Obtaining evidence that matter is made from sets of identical building blocks called atoms was the first step in our discovery that the universe has a fundamental quantum nature. The next step was the discovery that light and heat come in packets, which were first called *energy quanta* and later named photons and phonons (Einstein, 1907; Planck, 1901). The name photon is now used to describe the energy quanta for all parts of the electromagnetic spectrum, from radio to gamma rays, while phonons describe the packets of vibrational energy that run around inside solids and liquids, which include sound (Street, 2019). Roughly, the concentration of photons is the intensity or brightness of light, and the concentration of phonons is the temperature of a solid.

The interplay of photons, phonons, atoms, and their electrons determines the properties of all matter, including the materials at the heart of mobile phones, solar panels, and computers. This chapter explores the patterns of electron probability waves around atoms, called orbitals, and the way they spread out to encompass groups of atoms to make molecules, and spread out even further to become the conduction bands in metals that allow electrons to flow freely. The patterns of electron probability around atoms have a close resemblance to patterns of sound vibration in everyday objects such as musical instruments. We examine these patterns, easily studied in classroom activities, to help create an intuitive understanding of the basic structure of atoms.

The connection between electrons in atoms and standing waves in the macroscopic world arises because of Louis de Broglie's 1924 discovery of the waviness of matter (matter waves). We will discover how waviness limits our ability to "see" the smallest building blocks of matter and how it sets the size and shape of atoms. Looking at the beautiful complex patterns of electron waviness around atoms, we will develop an intuitive understanding of these patterns by relating them to similar patterns in large-scale objects from vibrating plates to the Sun. We will also explore the fundamental differences between electrons in atoms and photons that cause electrons to stack up in a *staircase* of increasing energy, while photons can happily coexist on any step of the staircase.

To explain the size of atoms and their energy processes, we will explore the two fundamental principles of quantum physics, the Heisenberg uncertainty principle, and the Pauli exclusion principle, which together with the properties of standing waves and the waviness of electrons allow us to give a qualitative explanation of the physics behind the special materials like semiconductors that underpin modern technology.

## SEEING THE VERY SMALL

Our world is revealed to us mainly by photons of light that scatter off matter and are detected when they give up their energy to receptors in our eyes. In daylight, the photons come from the Sun. At night and in dark places, we create the photons using electric lights. With magnifying lenses, we can see small things, but there is a limit. For example, we cannot see atoms even with a very strong magnifying lens.

The reason we cannot see atoms is that the wavelength of the photons we see with is roughly 1000 times bigger than an atom. Wavelength is a measure of the scale of waviness or fuzziness of the photons. Not surprisingly, photons are unable to make images of things that are smaller than their intrinsic fuzziness, which is their wavelength. For this reason, we cannot make an image of an atom using visible light.

So how can we see atoms? The answer is to use photons with a wavelength smaller than atoms, such as x-rays. Huge machines called synchrotron radiation sources are used to make powerful x-ray beams for seeing at the atomic scale.

However, you do not have to use photons to make images. Instead, particles like electrons can be used. De Broglie's law (see Box 18.1) tells us that the

**BOX 18.1: DE BROGLIE'S LAW**

Everything, whether photons or particles like electrons or protons or even large molecules, has a wavelength, which depends inversely on momentum. Wavelength, $\lambda = h/p$, where $p$ is the momentum, and $h$ is Planck's constant. Momentum is mass times speed when speed is low. Increasing momentum increases bulletiness or destructive ability, but reduces the wavelength to allow imaging on smaller scales.

The magnitude of Planck's constant is tiny: $6.6 \times 10^{-34}$ joules per hertz.

wavelength of photons, electrons, or anything depends inversely on their momentum. So to get the small wavelength you need for seeing small things, you need more momentum. Electrons are better than photons because their mass gives them much more momentum. An electron gun can fire a high-speed beam that can be conveniently focused using electric and magnetic fields, to create an *electron microscope*. Through these devices, we can clearly see down to the level of atoms. When we do, we see lines of atoms looking like fruit packed neatly in a box, as illustrated in Figure 18.1. Atoms, at about $10^{-10}$ metres or 100 picometres in size are the smallest things we can directly image.

18.1
Atomic scale image of tungsten selenide and the orderly arrangement of tungsten and selenium atoms except for some three-fold defects. (*Source: Lin, Y., Björkman, T., Komsa, H. et al. Three-fold rotational defects in two-dimensional transition metal dichalcogenides. Nat Commun 6, 6736 (2015). https://creativecommons.org/licenses/by/4.0/*)

Today, imaging at the atomic scale is crucial to many fields of research such as drug development and semiconductors for use in computers and phones. Both electron microscopes and x-rays are widely used. However, such imaging has many technical difficulties caused by De Broglie's law. As we teach it to primary school students, reducing the waviness (that is the wavelength) increases the bulletiness. At 100 picometres wavelength, both photons and electrons are highly destructive because they have so much more momentum. They very easily heat up or destroy the things being observed.

If you want to look at things even smaller than atoms, you must make the wavelength even smaller. This means that even more momentum is needed. But there is a catch! When you use extra momentum to make the wavelength smaller, you turn the beams into more and more destructive bullets. Instead of making images, you have to study the fragments of the collisions and use this information to reconstruct the inner structures.

In this way, the subatomic world was first uncovered by Ernest Rutherford in 1909, by observing how helium nuclei scattered off platinum foils. In the words of a Massachusetts Institute of Technology laboratory manual, "*It was as though bullets fired at a bale of cotton could occasionally ricochet backward. Such an observation might lead one to suspect rocks in the cotton*". The rocks in the cotton were the *nuclei* of platinum atoms. Rutherford determined that the nucleus was about 10,000 times smaller than the atom. Today, we know this factor is close to 50,000.

Atomic nuclei were first thought to be just tiny balls of protons, but it was soon realised that they had roughly equal numbers of neutrons. It took more momentum to examine atomic nuclei from inside, and even more to discover the *quarks* that protons and neutrons are made from, as well as the nuclear forces that hold nuclei together. To observe these tiniest things, we need huge machines capable of accelerating particles to give them the largest possible amount of momentum. The biggest such machine in the world is the Large Hadron Collider at CERN (The European Organization for Nuclear Research). In 2012, it discovered the Higgs Boson, the last of the 17 fundamental building blocks from which everything is made.

But we digress! We have learnt how waviness limits our ability to see atoms. We will now see how waviness determines the structure of atoms. For our purposes, we just need to know that tiny atomic nuclei strongly attract clouds of electrons until the number of electrons exactly matches the number of protons. The only part of the atom we normally "see" is this cloud of negatively charged electrons. We will go on to discuss the structure of these electron clouds.

### *Energy levels and standing wave patterns*

In the 19th century, people had realised that different elements like hydrogen, oxygen, and nitrogen give out a set of different sharply defined colours of light called spectral lines. Scientists looked at the spectrum of sunlight and saw

**BOX 18.2: ENERGY, FREQUENCY, WAVELENGTH, AND MASS**

When an electron jumps between two energy levels with energy difference $E$, it creates a photon of frequency $f$ (= $c/\lambda$ for wavelength $\lambda$), following Planck's formula, $E = hf$, where $h$ is Planck's constant.

Note that $E$, $f$, and $\lambda$ are all ways of specifying the photon energy. Also, because energy is equivalent to mass by $E = mc^2$, we can use energy, mass, frequency, and wavelength interchangeably. Physicists often quote mass in energy units and use the energy unit electron-volt, which is equal to $1.6 \times 10^{-19}$ joules. This can be confusing for students and teachers!

patterns of colours for the gases they knew, but they also saw a bright yellow line that did not match any of the known gases. They presumed there must be another element present on the Sun that was unknown on Earth. They called it *helium* after helios, the Greek name for the Sun. Twenty-eight years later helium was discovered on Earth. This discovery was a triumph for atomic physics, but it did not explain the origin of the spectral lines.

In 1911, Niels Bohr put forward his model of atoms in which electrons orbit the nucleus in separate orbits, each with its own energy. Atoms give out a photon when an electron jumps from a higher to lower energy level. From Planck's formula for the energy of photons, $E = hf$, we know that each photon energy corresponds to a different frequency and, correspondingly, to a different colour. Whole packets of energy had to be emitted or absorbed in an electron jump to explain the pure colours of light emitted by different gases. In this way, Bohr explained the pattern of colours produced by hydrogen atoms in terms of electrons jumping between different allowed energy levels (Box 18.2).

Once de Broglie had predicted matter waves, Bohr's theory evolved into the modern theory in which each electron energy corresponds to a particular *standing wave* pattern of an electron matter wave, electron waviness. We no longer have an orbit. Instead, we have a standing wave pattern called an *orbital* that has a fuzzy wavelike shape, but still has a fixed energy. The wave is a wave of probability that tells the chance of finding an electron at any particular location. The probability wave is like a heat map, used to study football matches, where the map is darker in the places where there is a high chance of finding a particular football player and pale in the places where the player is unlikely to go.

Electrons can jump from one orbital to another by either emitting or absorbing a photon. Absorbing involves an electron jumping to higher energy level while an electron must jump into a lower level orbital to emit a photon. Each orbital has a different three-dimensional pattern like those shown in Figure 18.2 and each has its own specific *energy level*.

The patterns of the orbitals in Figure 18.2 are not unlike the patterns of vibrations in macroscopic objects. Each orbital is defined by three numbers

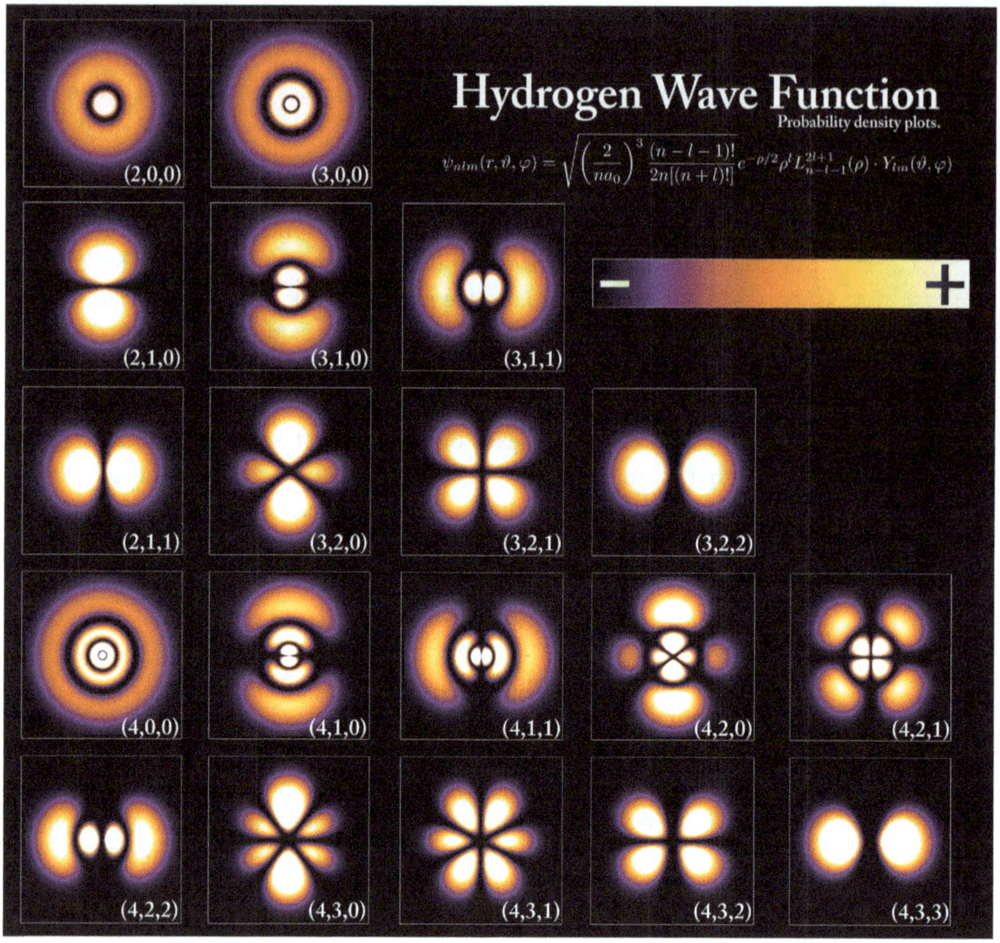


18.2

Orbitals are standing waves of electron probability that describe the patterns for where electrons are most likely to be found around the nucleus of an atom. They are fuzzy three-dimensional patterns, here shown in two-dimensional cross sections. In reality, they have additional structure in the third dimension. Like all standing waves, they can be defined by numbers, single integers for one-dimensional waves such as standing waves in a string, and three integers in three dimensions. For atoms these are called quantum numbers. The standing wave patterns are simplest for hydrogen but follow the same basic patterns for all atoms. Normally, hydrogen's single electron is in the 1, 0, 0 pattern unless it receives energy from photons – for heavier atoms, more and more of the complex patterns are filled because electrons can't share energy states as discussed below. The colour scale refers to the probability of an electron being found at that location. (*Source: PoorLeno / Public domain*)

called *quantum numbers* which are used to classify the patterns. In the next section, we will see how similar numbers classify patterns of vibration in macroscopic systems.

## STANDING WAVES AND THE QUANTUM WORLD

The orbitals of atoms can be understood more intuitively by treating them as a special case of the general phenomenon of standing waves which occur in

***Table 18.1*** **Physics is confusing because of many different names for similar phenomena. The table below gives names and descriptions for related phenomena in atoms (first five entries) and the standing waves illustrated in Figure 18.3.**

| *Similar concepts with different names* | *Description* |
|---|---|
| Wavefunction | A wave of quantum probability |
| Probability wave | A wave of quantum probability |
| Matter wave | Quantum probability wave for particles with non-zero rest mass |
| Orbital | A standing wave of quantum probability for an electron around an atom |
| Electron cloud | All the filled electron orbitals around an atom |
| Mode | Electrical or acoustic standing wave or wave pattern |
| Electromagnetic mode | Standing wave in a microwave oven etc. |
| Optical mode | Standing or travelling wave of light reflecting between mirrors and in optical fibres |
| Acoustic mode | Standing wave pattern for mechanically vibrating objects |
| Antenna pattern | Wave pattern created by a standing wave in a radio antenna |
| Ripple pattern | Antenna pattern for vibrating objects on water |
| Chladni figure | Acoustic mode pattern made by sand on a vibrating object such as a metal plate |

almost all fields of physics including optics, acoustics, and radio science. Each field has evolved separate language to describe standing wave phenomena. Acoustic standing waves are ideally suited to activity-based learning, and when they are connected to other manifestation of standing waves, from the scale of orbitals to the scale of the Sun, students can obtain a broad and unified understanding across many areas. Table 18.1 gives a glossary of names and descriptions of different standing wave phenomena, while Figure 18.3 gives some examples. It shows standing waves, from optical standing waves created when laser light is reflected back on itself, to sound waves in metal plates, to probability waves for electrons in hydrogen, and one created when electrons reflect off atoms in a metal surface. The bottom left image is created by a bee flapping its wings on a pond.

Figure 18.3 is a tiny sample of the general phenomenon of standing waves: patterns created when waves of the same wavelength pass through each other. Any process that causes waves to superimpose on themselves creates standing waves. Unlike travelling waves that propagate, standing waves *stand* in one location.

One-dimensional standing waves, like waves in a string or wire, are easy to conceptualise. (See Box 18.3 for classroom activities on standing waves.) Every acoustic musical instrument involves standing waves. One-dimensional standing waves are formed in violin strings and in the air columns of flutes,

18.3
A gallery of standing waves. The top three rows show three different standing waves of light and sound called optical and acoustic modes (left and right); the next two are electron probability distributions for hydrogen orbitals. The bottom row shows the antenna pattern of water ripples created by a bee buzzing on the surface of a pond, and a gold reproduction of electron standing waves within a ring of single atoms on a metal surface observed using a scanning tunnelling microscope. Author's images except the electron standing wave which is by J Voss-Andreae/public domain. (*Source: Author's images except the electron standing wave which is by J Voss-Andreae/public domain.*)

while drums and cymbals have two-dimensional standing wave patterns similar to those shown in Figure 18.3.

The classification of standing waves is naturally very similar to the classification of orbitals because orbitals are simply standing waves of electron waviness. Typical classifications in two dimensions are pairs of numbers that count the number of nodes in each dimension. In three dimensions, at least three numbers are needed. These are like the quantum numbers for the orbitals discussed in Figure 18.2.

Three-dimensional standing waves are harder to visualise. The largest scale in which three-dimensional standing waves are observed is in the Sun, as shown in Figure 18.4. The Sun is continually vibrating in many complex standing wave patterns caused by pressure fluctuations. Figure 18.4 shows just one of the thousands of different standing wave patterns, with the Sun cut away to see the pattern penetrating towards the centre.

Standing waves generate propagating waves, like the sounds we hear from musical instruments, the ripples emanating from the bee in Figure 18.3, and the

**BOX 18.3: CLASSROOM ACTIVITIES ON STANDING WAVES**

With a rope across the classroom, it is easy to shake the rope and observe standing waves. We find *anti-nodes* with large vibration and *nodes* with small vibration. We can count the numbers of nodes: it might be zero, one or two or more depending on your skill with the rope. These are the values of the *quantum number* for the rope. The higher the quantum number, the higher the frequency of vibration. A good classroom activity is to compete for the highest quantum number anyone can create and try to estimate the frequency for each quantum number.

Two-dimensional standing waves are easily observed in a thin metal plate sprinkled with sand and in water ripples in round baking dishes partly filled with water. The metal plate must be connected to a loudspeaker to create a set of patterns like those shown in Figure 18.3, due to sand collecting on nodal lines. Search *Chladni Figures* on YouTube for examples and instructions. For water in baking dishes, place the dish on a cushion, and experiment with tapping. With sky reflections, the patterns are beautiful in slow-motion videos.

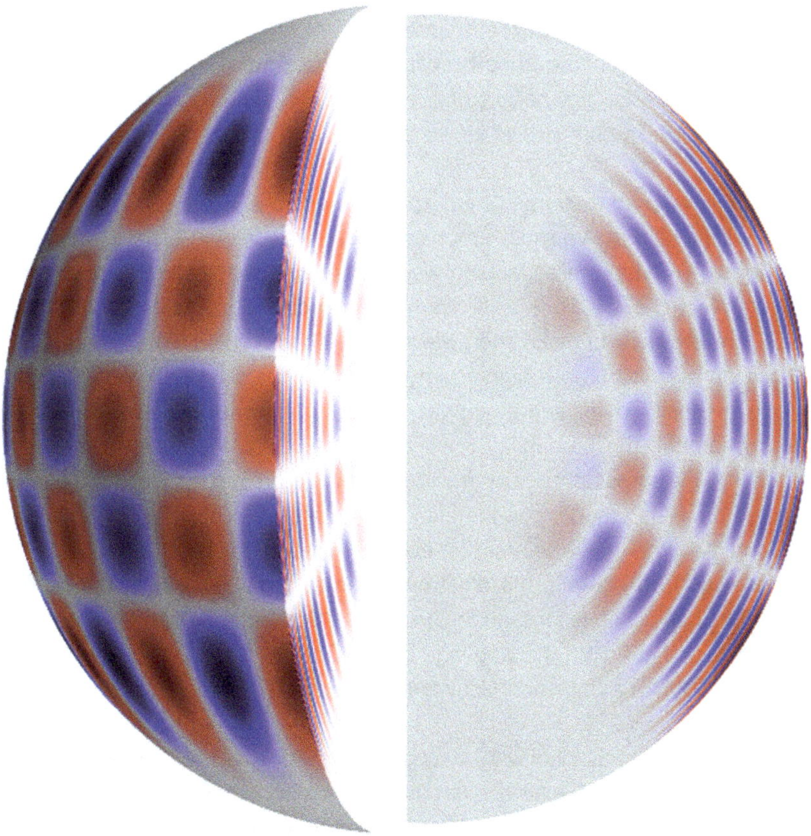

18.4
Simulated three-dimensional pattern of a vibrational mode of the Sun. Solar physicists observe the frequency of modes like this and use them to study the Sun's interior because the vibration pattern extends deep to the centre of the Sun. This particular mode is labelled 14, 20, 16. (*Source: Warrickball / CC BY-SA (https://creativecommons.org/licenses/by-sa/4.0)*)

signals that emanate from radio and TV antennas. The pattern of the standing wave is reflected in the pattern of wave emission, which is called the *antenna pattern* for radio antennas. This determines the strength of the emitted signal in different directions. The antenna pattern for the ripples created by the bee in Figure 18.3 has directions where emission is strong and others where it is weak. Equally, the shape of atomic and molecular orbitals determines the probability of emission of photons in particular directions.

## ENERGY LEVELS AND QUANTISATION

We have seen that standing waves occur on all scales, from the quantum world of atoms and photons to the macroscopic world of our everyday experience. And they generate propagating waves and lose energy like the dying note of a piano. In our experience, the loss of energy from a vibrating object seems continuous, just like water and air seem continuous. However, this is an illusion.

It is just because the quanta are so tiny, whether they are atoms or photons or phonons, that most things *seem* continuous. Look at a clear, sharp image on a large TV screen: then go up close and see that the image is made of individual dots called pixels. This is like the transition from the macroscopic world to the quantum world. Most of the time we can be perfectly happy about the smooth, crisp TV images, but if we want to understand how the TV works, we need to know about the *pixels* in which TV images are quantised. The physics of atoms allows us to understand how the world works!

We have seen that orbitals are three-dimensional standing waves of electron waviness, but we must emphasise that waviness is always *probability* waviness. So orbitals are standing waves of quantum probability that describe *where* electrons are likely to be found. But probability also relates to *when* electrons are likely to make energy-level transitions.

If an electron is in a high-energy orbital, it has a *probability* of jumping into a particular lower energy orbital. If another electron occupies that lower energy orbital, the probability is zero because two electrons can never be fitted into the same orbital, as we will discuss in more detail below. If the lower energy orbital is empty, then the electron has a certain non-zero probability of making the jump that is different for every transition. This is called the *transition probability*. Nothing tells the electron when to jump (except in the case of stimulated emission discussed in Chapter 4). Some transitions have very low transition probability. In the language of radioactivity, they have a long *half-life*. (See Box 18.4 for a classroom activity and Chapter 17 Gold: Einstein's Metal for discussion of half-life).

This idea of probability wave or quantum waviness is the one central idea in quantum physics that we have to get used to. This becomes even more apparent when we consider the role of quantum uncertainty.

**BOX 18.4: GLOWING IN THE DARK: AN ACTIVITY FOR OBSERVING THE PROBABILITY OF ELECTRON TRANSITIONS**

Glow in the dark stars such as those illustrated in Figure 18.5 are often found in children's bedrooms, and similar paper and paint are easily available. You can write on this stuff with low-cost blue or ultraviolet LED pen light and see it glowing green for many minutes.

Blue light consists of high-energy, low-wavelength photons around 400-nm wavelength. These photons have enough energy to make some electrons jump into a high energy orbital. The same electrons then make a transition in two steps, the last one being a low probability transition back to their original orbital. The energy difference of this step corresponds to a green wavelength of about 530 nm. The low probability means that it has quite a long half-life so that the paper glows in green photons for a long time.

Using eyes alone, ask students to estimate how brightness decays with time, using an arbitrary 1–10 scale of brightness.

Then use a mobile phone app to measure the brightness of a glowing spot over time and plot the decay (which should be exponential).

Compare the graphs and estimates for the half-life.

18.5
Glow in the dark stars illustrate the probability of electron transitions. (*Source: authors*)

## HEISENBERG PLUS PAULI: WHY ATOMS HAVE ORBITALS

The oscillation of electrons in all atoms is explained by the *Heisenberg uncertainty principle* which states that the product of the uncertainty in the position of an electron (or anything else!) times the uncertainty in its momentum must be greater than or equal to Planck's constant divided by $4\pi$. (Heisenberg's discovery is described in Chapter 4 and an activity for teaching the uncertainty principle is given in Chapter 7). Electrons can never be stationary because that would mean zero momentum uncertainty and hence infinite position uncertainty. You can only have momentum *uncertainty* if there *is* momentum. This means electrons must have speed, and that means their position must also be changing.

The uncertainty principle is closely connected to de Broglie's law, which states that everything has waviness. The electron's position is fuzzy because the wave is a probability wave: there are chances of it being in different places. But its speed is also fuzzy. The more I try to force an electron into a tiny box, the more it struggles to get out by jiggling faster, that is, it increases its momentum variation.

When you put an electron near a tiny atomic nucleus that provides a strong attractive force, the electron cannot be confined very close because the tiny position uncertainty would create huge momentum uncertainty. It has to find a certain balance of position uncertainty and momentum uncertainty: this is called the *ground state* and is the lowest energy that an electron can have in an atom. It is the simplest orbital, shown at the top of Figure 18.2. All the others have more nodes and higher energy similar to the more complex patterns in Figure 18.3.

The second crucial rule for the behaviour of electrons is called the *Pauli exclusion principle.* This is the rule that says electrons can never share the same orbital. It forces electrons to stack themselves up in a so-called energy staircase which combined together is the cloud of electrons around atoms.

The Pauli exclusion principle is connected to the mysterious property of quantum spin, which is measured in units of Planck's constant divided by $2\pi$. Every electron is like a spinning top, except that for an electron its spin can have only two values, plus 1/2 and minus 1/2. One value means clockwise rotation, the other anticlockwise. Electrons are unable to *not spin.* Imagine electrons as tiny identical spinning things with a fixed spin speed and a fixed electric charge. The two values of spin mean that each orbital pattern shown in Figure 18.2 represents two orbitals, one for each spin value. If you know cricket or table tennis, you will know how spin is important. For electrons, it is unavoidable.

Because standing waves can only be made with whole wavelengths around a ring only certain wavelengths are allowed, so the energy levels are quantised. The Pauli exclusion principle disallows multiple electrons in the same orbital. Consequently, all the electrons are forced to be in different orbitals (or standing wave patterns) which generally correspond to different energy levels. This means that the more electrons you have in an atom, the higher the energy staircase. As a result, the electrons in an atom are always stacked up like a class of children sitting on a long narrow staircase, with one child on each step. The bottom step is

**BOX 18.5: STAIRCASE GAME FOR THE PAULI EXCLUSION PRINCIPLE**

Most students are selected to be electrons for a staircase of atomic orbitals of an atom. A few students are chosen to be lasers, able to emit (throw) photons in the form of red or blue balls.

Decide that blue photons have energy equal to four steps, while red balls only have one step of energy. Ideally, the number of "electrons" should match the number of steps of the staircase, and there should be a level area above which represents free space. A red photon can be absorbed (caught) thereby raising an electron one level, but only if the level above is empty. An electron without a vacant step above *must not* catch a red photon: they must let it bounce off. Blue lasers raise electrons four levels but only if the fourth level above is empty. If electrons are *within* four steps of the top, a blue laser ionises and contributes its remaining energy as kinetic energy: the ionised electron runs around on the landing.

The aim of the game is for the red and blue lasers to *fully ionise* the atom...that is to raise all the electrons to the upper landing where the electrons are free. An electron must not catch photons that cannot be absorbed.

A ten-level atom requires 55 units of energy, requiring 15 red balls and ten blue balls if it is to be fully ionised without excess kinetic energy.

the ground state. The bigger the class, the higher the last filled step. A classroom game based on this idea is described in Box 18.5.

The Pauli exclusion principle explains the structure of atoms and, astonishingly, also explains white dwarf stars and neutron stars. However, there is a whole domain in which it does not apply as we will see below.

## FERMIONS AND BOSONS

It is very important to know that the Pauli Exclusion Principle *only* applies to electrons and other spin-half particles like neutrons and protons that collectively are called *fermions*. It arises because of the so-called anti-symmetry of the probability waves of particles with half-integer spin.

The behaviour of the other class of particles, *bosons*, is quite different. In musical instruments, the standing waves are waves of phonons, the quanta of sound vibration. In radio antennas or lasers, the standing waves are photons. Both phonons and photons have integer spin: the photon has a spin of one, and the phonon has zero spin (still an integer!). This difference between integer spin and half-integer spin is dramatic. The rule for fermions is *never share*. The rule for bosons is *share as much as you like*.

While electrons can have only one electron in each orbital, you can have as many phonons or photons as you like sharing the same standing wave. For the same standing wave pattern, you can increase the total energy by simply

**BOX 18.6: THE MATHEMATICS OF FERMIONS AND BOSON**

Two rather simple formulae that differ only by a plus or minus one describe the dramatic difference between fermions that have half-integer spin and bosons that have integer spin. Here, we will not show these formulae, which apply to a "gas" of particles. At zero temperature, the boson formula resolves to 1/(1 – 1), while for fermions it resolves to 1/(1+1). The minus 1 causes the first to be equal to infinity because one divided by zero equals infinity. This says that an infinite number of bosons can reside in any state. The second formula resolves to ½. With times 2 not shown, this says that just 1 particle can reside in any state. You can easily find the formulae called the Bose-Einstein and the Fermi-Dirac distribution functions online.

adding more photons or phonons of the same frequency. For example, you can trap photons between not quite perfect mirrors in an optical standing wave as illustrated in Figure 18.3 (and later in Figure 18.7). If you start with *N* identical photons, you can observe them leaking out in a random stream, one photon at a time. The energy level of the standing wave is *N* times the photon's energy *hf*, which you can determine by counting the photons as they leak out. Similarly, if you vibrate a metal plate more loudly, you have put in more phonons, its energy level is higher and its vibration is bigger. Box 18.6 explains the mathematics.

Despite the complete difference between how electrons and photons distribute themselves, the concept of energy levels (like the energy staircase in atoms) is quite a universal concept in physics. For photons and phonons, energy is directly connected to frequency by Planck's formula, $E = hf$. For macroscopic standing waves, every pattern corresponds to a particular frequency, and for orbitals, an energy level is associated with each different standing wave pattern. For example, a guitar string can be plucked to excite various harmonics corresponding to more nodes and antinodes, with each having a higher frequency.

For the standing waves in the top three panels of Figure 18.3, each pattern corresponds to a different frequency. Because frequency defines energy, each standing wave pattern represents an energy level. Hence, an energy staircase still exists for bosons (photons and phonons) just as it does for fermions (electrons). The difference for bosons is that I can put as many as I like on any step of the staircase. At the quantum level, the loudness or the brightness (of a particular standing wave pattern) is described by the number of phonons or photons, each with energy $E = hf$. Our eyes can barely detect single photons but the softest sound we can hear has countless billions of phonons.

## ALL BOUND SYSTEMS HAVE ENERGY LEVELS

Energy levels are related to the allowed frequencies, wavelengths, or colours of photons from different atoms, but they too are part of a much more general phenomenon: they arise because a system is *bound*, in other words, it has boundaries. For example, when electrons are bound to atoms, they have certain energy levels related to the requirement of probability waves to be smooth and unbroken around a closed path. This leads to the specific wavelengths of light emission from each type of atom associated with electrons jumping from one orbital to another. Likewise, photons can only be absorbed when an available energy transition matches the energy of the incoming photon.

However, as soon as electrons are unbound, like electrons floating around in space, they can move at any speed and have any amount of kinetic energy because their matter waves do not trace closed paths. A photon that interacts with such an electron can give up any fraction of its energy, just as billiard balls do when they collide.

Metals and semiconductors are very important examples of bound systems in which the electron probability waves overlap so that the electrons are shared between atoms. Metals such as copper, iron or aluminium share their outer electrons with adjoining atoms. The shared electrons are able to travel through a landscape of regularly spaced atoms. The atoms themselves provide repulsion because the electrons in the atoms repel approaching electrons. You could imagine an electron in a metal as being a bit like a billiard ball rolling on a billiard table covered in lots of little uniformly spaced hills. A ball rolling on this regular bumpy landscape of mounds and valleys created by billions of atoms, as illustrated in Figure 18.6, has recurring patterns of motion including various sized loops and circuits around different numbers of mounds.

Like in the case of orbitals, when an electron moves in a closed loop, its waviness has to be continuous – you cannot have discontinuous breaks in any

18.6
This image of the attractive forces from atoms was measured by an atomic force microscope. It shows a regular hilly landscape of attraction and repulsion due to the pattern of atoms. Electrons move around the hills in various loops that are like extended orbitals. (*Source: Image courtesy Nanoscience Instruments.*)

wave that is able to join back onto itself. The situation is similar to electrons orbiting the nucleus of atoms, but now they are taking much bigger circuits throughout the metal. Because the waves must be continuous, you can only have integer numbers of wavelengths in any patterns. This creates quantised energy levels corresponding, by de Broglie's formula, to certain momentum values. These correspond to energy levels in metals that are referred to as the *conduction band.*

In semiconductors like silicon, which are so important to modern technology, there are few shared electrons unless impurities are added that contribute shared electrons from their outermost orbitals. This method, called doping, provides the way to engineer the electron's bumpy landscape to create special properties that are essential in making the components of mobile phones, computer chips, and solar panels.

In solar panels, the energy bands are optimised so that photons from the Sun scatter electrons to higher energy, where they provide the voltage and current of electrons for solar energy systems.

Also, in these engineered bumpy landscapes electrons can collide with phonons. This electron-phonon scattering enables the creation of electronic thermometers and solid-state coolers. In optical devices, standing waves of photons can be made to interact with standing waves of phonons, as illustrated in Figure 18.7. These provide new ways of using quantum physics.

We have seen that energy levels arise in all bound systems in which particles (mainly electrons and photons) are brought back to the same place by orbiting or reflecting or merely navigating a landscape of hills and valleys created by atoms in a metal. For electrons, the binding can be a central attraction of electrons by the nucleus, or in the case of metals and semiconductors, boundaries created by the repetition of multiple repulsive "bumps". In semiconductors, these principles are used to create the technology that powers the modern world. It all comes directly from Einsteinian physics.

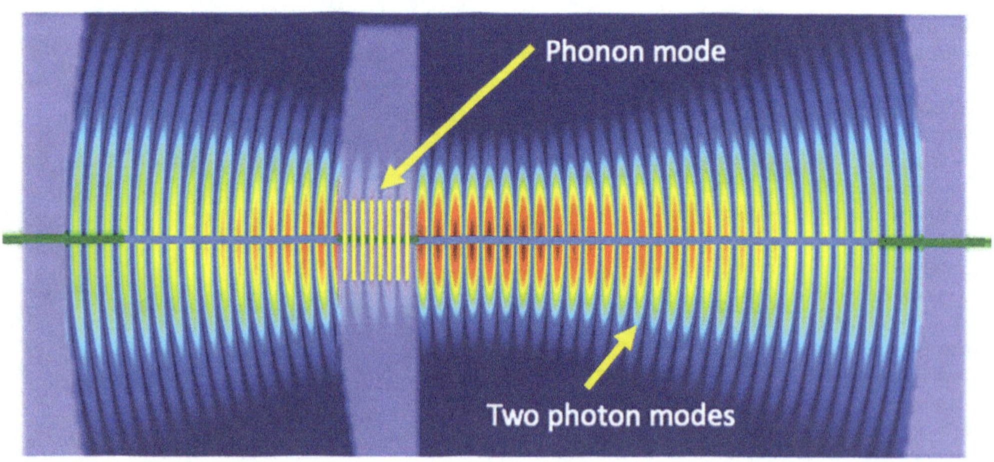


18.7
This image shows an apparatus designed to make photons and phonons interact and exchange energy. Curved mirrors at each end reflect photons back and forth to make an optical standing wave while a quartz lens reflects phonons back and forth in an acoustic standing wave. Photons can either absorb a phonon or create a new phonon, thereby gaining or losing some energy, in both cases changing their frequency. New sensors and devices are being developed based on manipulation of quantum phenomena like these. (*Source: authors*)

## CONCLUSION

The world of atoms and solids is the world around us, but the tiny size of atoms and quantum rules that they obey is normally considered beyond the reach of school science. Here I have tried to create an intuitive picture of the world of atoms suitable for middle and secondary school students, by emphasising connections between electron orbitals and standing waves in the macroscopic world.

First, we examined the physics of seeing tiny things and then discovered how the same physics creates orbitals in atoms. Then we explored the similarity of the patterns of orbitals and macroscopic standing waves in different situations. We saw that in both cases the pattern of vibration is linked to a unique frequency or energy level. In the quantum case, we emphasised that the patterns describe the probability of detecting a particle (e.g., a photon or electron) at a particular location.

We also examined the differences between standing waves made by electrons, photons, and phonons. The differences come about because photons and phonons have integer spin and obey entirely different rules from electrons that have half-integer spin. Spin divides the universe into two classes: the fermions that refuse to share energy levels, each needing one step on an energy staircase. The bosons, by contrast, share freely: they are happy to crowd onto the bottom step.

We considered what happens when electrons can move from atom to atom inside metals and saw that the same physics of electron probability waves leads to conduction bands in semiconductors that can be manipulated.

Most modern electronic technology depends on our ability to understand matter at the atomic scale and then manipulate its properties to create the devices in which photons, phonons, and electrons interact. From this physics, we make solar cells and cameras where photons deliver electrons, LEDs where electrons deliver photons, thermometers where phonons affect electrons, and much more. It all depends on the physics of atoms.

## FURTHER READING

Chapter 15: Introducing quantum physics with toy photons
Chapter 16: Teaching quantum physics to middle and high-school students using phasor wheels
Chapter 17: Gold: Einstein's metal

## REFERENCES

Einstein, A. (1907). Plancksche theorie der strahlung und die theorie der spezifischen wärme [Planck's theory of radiation and the theory of specific Heat]. *Annalen Der Physik*, *22*, 180–190.

Planck, M. (1901). On the law of distribution of energy in the normal spectrum. *Annalen Der Physik*, *4*, 533. Retrieved from papers2://publication/uuid/2BC0A8A4-A187-4E88-A6C2-AE23E82CFB80.

Street, R. A. (2019). Physics of quantum mechanics – why are there phonons? *ArXiv E-Prints*, 1–5. Retrieved from https://arxiv.org/pdf/1910.10020.pdf.

# Section 3

# Introducing Einsteinian physics around the world

# 19 Designing learning resources and investigating student motivation and learning in general relativity and quantum physics in Norway

*Ellen Karoline Henriksen*

**WHAT?**

This chapter provides an overview of ReleQuant, a physics education project in Norway that has developed digital, research-based learning resources in general relativity and quantum physics for upper secondary physics students.

**FOR WHOM?**

This chapter is intended for readers who are interested in the development of digital learning resources and curriculum development in Einsteinian physics, and also, for those who wish to learn more about secondary school students' learning processes and motivation in general relativity and quantum physics.

## INTRODUCTION

Three students discuss the thought experiment "Schrödinger's cat":

*Student 1:* I share Einstein's view that the cat is not both dead and alive at the same time before we open the box, it is, or the cat is one or the other, it is just that we don't know which state it is in.

*Student 2:* But isn't that what makes, the fact that we don't know, what makes it in a way both dead and alive? (...) When we are standing outside the box, then it is dead and (...) I don't know, alive, that doesn't fit for me.

*Student 3:* It has to do with the fact that that radioactive nucleus was both in state, two different states; emitted radiation and not emitted radiation, that thing that would...that poison thing. And that once you measured, it was then that it sort of fell down to one of those states. That was the point.

The above exchange took place between three Norwegian upper secondary physics students working with learning resources in quantum physics developed through project ReleQuant. We take this exchange as a sign that students are able to engage with modern physics ideas such as superposition of quantum states and the role of measurement, and that students can develop and deepen their understanding through such discussions. In this chapter, I describe why and how the ReleQuant learning resources were developed and what we have found out about students' motivation, understanding, and learning by studying their work with the resources.

## CHALLENGE: TEACHING UNFAMILIAR TOPICS WITH UNFAMILIAR LEARNING GOALS

In 2006, general relativity became part of the Norwegian national curriculum for upper secondary physics. Also, the learning goals in quantum physics were revised. Both these topics of Einsteinian physics were now to be treated qualitatively and with a view to philosophical and epistemological implications. Students taking the optional physics subject in their final year of school (age 18–19) were expected to (NDET, 2006):

- (...) give a qualitative description of the general theory of relativity;
- give an account of Einstein's explanation of the photoelectric effect and give a qualitative account of how results from experiments with the photoelectric effect, Compton scattering, and the wave nature of particles represent a break with classical physics;
- give an account of Heisenberg's uncertainty relations, describe the phenomena "entangled photons" and give an account of their epistemological consequences.[1]

These learning goals represented a considerable challenge for physics teachers for several reasons. First, many of the teachers had not encountered general relativity or quantum entanglement as part of their physics training. Second, the emphasis on qualitative descriptions and philosophical aspects was unfamiliar and very different from the typical physics instructional approach (Bøe et al., 2019) where the teacher communicates the "correct physics content knowledge" to students, followed by individual problem-solving with correct answers listed in the back of the book. Given these new curricular aims, teachers had to learn both new physics content and new ways to arrange for students' learning. Learning goals now include the ability to qualitatively discuss quantum physics concepts and the implications of quantum physics for what we can know about physical entities and processes on the subatomic scale.

With these challenges as a starting point, we established project ReleQuant (www.mn.uio.no/fysikk/english/research/projects/relequant) with the dual aim of 1) developing research-based learning resources and 2) conduct-

1 The official English translation says "cognitive consequences"; however, in our view "epistemological" is a better translation, since this term here refers to the nature, origin, and limits of our knowledge of the natural world.

ing science education research to map students' understanding and learning (Bungum et al., 2015; Henriksen et al., 2014). A third goal of the project (which is not detailed here) was to provide an arena for professional development for physics teacher students as well as practising physics teachers that were partners in the project.

### *Developing the learning resources "quantum physics" and "general relativity"*

As a methodological frame for developing the learning resources, we chose design-based research (Anderson and Shattuck, 2012), also known as educational design research (Reeves and McKenney, 2013). Design-based research allows the researchers to systematically develop learning resources based on research through several iterations of classroom trials and in close collaboration with teachers. In the case of ReleQuant, we developed two learning resources for publication on www.viten.no/eng/, the national open access Norwegian platform for web-based learning resources in science for grades 8–13. Combining curricular aims, previous research on students' understanding and current perspectives on students' learning and motivation, the project team decided that the resources should (Henriksen et al., 2014) draw on sociocultural theories of learning; address documented conceptual challenges for students; invite students to express their understanding orally and in writing; allow students to explore digital animations and simulations of phenomena; illustrate how scientific knowledge develops through discussion between scientists; illustrate how science is applied in a range of real-world contexts; inspire reflections on quantum physical concepts and the status of our knowledge about the subatomic world; support teachers' varied use of communicative approaches; and support teachers to assess their students.

The resource "*Quantum Physics*" (Bungum et al., 2015) was launched in October 2016 and "*General Relativity*" (Kersting et al., 2018) in January 2018 (Figure 19.1), both after three to four rounds of classroom data collection in upper secondary schools in the greater Oslo area in south-eastern Norway. Both resources contain writing tasks and discussion tasks for students to work individually or in groups; text and images illustrating central principles; video clips showing scientists explaining concepts or discussing philosophical aspects; applications and real-world examples like GPS positioning and X-ray imaging, and animations and simulations for students to work with (e.g., for exploring the photoelectric effect), see Figures 19.2 and 19.3.

### *Physics education research in the ReleQuant Project: What did we find out?*

The research group used a range of different data sources in the research and development process in ReleQuant. The platform *www.viten.no* allows collection of students' written responses to questions such as "What do you

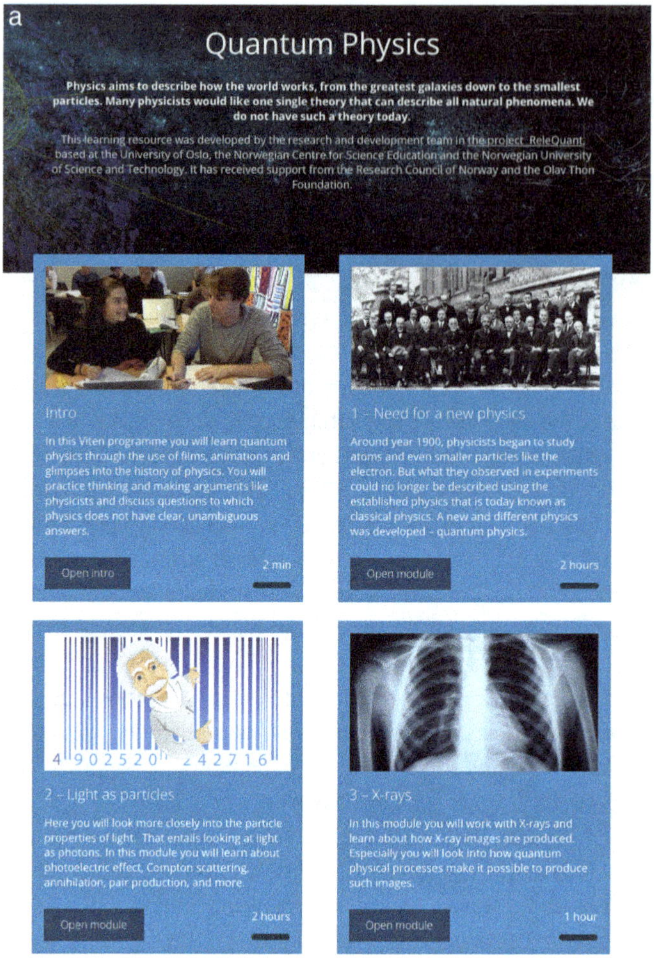


19.1
Screenshots of the front pages of the learning resources: (a) "Quantum Physics" and (b) "General Relativity". Each resource has three to five modules designed for students to meet learning goals in the national physics curriculum. Both resources are available at www.viten.no/eng. (*Source: Copyright by Project ReleQuant*)

know about gravity?". Recordings of small-group discussions were collected by inviting students to use the recording function on their own smartphones and send the sound files to the teacher after the lesson. We also conducted group interviews with groups of four to six students to explore their understanding, self-experienced learning and motivation, and their challenges while working with the resources. Classroom and video observations and informal interviews, as well as focus groups with teachers, were also part of the ReleQuant data material.

Three categories of results are presented here: 1) results regarding the development process; 2) results regarding students' understanding of quantum physics; and 3) results regarding students' understanding of general relativity. Notably, two of the short research papers that the ReleQuant group has published are based on research conducted by master students in their final year of a five-year physics teacher education programme.

19.1
*(continued)*

19.2
An example of the many discussion tasks inviting students to explore central ideas in Einsteinian physics. From www.viten.no/eng – General Relativity. (*Source: Copyright by Project ReleQuant*)

## Discuss in pairs

Are you moving in spacetime right now? Do you follow a geodesic curve through spacetime?

According to Newton, why do you experience a pull towards the ground? And according to Einstein?

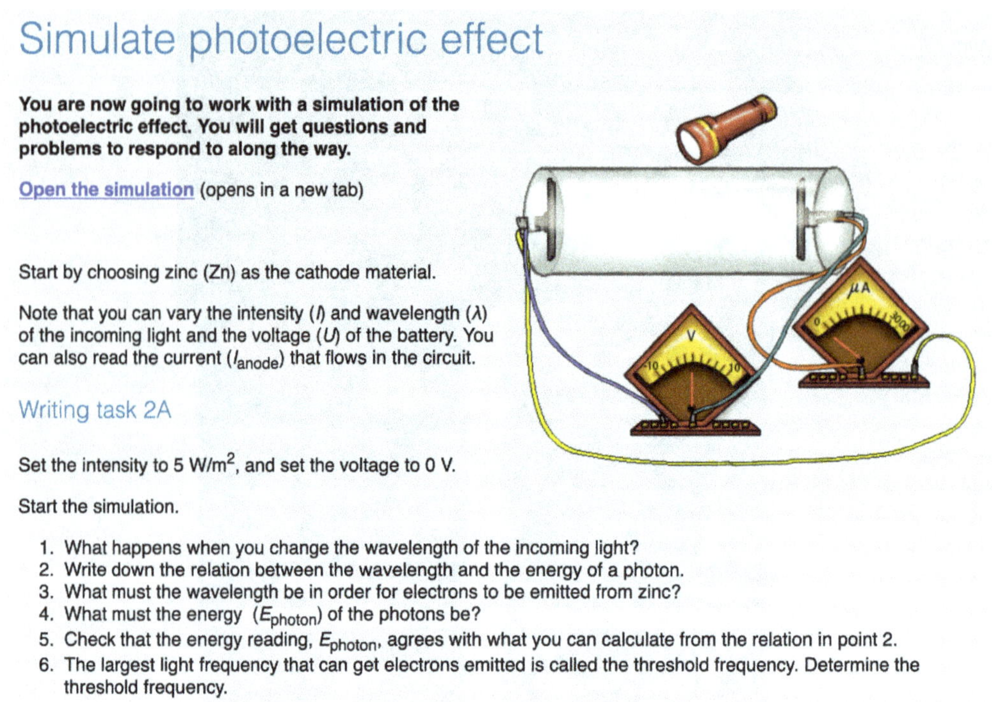

# Simulate photoelectric effect

**You are now going to work with a simulation of the photoelectric effect. You will get questions and problems to respond to along the way.**

Open the simulation (opens in a new tab)

Start by choosing zinc (Zn) as the cathode material.

Note that you can vary the intensity ($I$) and wavelength ($\lambda$) of the incoming light and the voltage ($U$) of the battery. You can also read the current ($I_{anode}$) that flows in the circuit.

## Writing task 2A

Set the intensity to 5 W/m$^2$, and set the voltage to 0 V.

Start the simulation.

1. What happens when you change the wavelength of the incoming light?
2. Write down the relation between the wavelength and the energy of a photon.
3. What must the wavelength be in order for electrons to be emitted from zinc?
4. What must the energy ($E_{photon}$) of the photons be?
5. Check that the energy reading, $E_{photon}$, agrees with what you can calculate from the relation in point 2.
6. The largest light frequency that can get electrons emitted is called the threshold frequency. Determine the threshold frequency.

19.3
A simulation of the photoelectric effect lets students work with the idea of "light as particles" in the Quantum Physics learning resource at www.viten.no/eng – Quantum Physics. Simulation adapted from phet.colorado.edu. (*Source: Copyright by Project ReleQuant*)

### *Developing "Einsteinian Physics" learning resources in a design-based research process*

Henriksen et al. (2014) outlined the aims and the thinking guiding the development in ReleQuant, whereas Bungum et al. (2015) described the first cycle of development of the quantum physics resource. Part of the development work that needed to be done was to "translate" curricular aims into more specific learning aims, and to develop learning resources and tasks and activities. One example concerns the curricular aim to "give a qualitative account of how results from experiments with the photoelectric effect, Compton scattering, and the wave nature of particles represent a break with classical physics". How, specifically, can we say that quantum physics breaks with central assumptions inherent in classical physics? This was a concern found both among our collaborating teachers and in previous research (Bungum et al., 2015). In the quantum physics resource, we decided to focus on three aspects of the break with classical physics: quantisation versus continuity; classical determinism contra statistical description, and local reality versus nonlocality (the latter is related to the phenomenon of entangled photons, which is mentioned in the curriculum).

In the case of quantum physics, this process of translating curricular aims was seamlessly built into the design-based research process. In the case of general relativity, we combined the Model of Educational Reconstruction (Duit et al., 2012) with the design-based research framework to formulate suitable learning goals in line with the rather vague curricular aim (Kersting, 2019b; Kersting et al., 2018).

### *Students' understanding of quantum physics*

Learning quantum physics is challenging, but also fascinating for students since it breaks with their intuition and everyday experiences, and presents them with intriguing phenomena like quantum uncertainty, entanglement, and wave-particle duality. The ReleQuant group has explored several aspects of learning quantum physics. Henriksen et al. (2018) addressed students' interpretation of the wave-particle duality for light and found that some students had an "uncritical duality" notion, accepting without question the two ontologically different descriptions of light, while others tried to reconcile the quantum description with classical notions. From students' discussions about the nature of light, it appeared that many seemed to implicitly take a realist view on the description of physical phenomena, contrary to the Copenhagen interpretation, which is prevalent in textbooks. Looking more closely at learning processes in the classroom, Bungum et al. (2018) analysed characteristics of student small-group discussions in quantum physics and their potential for promoting learning. They concluded that small-group discussions could support students' understanding in quantum physics through prompting students to 1) articulate conceptual difficulties; 2) deepen their understanding through the exchange of views; and 3) formulate new questions. The paper also pointed out the very central role of the teacher in supporting students' learning through scaffolding their discussions.

Two short papers report on students' understanding of specific quantum physics concepts. Myhrehagen and Bungum (2016) described how students interpreted the Schrödinger's cat thought experiment and pointed out how this historical example may provide a good entry point for exploring some central quantum physical concepts in depth. The role of *observation* in quantum physics is different from that in classical and daily contexts: observation cannot be regarded as passive "looking", but rather as an *interaction* that determines the state of the quantum system that is observed. Huseby and Bungum (2019) investigated students' understanding of quantum observation and found that many students face challenges in understanding quantum physics since they hold on to a view of observation as passive looking. Therefore, it was recommended that observation as interaction needs to be discussed explicitly in upper secondary physics. Since the learning goals and learning activities with the ReleQuant resources call for students' roles that differ from those in traditional physics classrooms, we have also investigated how such an innovative approach was received by the students. Bøe et al. (2019) pointed out that students were motivated by historical and philosophical aspects of physics, but they did not see these aspects as learning goals in their own right since students were enculturated into a traditional physics learning culture where doing calculations and getting correct answers was most highly valued. The lack of correct answers frustrated students because it took away their usual source of mastery experiences and self-efficacy. As the authors remarked, learning activities, learning goals, and assessment forms need to be better aligned and clearly articulated when innovative learning approaches are introduced.

*Students' understanding of general relativity*

Conceptions of general relativity among pre-university students are an understudied area in science education research, and the ReleQuant project has contributed to exploring this field. Kersting et al. (2018) described the development of the general relativity learning resource and also presented results concerning students' understanding of concepts like the principle of equivalence and relativistic phenomena such as gravitational frequency shift and time dilation. The paper presented empirically supported design principles for designing learning resources for general relativity in upper secondary school. A specific challenge for students working with general relativity is interpreting four-dimensional spacetime. Kersting and Steier (2018) investigated how the prevalent rubber sheet analogy of spacetime was interpreted by students. The authors also offered advice on teaching general relativity based on the identified strengths and weaknesses of this analogy. Kersting (2019a) further explored the possibilities and pitfalls inherent in different models and metaphors for curved spacetime. She presented a model constructed to help upper secondary students visualise how warped time gives rise to the phenomenon of gravity, and suggested four specific instructional strategies on how to use this model productively in the physics classroom.

The ReleQuant group has also investigated the role of students' imaginative processes and their use of gestures as well as artefacts when trying to make meaning of general relativity (Steier and Kersting, 2019, Steier et al., 2019). These authors suggested to actively use "imaginative activities" in instruction, inviting students to use thought experiments, construct new metaphors, and use gesture and artefacts to illustrate abstract ideas in science.

## CONCLUSION

At the time of writing this chapter, the ReleQuant project is in its finishing phase in the sense that the funding period has ended, students have submitted their master and doctoral theses, most of the reports from the project have been published or submitted, and the two learning resources are online and working. However, the project continues as part of the Einsteinian Physics Education Research (EPER) collaboration. Through this collaboration, the Norwegian group will continue to explore ways of letting students of different ages meet central ideas in general relativity as well as quantum physics.

An interesting development in our local Norwegian context is that the national curriculum is again in the process of change, and upper secondary physics students have new learning goals from 2021. The new curriculum retains general relativity and quantum physics, but with somewhat weaker emphasis on epistemological and philosophical implications. We believe that the knowledge base that has been developed through the ReleQuant project and its international collaboration partners in EPER will be a central resource in the process of developing teaching approaches and implementing the Einsteinian physics components in the new curriculum. Internationally, there is a trend towards integrating Einsteinian physics perspectives into the curricula of an increasing number of countries (Stadermann et al., 2019). ReleQuant and the EPER collaboration will be central discussion partners and offer teaching approaches and learning resources attuned to different age groups, school systems, and curricular frameworks, thus letting young people around the globe share our best understanding of physical reality.

## FURTHER READING

Chapter 14: Using the language of gravity to teach about space, time, and matter in general relativity
Chapter 16: Teaching quantum physics to middle and high-school students using phasor wheels
Chapter 23: Towards a study module on general relativity in the Netherlands

## ACKNOWLEDGEMENT

This research has been supported by the Research Council of Norway (project number 246723), the Olav Thon Foundation, and the Norwegian Centre for Science Education. We are grateful to participating teachers, students, and colleagues in the ReleQuant research group for their contributions and collaboration.

## REFERENCES

Anderson, T. & Shattuck, J. (2012). "Design-based research: A decade of progress in education research?" *Educational Researcher*, 41, 16–25.

Bungum, B., Bøe, M. V. & Henriksen, E. K. (2018). "Quantum talk: How small-group discussions may enhance students' understanding in quantum physics." *Science Education*, 102, 856–877.

Bungum, B., Henriksen, E. K., Angell, C., Tellefsen, C. W. & Bøe, M. V. (2015). "ReleQuant–Improving teaching and learning in quantum physics through educational design research." *Nordic Studies in Science Education*, 11, 153–168.

Bøe, M. V., Henriksen, E. K. & Angell, C. (2019). "Actual vs. implied physics students: How students from traditional physics classrooms related to an innovative approach to quantum physics." *Science Education*, 102, 649–667.

Duit, R., Gropengiesser, H., Kattmann, U., Komorek, M. & Parchmann, I. (2012). "The model of educational reconstruction – A framework for improving teaching and learning science." *Science Education Research and Practice in Europe*.5, 13–37.

Henriksen, E. K., Angell, C., Vistnes, A. I. & Bungum, B. (2018). "What is light? Students' reflections on the wave-particle duality of light and the nature of physics." *Science & Education*, 27, 81–111.

Henriksen, E. K., Bungum, B., Angell, C., Tellefsen, C. W., Frågåt, T. & Bøe, M. V. (2014). "Relativity, quantum physics and philosophy in the upper secondary curriculum: Challenges, opportunities and proposed approaches." *Physics Education*, 49, 678–684.

Huseby, A. & Bungum, B. (2019). "Observation in quantum physics: Challenges for upper secondary physics students in discussing electrons as waves." *Physics Education*, 54, 065002.

Kersting, M. (2019a). "Free fall in curved spacetime – How to visualise gravity in general relativity." *Physics Education*, 54, 035008.

Kersting, M. (2019b). *General Relativity in Secondary School: Research-Based Development of Learning Resources and Analyses of Students' Conceptual Understanding Using the Model of Educational Reconstruction.* PhD thesis, University of Oslo.

Kersting, M., Henriksen, E. K., Bøe, M. V. & Angell, C. (2018). "General relativity in upper secondary school: Design and evaluation of an online learning environment using the model of educational reconstruction." *Physical Review Physics Education Research*, 14, 010130.

Kersting, M. & Steier, R. (2018). "Understanding curved spacetime." *Science & Education*, 27, 593–623.

Myhrehagen, V. H. & Bungum, B. (2016). "From the cat's point of view': Upper secondary physics students' reflections on Schrödinger's thought experiment." *Physics Education*, 51, 055009.

NDET. (2006). *Physics - programme subject in programmes for specialisation in general studies* [Online]. Available: https://www.udir.no/kl06/FYS1-01?lplang=http://data.udir.no/kl06/eng [Accessed 01.10.2019].

Reeves, T. & McKenney, S. (2013). "7 things you should know about educational design research."' *Educause* [Online]. Available: http://www.educause.edu/library/resources/7-things-you-should-know-about-educational-design-research.

Stadermann, H., van den Berg, E. & Goedhart, M. (2019). "'Analysis of secondary school quantum physics curricula of 15 different countries: Different perspectives on a challenging topic." *Physical Review Physics Education Research*, 15, 010130.

Steier, R. & Kersting, M. (2019). "Metaimagining and embodied conceptions of spacetime."' *Cognition and Instruction*, 37, 145–168.

Steier, R., Kersting, M. & Silseth, K. (2019). "Imagining with improvised representations in CSCL environments." *International Journal of Computer-Supported Collaborative Learning*, 14, 109–136.

# 20 Towards a comprehensive general relativity course for secondary school

*Ute Kraus and Corvin Zahn*

**WHAT?**

This chapter describes a secondary school course in general relativity that makes extensive use of hands-on activities while being restricted to elementary mathematics. The course is made up of short, independent modules that focus on phenomena and on the theory's geometric concepts.

**FOR WHOM?**

This chapter is intended for readers who are interested in curriculum design and development in Einsteinian physics, and also, for those who wish to teach general relativity at the middle and upper secondary school level.

## INTRODUCTION

Encounter general relativity: Navigate through a wormhole or build a do-it-yourself black hole. Visualisations and models can make relativity accessible and at the same time, provide a sound physical description. These are the tools used extensively in a general relativity course under construction that will be made available as an open educational resource.

In Germany, like in many other countries, general relativity is in the process of making its way into secondary schools. To facilitate the teaching of relativity, the physics education group at Hildesheim University develops new media and teaching approaches and makes them available as open educational resources, both in German (tempolimit-lichtgeschwindigkeit.de) and English (spacetime-travel.org).

A part of this work is the ongoing development of a course in general relativity. The course provides teaching units on two levels, targeting students from grade 9 (age 15 years) and students from grade 11 (age 17 years), respectively. This course is an open-ended project with the long-term strategy to evolve into a comprehensive course covering a broad range of relativistic topics. A flexible course structure allows teachers to select units and build their own programmes.

The focus of this course is geometry. Its key objective is to explicitly describe the geometries of relevant spacetimes (e.g., black holes, compact stars, and the expanding universe) and show how geometry gives rise to the gravitational phenomena observed in these systems. Two distinctive features characterise the course. One is the use of visualisations from the first-person point of view ("What should we see when falling into a black hole?"). The other is the use of a specific class of geometric models (so-called sector models) that represent curved spaces or spacetimes true to scale. With these tools, we describe phenomena and geometries in a sound and rigorous way without using advanced mathematics. This chapter introduces the tools, describes the strategy of the course, outlines its scope, and provides summaries of a few teaching units.

## CONTENT AND STRUCTURE

To ascertain the topics that teachers are expected to teach in general relativity at the secondary school level, we reviewed current school curricula (see Kraus and Zahn, 2019b for a tabular summary of data). The conclusions from this review (that includes the curricula of the 16 German states and selected international curricula) can be summarised as follows:

- No broad consensus about what should be taught about general relativity at school has emerged by the present time. Curricular specifications range from very general (e.g. fundamental aspects of general relativity) to quite specific propositions (e.g., the equivalence principle and the three classical tests in the solar system). Some curricula focus on the fundamentals, others on applications (e.g., black holes, cosmology). Among those curricula stating specific topics, no single topic was found that is universally included.
- The dedicated teaching time, where specified, is typically short. In the German curricula that allot time for general relativity (four out of 16 states), the teaching time ranges from four to eight lessons (à 45 minutes), typically in grade 11 (age 17 years).
- In many German states, topics are at the discretion of the teachers or the schools for a certain amount of teaching time or certain classes. Examples are classes emphasising scientific methods (e.g., "Seminarfach"). This can be a possibility to introduce relativity at greater length (such as half a school year with two lessons à 45 minutes per week).

Based on these findings, we choose a course structure that can cater to different needs regarding content and teaching time. The course is being developed in two parts, one short and focussed on the core ideas, the other extended and structured for flexible use.

The short course provides a self-contained first introduction to the gist of how general relativity describes gravity. This includes the following three core points: 1. Gravity means that spacetime is curved. 2. The geometry of the curved spacetime determines the motion of particles and the propagation of light under the influence of gravity: Light signals and free particles follow geodesics (that is, locally straight paths). 3. The sources of gravitation determine the geometry of the curved spacetime: the Einstein field equations express this relation as a link between the properties of matter (density, pressure, and others) and the curvature of spacetime. The short course is organised into two units: one unit introduces geodesics and their role in general relativity; the other unit introduces curvature in two and three dimensions and its role in general relativity.

The extended course covers a broad spectrum of topics ranging from the geometric concepts of the theory to the study of specific spacetimes and the related astrophysical phenomena. As the extended course grows, we seek to cover the same topics as introductory university textbooks, but we rephrase the mathematical analysis with model-based activities and visualisations. The extended course will provide many teaching units from which teachers can build their own courses to suit their curricula and their time schedules. Details on selected teaching units are given below.

## CHALLENGES AND TOOLS

Teaching general relativity at school involves at least two major challenges: first, relativistic phenomena are not a part of daily life, and we cannot demonstrate them in the school laboratory. With a complete lack of experimental demonstrations and direct personal experience, how are students to get familiar with the phenomena? Second, the theory uses advanced mathematical tools that are not available at school. How can we rigorously formulate the theory without the specific mathematics? In our course, we address these issues by using two types of purpose-built tools: computer visualisations to provide a virtual experience of the phenomena, and sector models for representing curved spacetimes. In this section, we describe both types of tools, giving some examples that illustrate their roles in the course.

### *First-person visualisation*

To provide a virtual personal experience of relativistic phenomena, we create computer simulations. They visualise relevant scenes from a first-person point of view. Figure 20.1 gives an example. It shows what we should see if we were to travel into the vicinity of a black hole and then take a look around (Kraus, 2005): the black hole appears as a black disc because there is an inner region from which no light escapes. Outside the black disc, the Milky Way in the background appears distorted. By analysing the distortions, we can study the consequences of gravitational light deflection by a spherical mass. We are also developing interactive simulations that allow students to experiment with the scene. An example of a virtual experiment of this kind is a simulation that "positions" a black hole in between a camera and a filmed student. The camera image is transformed to

20.1
Computer simulations visualise relativistic phenomena: a black hole seen from nearby, with the Milky Way in the background. (*Source: Copyright by Ute Kraus, Corvin Zahn*)

simulate light deflection by the black hole, and it is displayed on the computer screen in real-time. This allows students to experiment with the effects of light deflection in a systematic way. For videos, teaching material, and more information see the Spacetime Travel website at spacetimetravel.org.

### *Sector models*

This course makes extensive use of sector models to represent curved spacetimes and explore their geometry. A sector model is a true-to-scale model of a space or a spacetime. It can be a physical model built from paper or cardboard cut-out sheets, or it can be created by three-dimensional printing. It can also be a digital model manipulated with drag and drop. The theoretical basis for sector models is the description of spacetimes in the Regge calculus (Regge, 1961). Further we describe how to build a sector model and how to use it to explore a geometry. For this introduction, we first consider a curved surface. We then show how the procedures are extended to three-dimensional space and to spacetime.

To build the sector model of a curved surface, first, subdivide the surface into small pieces of area. Figure 20.2a shows this step using a saddle as an example. Then, approximate each piece by a small *flat* piece of area, a sector, and assemble the sectors in the plane: this is the sector model (Figure 20.2b). In principle, the sector model is a "cut-up" version of the surface. It involves an approximation (replacing small curved quadrangles with small flat quadrangles); this approximation can be made as precise as desired by using more and smaller sectors.

Now, turn to the applications of the sector model. When we explore a geometry, there are two fundamental properties that we study most often in our course: curvature and geodesics. Here is how to use a sector model to test for curvature: join neighbouring sectors. If the surface in question is a plane surface,

20.2
A two-dimensional sector model: (a) a saddle is subdivided into quadrangles; (b) the sector model – each quadrangle is approximated by a *flat* quadrangle, a sector; and (c) test for curvature. (*Source: Copyright by Ute Kraus, Corvin Zahn*)

20.2
*(continued)*

all the sectors fit perfectly, like the pieces of a jigsaw puzzle. This perfect fit indicates zero curvature. Figure 20.2c shows the test for the saddle: three sectors have been joined, and now there is not enough space left for the fourth (bottom left). The only way to join the fourth sector to one of its neighbours would be to let sectors overlap. This overlap indicates curvature, and it also tells us the sign: negative curvature in this case. In the case of positive curvature, there would be more than enough space left for the last sector. We could then join it to a neighbour, and a gap would remain. We can go on to determine the magnitude of the curvature by measuring the size of the gap or the overlap and the size of the sectors; details can be found in Zahn and Kraus (2014).

The second fundamental application of sector models is the construction of geodesics. A geodesic on a curved surface is the generalisation of a straight line in the Euclidean plane. It is a line that is straight in an intrinsic kind of way in the sense that it keeps its direction at each point, that is, does not bend or kink to either side. When we apply this definition to lines on a sector model, we find that straight lines on the sector model are geodesics on the curved surface. How to use a sector model as a tool for constructing geodesics is discussed in more detail in Chapter 12 of this book. Chapter 12 also provides several additional examples of sector models. In the extended course, the use of sector models goes beyond curvature and geodesics; there it facilitates the discussion of other aspects like parallel transport and geodesic deviation.

For teaching relativity, we extend the sector model approach from curved surfaces to the curved spaces and spacetimes that we wish to explore. We can,

| | | | |
|---|---|---|---|
| 8:40 – 9:25 am | Mathematics *Ms Becker* | Geography *Mr Richter* | Latin *Ms Schulz* |
| 8:35 – 8:40 am | break | break | break |
| 7:50 – 8:35 am | German *Ms Meyer* | English *Mr Weber* | History *Mr Schmidt* |
| | Room 101 Class 9A | Room 102 Class 9B | Room 103 Class 10A |

20.3 Spacetime sectors in daily life: this schedule of a German secondary school is a sector model of Minkowski spacetime. (*Source: Copyright by Ute Kraus, Corvin Zahn*)

for instance, study the three-dimensional curved space of a star. Creation and use of the sector model are translated from two to three dimensions: space is subdivided into small volumes, each small volume is approximated by a small volume of Euclidean space, resulting in a sector model made up of three-dimensional blocks. For more details on sector models for spaces, see Zahn and Kraus (2014).

To study the motion of particles, we need to include time. In spacetime models, either one or two spatial dimensions are suppressed, so that a sector model represents either a line in space plus time or a surface in space plus time, giving a total of two or three dimensions, respectively. Spacetime sectors are something that we are actually quite familiar with in everyday life. Take the schedule of a school, as shown in Figure 20.3: the school building (space) is subdivided into classrooms, the school day (time) is subdivided into lessons, "the lesson from 8:40 to 9:25 am in classroom 101" is a piece of spacetime – a spacetime sector. The creation of the sector models carries over from space to spacetime, and the use of the tools remains the same: we can test for curvature by joining blocks, and straight lines on the sector model are geodesics of the spacetime. The handling of spacetime sectors is somewhat more involved and requires some basic knowledge in special relativity. In practice, we use digital models with support features that make sector manipulation quick and easy. Details on the creation and the use of spacetime sector models are given by Kraus and Zahn (2019a).

As this short overview shows, sector models are built in a transparent way and make notions of non-Euclidean geometry accessible. This novel tool, with its great potential for teaching geometry, provides the basis for our comprehensive general relativity course for secondary school. Sector models are available online as cut-out sheets for DIY physical models at spacetimetravel.org/list_of_sectormodels; web applications for digital models will become available on the Spacetime Travel website in due course.

## EXAMPLES AND OUTLOOK

To give a flavour of how the units of this course can be used in class, we provide summaries of a few units and describe the student activities that they include. We also provide references to more detailed descriptions.

***Short course: Curvature and geodesics***

As described in the first section, the two units of the short course combine to give a concise introduction to the main ideas making up the description of gravity by general relativity. The two units can be used in either order or can also be used independently. They require some previous knowledge in geometry, including properties of straight lines, triangles, and circles. We have elaborated the short course units for grade 9 students (age 15 years). The typical teaching time would be 90 minutes each. Our experience indicates that teachers can adapt them for classes between grade 8 and grade 13 (age 14 to 19 years) by adjusting the amount of hands-on work and by allowing for more or less time, respectively.

*Gravity is geometry*

In this unit, students study the curvature of surfaces and three-dimensional spaces. They get to know criteria for positive, negative, and null curvature and apply them both to the surfaces of different artefacts (e.g., an egg, a torus, a cylinder) and the surfaces described by their sector models. The sector models of surfaces are provided as paper models. Students then work with three-dimensional sector models provided as blocks made of cardboard, and they apply curvature criteria to these curved spaces. The comparison between a vacuum space (black hole) and a space filled with high-density matter (neutron star) illustrates the relation between curvature and matter stated in Einstein's field equations. This unit is described in more detail by Zahn and Kraus (2014) and Kraus and Zahn (2019c).

*Light deflection near a black hole*

In this unit, students are introduced to geodesics and how to construct them. They carry out the construction first on a sphere (a writeable globe), then on the sector model of a sphere, and finally on the sector model of the equatorial plane of a black hole. The exploration of geodesics near a black hole leads to a discussion of gravitational light deflection, and gravitational lensing is explored with interactive visualisations. This unit is described in more detail in Chapter 12 of this book and by Zahn and Kraus (2019).

***Extended course***

The extended course contains the short course material and extends it broadly in two directions: the discussion of geometric concepts is extended (e.g., going from space to spacetime) and is deepened (e.g., quantitative treatment of curvature). And the course lets students explore more spacetimes, from black holes to the expanding universe.

The extended course contains units on two levels: a basic level for students from grade 9 (age 15 years) and an advanced level for students from grade 11 (age 17 years). Here, we give two examples of units at the advanced level.

### *From the metric to the geodesic*

In this unit, students are introduced to several metrics and learn to create the sector model of a two-dimensional space from scratch. They create both sector models used in the short course unit "light deflection near a black hole". As an outcome of studying this unit, students can explore two-dimensional curved spaces on their own: when given a metric, they can find geodesics, and they can determine the curvature. Mathematical knowledge required for this unit includes the Pythagorean theorem. The unit is described in detail by Kraus and Zahn (2016).

### *Vertical free fall: Einsteinian vs. Newtonian description*

Students use a spacetime sector model to explore vertical free fall. The Einsteinian description (object follows a geodesic in spacetime) is compared with the Newtonian description (object is accelerated by a gravitational force). This unit includes some use of special relativity. It is described in more detail by Kraus and Zahn (2019a).

## OUTLOOK

Current work is extending the course with new sector models, new visualisations, and additional teaching units. These allow exploring more spacetimes, including neutron stars, the expanding universe, wormholes, and gravitational waves. The scope of the sector model approach and, thereby, the scope of this model-based course is finally set by what can or cannot be represented in the form of sector models. The major limitation is the need to build the models in either two or three dimensions. Therefore, in the case of a four-dimensional spacetime, what we can represent are two- and three-dimensional subspaces. In some cases, when the spacetime is sufficiently symmetric, two- and three-dimensional subspaces may, in combination, give the full picture. Examples include the non-rotating black hole, a spherically symmetric, and static spacetime, and the expanding universe, being homogeneous and isotropic.

## CONCLUSION

We have outlined the principles of a secondary school general relativity course that is currently under development. This chapter is one of several in the present book discussing the question of how to teach general relativity at school from different perspectives. The focus of the work presented here is on a detailed and in-depth description of the geometric concepts of general relativity that shows the nuts and bolts of the working of the theory. This focus is made possible by a novel approach making use of the description of curved spacetimes with sector models. Since there is a clear correspondence between the model-based and the mathematical description, these two can be seamlessly integrated. Students who go on to university physics can make this link and build on the geometric insight developed in this model-based course.

## FURTHER READING

Chapter 12: Introducing the geometric concepts of general relativity with sector models
Chapter 23: Towards a study module on general relativity in the Netherlands
Chapter 24: Introducing general relativity without special relativity – Classroom experience from the Czech Republic

## REFERENCES

Kraus, U. (2005). "Destination Black Hole." spacetimetravel.org/reiseziel.

Kraus, U. & Zahn, C. (2016). "Teaching gravitational light deflection: A short path from the metric to the geodesic." spacetimetravel.org/aur16.

Kraus, U. & Zahn, C. (2019a). "Sector models—A toolkit for teaching general relativity: III. Spacetime geodesics." *European Journal of Physics*, 40, 015602, online version on spacetimetravel.org/sectormodels3.

Kraus, U. & Zahn, C. (2019b). "The status quo of general relativity in school curricula." teaching-relativity.org/GRCurriculum.

Kraus, U. & Zahn, C. (2019c). "Teaching general relativity with sector models: the field equations." *Journal of Physics: Conf. Series*, 1286, 012025.

Regge, T. (1961). "General relativity without coordinates." *Il Nuovo Cimento*, 19, 558–571.

Zahn, C. & Kraus, U. (2014). "Sector models—A toolkit for teaching general relativity: I. Curved spaces and spacetimes." *European Journal of Physics*, 35, 055020, online version on spacetimetravel.org/sectormodels1.

Zahn, C. & Kraus, U. (2019). "Sector models—A toolkit for teaching general relativity: II. Geodesics." *European Journal of Physics*, 40, 015601, online version on spacetime-travel.org/sectormodels2.

# 21 Einstein-First

## *Modernising the school physics curriculum in Western Australia and beyond*

*Carolyn Maxwell, David Blair, Jyoti Kaur, Rahul Choudhary, and Marjan Zadnik*

**WHAT?**

This chapter provides an overview of Einstein-First, a physics education project in Western Australia that aims to modernise the school science curriculum using engaging hands-on learning resources.

**FOR WHOM?**

This chapter is intended for readers who are interested in the integration of Einsteinian concepts across science subjects and other learning areas, and also, for those who wish to learn more about curriculum development in Einsteinian physics.

### INTRODUCTION

Some university physics lecturers are known for saying, "Forget what you learnt at school". Most physicists and educators believe that Einsteinian physics is too difficult for young people to learn and, as a result, should be left until university studies. This means that only physics graduates have the opportunity to learn Einsteinian physics. Everybody else learns little more than school children did 100 years ago and that means almost no post-1900 concepts or applications!

The Einstein-First project aims to change this, in line with our belief: *every child has the right to share our best understanding of physical reality.*

Einsteinian physics underpins the modern technologies and devices that we use and depend on today, for example, mobile phones, computers, GPS, solar power, lasers, and medical apparatus. It also explains some of the complex challenges humans currently face, such as global warming. We believe that an education system that starts with fundamental Einsteinian concepts and uses them to develop an understanding of technologies relevant to everyday life is also likely to raise standards of scientific literacy, improve gender equity, and improve student attitudes to Science, Technology, Engineering, and Mathematics (STEM) subjects. Research already conducted by the Einstein-First project has shown that teaching Einsteinian physics concepts at an early age improves both the gender gap in physics learning and student attitudes to physics (Kaur et al., 2018; 2020).

***Table 21.1*** **Fundamental tension between school physics and Einsteinian physics**

| *Schools today* | *Modern understanding* |
|---|---|
| Rigid Euclidean space | Space is flexible and curved |
| Time is absolute | Relative spacetime |
| Light is a wave, bullets are particles | Everything has "waviness" and "bulletiness" |
| Energy is massless | Mass is a form of energy: $E = mc^2$ |
| Newtonian determinism | Quantum probability and uncertainty |
| Gravity: instantaneous force | Gravity: spacetime curvature, speed $c$ |

## BACKGROUND TO THE PROJECT

Today, there is consensus amongst physicists that Einsteinian physics is our best understanding of the universe. The many practical applications discussed in this book demonstrate the relevance of Einsteinian physics. One of the most recent applications is the development of gravitational wave detectors leading to the discovery of gravitational waves emanating from coalescing black holes. Whether we care about the truth or we care about educating people to be literate in our modern technological world, Einsteinian physics is fundamental.

The current fundamental tension between Einsteinian physics and the physics students learn at school is shown in Table 21.1.

A team of physicists and educators at The University of Western Australia began to question the fundamentals of science teaching in schools in 2013. Pilot studies with Year 6 and Year 7 students (Pitts et al., 2014) quickly demonstrated that 11- and 12-year-old children relished learning Einsteinian concepts, and since that time numerous studies (Choudhary et al., 2018, Kaur et al., 2020) have demonstrated that the concepts of Einsteinian physics are within reach of children as young as eight years. A summary of the findings of studies carried out by our project is given in Box 21.1.

News of the project spread internationally, leading to an international workshop supported by Norway and Australia at the Gravity Discovery Centre, 90 km north of Perth, Western Australia, in 2016 and a German-funded international workshop on teaching general relativity in 2019. The 2016 workshop led to the creation of the Einsteinian Physics Education Research Collaboration to combine international resources to create an Einsteinian-oriented curriculum for use in schools internationally. The current Einstein-First project is an outcome of this collaboration. In February 2020, the project team in Western Australia hosted the Inaugural Einstein-First International Workshop: Teaching Einsteinian Physics at Schools, at The University of Western Australia and the Gravity Discovery Centre. Speakers from Australia, Norway, Germany, Italy, Brazil, and Israel presented their research and findings; we conducted workshops on our proposed primary and secondary curricula as well as virtual reality applications; and the team held a collaboration meeting involving many representatives from our international research partners.

**BOX 21.1: OUTCOMES FROM SEVEN YEARS OF TRIALLING WITH STUDENTS AGES 8-16**

1. Children accept Einsteinian concepts with ease (Kaur et al., 2017; Pitts et al., 2014).
2. Children are aware that they learn "old stuff" at school (Blair, 2019).
3. Most children are aware of Einstein and want to learn Einsteinian concepts (Kaur et al., 2018).
4. Children are extremely motivated by the activity-based Einsteinian curriculum (Choudhary et al., 2019; Kaur et al., 2018).
5. Teachers enjoy teaching using our activity-based learning, especially those with a minimal physics background (Kaur et al., 2018).
6. Evidence of gender levelling: girls improve more than boys (Kaur et al., 2020).
7. Positive responses from teachers attending our curriculum consultation workshops and from parents (Foppoli et al., 2019).
8. Strong evidence that all students, independent of aptitude, accept the core concepts (Kaur et al., 2017).
9. Published histograms from multiple interventions demonstrate very impressive improvements in both attitude and understanding (Kaur et al., 2020).
10. Long term retention demonstrated over three years (Kaur et al., 2018).

## THE AIMS OF THE EINSTEIN-FIRST PROJECT

The Einstein-First project was named to emphasise the importance of introducing the Einsteinian language of reality *at the very beginning of education.* Our goal is to clarify concepts, find unifying principles, use straight-forward, simple language and teach the terms, ideas and principles of Einsteinian physics at an early age via hands-on, activity-based learning.

The Einstein-First project is a five-year project part-funded by the Australian Research Council. It aims to trial and evaluate a seamless progression of learning Einsteinian physics through primary and secondary school (Years 3 to 10) over five years. Research-informed teaching and learning materials and assessment instruments will be trialled in partner schools by teachers who have been trained by the project team.

## PRINCIPLES BEHIND OUR MODERNISED PHYSICS CURRICULUM

### *Seamless school curriculum*

In the longer term, our project aims to integrate Einsteinian physics concepts into the curriculum for all students from the start to the end of schooling –

Kindergarten/Pre-primary to Year 12. However, over the next five years, our project will trial our modernised curriculum from Year 3 to Year 10 only. Currently, in Australia, the Year 10 science curriculum contains some cosmology and the Year 11 and 12 physics and chemistry curricula include Einsteinian physics concepts.

### *Spiral curriculum*

The Einstein-First project has adopted Jerome Bruner's spiral curriculum approach, which uses Piaget's Constructivist approach to learning. According to Bruner (1960), "*Any subject can be taught effectively in some intellectually honest form to any child at any stage of development*" (italics added). In short, this approach seeks to teach ideas at a simple level first then revisit concepts at increasingly complex levels later on. We aim for the first pass of the spiral to occur in primary school. The concepts taught then are later developed semi-quantitatively in middle school, preparing for a third deeper quantitative pass in senior high school. The Australian Curriculum already has strong elements of a spiral curriculum design, but our project will strengthen this spiral nature through careful choice of content and learning activities.

### *Engaging, experiential hands-on learning activities*

The Einstein-First project is premised on activity-based learning. We emphasise hands-on activities, but also make use of digital learning resources and other web-based resources when appropriate. We use simple, cost-effective equipment such as nerf guns, balloons, toy cars and balls of varying masses, a lycra sheet stretched over a frame for curved space and gravity models, phasor wheels constructed from straws and rulers, a camera obscura made from a shoebox, upturned woks and magnets, and slinky springs. Digital resources include existing online videos, videos filmed by the Einstein-First team, as well as virtual reality headsets and software. See Chapters 15 and 16 for detailed descriptions of selected activities that have been used already in primary and middle school lessons conducted by Einstein-First team members Tejinder Kaur and Rahul Choudhary.

### *Reinforcement of overarching themes*

The following six overarching themes frame the Australian science curriculum: patterns, order and organisation; form and function; stability and change; scale and measurement; matter and energy; and systems (ACARA, 2020). Throughout our proposed modern physics curricula, we keep in mind the dualities inherent in Einsteinian physics concepts, which are related to the Australian science curriculum's overarching themes:

1. Two properties: everything has "waviness" and "bulletiness"
2. Two forms of energy: mass and frequency (all energy can be expressed as $E=mc^2$ or $E=hf$)

3. Two forms of matter: matter and antimatter
4. Two types of fundamental building blocks of matter: bosons, which are force carriers (including photons); and fermions, which include baryons (protons and neutrons) and leptons (electrons)
5. Two processes: fission and fusion
6. Curved space and flat space
7. Stars and black holes
8. Euclidean and non-Euclidean geometry

We aim to introduce all these concepts in the primary curriculum (Years 3 to 6) and then develop them more fully in the secondary curriculum (Years 7 to 10).

### *Integration of Einsteinian concepts across science subjects and other learning areas*

Wherever possible, Einsteinian physics concepts will be integrated beyond physics lessons and into the teaching of chemistry, Earth and space science, and biological science. They will also be integrated across associated learning areas, including mathematics, arts, technologies, humanities and social sciences, and english.

At the primary level, we encourage the integration of mathematics and arts with Einsteinian physics concepts to allow the Einsteinian ideas to be explored across the curriculum. Besides mathematics, we believe it is crucial to include history, art, and drama. We believe that linking elements of mathematics, science, and arts curricula can help to increase these subjects' relevance for students, and foster better student engagement. For example, the concept of photons in light (physics) can be introduced using toy photons and then be linked to learning about the mathematics of arrows (mathematics). This learning could be complemented further by acting out a role-play about light (drama).

We propose the following: history should include the history of science and mathematics. Reading should include stories about science and scientists. Drama should include role-plays about the discovery, debating, and overthrow of ideas. Music should include science songs as singing provides a memorable and enjoyable way of learning.

The Einstein-First team has developed several role-plays in which scientists through the ages present their ideas. The team has successfully trialled these role-plays with students ranging in age from eight to 15 years, and the role-play scripts are available on our website (www.einsteinianphysics.com). We also propose developing role-plays on topics such as Australian History and the Size of the Universe; The Discovery of Zero; the Discovery of Radio; and The History of Probability.

### *Teacher involvement*

The Einstein-First project has actively involved school teachers in the development of its primary and middle school curricula. Consultation workshops with teachers and subject specialists have been conducted since May 2019 to assist the project team in refining curriculum topics, the depth, breadth and sequence of concepts, and activities.

A key aspect of the implementation of an Einsteinian curriculum is teachers' professional development. The project team has begun the process of training teachers from our partner schools, who will implement and test the new curriculum. We have adopted the strategy of defining the curriculum by activities to take into account that many teachers have a minimal science background. For example, the activity "rolling balls on a stretched lycra sheet" uncovers the concept that gravity is associated with curved space. The concepts themselves can seem intimidating, but the activities clarify them.

Particularly at the primary school level, where teachers are usually generalists, directing teachers to a range of background resources encourages engagement with Einsteinian concepts, as well as integrating these ideas into different subject areas. Many physical and digital resources are available including children's books on Einstein, videos illustrating Einsteinian concepts, and amazing images of gravitationally lensed galaxies taken by the Hubble space telescope. These resources make the concept of curved space just one click away on the internet.

Another essential part of the project will be the assessment of our ability to educate teachers effectively regarding Einsteinian physics concepts. In the long term, beyond the participation of our partner schools in the research trials, we need to ensure that we have suitable material and adequate training procedures so that teachers develop sufficient confidence to be able to implement the proposed curriculum.

### *Engagement with education and curriculum authorities*

Engagement with curriculum and assessment authorities at both state and national levels and funding support for broad-scale teacher professional development are crucial if, as hoped, the curriculum is to be implemented across Australia in future. Through our partnerships with science teachers' associations, we aim to promote widespread understanding of the project to ensure national awareness of the need for a revised curriculum. We propose a national review process conducted by a panel drawn from professional organisations and curriculum authorities, which will use the project's research results to determine how best to incorporate Einsteinian physics into the national curriculum. Hopefully, the research and the review will also contribute to the worldwide introduction of Einsteinian science at school.

## EINSTEIN-FIRST PRIMARY SCHOOL CURRICULUM

Analysis of the current primary school curriculum shows many ways in which the traditional Newtonian description of reality is implicitly embedded in learning. Relatively simple modifications are needed at the primary school level to

***Table 21.2* Einsteinian physics themes and topics in our proposed primary school curriculum**

| *Year* | *Themes* | *Physical sciences* | *Earth and space sciences* | *Chemical sciences* |
|---|---|---|---|---|
| Year 3 | **Energy, gravity, spacetime, and matter** | Heat | The Sun, Earth, and Moon | States of matter |
| Year 4 | **Energy and matter** | Forces | The atmosphere and climate change | Properties of materials |
| Year 5 | **Light and the solar system** | Light | The solar system | Atomic structure and matter |
| Year 6 | **Electricity generation and transfer** | Alternative energy sources | – | Chemical cells |

open children's minds to the new reality, thereby creating expectations of a progression of learning that will equip them with scientific literacy relevant to the modern world.

With each new scientific paradigm has come new mathematics. Newtonian physics needed calculus, while Einsteinian physics needs curved space geometry, vector mathematics, and statistics. At primary school, it is most important to open children's minds to the idea that mathematics is more than numbers, and that it includes the mathematics of arrows, experimental geometry on curved and flat spaces, and statistics. The basic principles of these forms of mathematics can be introduced conceptually at an early age through activities, graphs, models, and games at the primary level, as preparation for further development at high school. We do not propose introducing any of the sophisticated mathematics that is needed only for specialists at the tertiary level.

We are developing our proposed curriculum from Kindergarten/Pre-primary onwards, although the current five-year research project we are undertaking includes trials from Year 3 to Year 10 only. Teaching resources, developed with teachers and Pre-primary specialists, involving simple, hands-on activities and suggested digital resources will nevertheless be provided to participating teachers of Kindergarten to Year 2 students in our partner schools.

The Einsteinian physics themes and topics covered in our proposed primary school curriculum, across the physical sciences, Earth and space sciences and chemical sciences strands, are shown in Table 21.2.

## EINSTEIN-FIRST MIDDLE SCHOOL CURRICULUM

The Einsteinian physics themes and topics covered in our proposed middle school curriculum are set out in Table 21.3. Note that the topics are too numerous to reproduce fully here, so only a few examples are given.

***Table 21.3*** **Einsteinian physics themes and topics in our proposed middle school curriculum**

| *Year* | *Themes* | *Examples of topics* |
|---|---|---|
| Year 7 | **Gravity and spacetime** | Spacetime; Einstein's happiest thought; Einstein's gravity |
| Year 8 | **Energy and matter 1** | Energy; black holes; terminal velocity; photons |
| Year 9 | **The rules of the quantum world** | Photons and matter; Planck's formula; quantum entanglement; the photoelectric effect |
| Year 10 | **Energy and matter 2** | Energy conversion; fusion and fission; the structure of atoms and nuclei; the structure of matter |

## CONCLUSION

Research conducted by the Einstein-First project team over the past eight years has demonstrated that Einsteinian concepts can be taught successfully to young children. The next step is the training of teachers and collaborating with them to refine and implement our primary and secondary school Einsteinian physics curricula. Data obtained from these trials will be evaluated by both Western Australian and national curriculum and assessment authorities with a view to incorporating Einsteinian physics into the Australian Curriculum in future.

Our work, along with others', has emphasised the need for a paradigm shift in school physics education in Australia and worldwide. The Einstein-First project aims to be at the forefront of this shift.

## FURTHER READING

Chapter 8: Pushing the boundaries of Einsteinian physics education using virtual reality technology
Chapter 15: Introducing quantum physics with toy photons
Chapter 18: Patterns and atoms: The structure of atomic matter

## ACKNOWLEDGEMENTS

Australian Research Council: This work was supported by the Australian Research Council Linkage Project.

We wish to thank all our partner investigators who helped to make the project possible: Dr Stephen Humphry (UWA); Prof David Treagust (Curtin University); Prof Li Ju (UWA); Mr David Wood (STAWA); Dr Elaine Horne; Prof Susan Scott (ANU); Prof Grady Venville (ANU); Ms Sandra Halley (DET, WA); Ms Glenda Leslie (AISWA); Mr Richard Meagher; Dr Ron Burman; Dr Robin Groves; Ms Jan Devlin; Mr Don Pridmore; Mr Shachar Boublil.

We also gratefully acknowledge our partner organisations: Gravity Discovery Centre Foundation; Department of Education, Western Australia; Association of Independent Schools of Western Australia; Science Teachers Association of Western Australia; Mount Lawley Senior High School; University of Oslo; Seoul National University; University of Hildesheim; University of Glasgow; and Beijing Normal University.

## REFERENCES

Australian Curriculum, Assessment and Reporting Authority (ACARA) (2020). https://www.acara.edu.au/curriculum Retrieved on May 18, 2020.

Blair, D. (2019). "Putting Einstein first: It's time to stop lying to our children about physics." Ockham's Razor ABC Science. https://www.abc.net.au/news/science/2019-12-13/time-to-stop-lying-to-our-children-about-physics/11789858.

Bruner, J. (1960). *The Process of Education*. Cambridge, MA: The President and Fellows of Harvard College, p. 33.

Choudhary, R., Foppoli, A., Kaur, T., Blair, David G., Zadnik, M., Meagher, R. (2018). "Can a short intervention focused on gravitational waves and quantum physics improve students' understanding and attitude?" *Physics Education*, 53, 065020.

Choudhary, R., Kraus, U., Kersting, M., Blair, D., Zahn, C., Zadnik, M., Meagher, R. (2019). "Einsteinian physics in the classroom: Integrating physical and digital learning resources in the context of an international research collaboration." *The Physics Educator*, 1(4), 1950016.

Foppoli, A., Choudhary, R., Blair, D., Kaur, T., Moschilla, J., & Zadnik, M. (2019). "Public and teacher response to Einsteinian physics in schools." *Physics Education*, 54, 015001.

Kaur, T., Blair, D., Moschilla, J., Stannard, W., Zadnik, M. (2017). "Teaching Einsteinian physics at schools: Part 3, review of research outcomes." *Physics Education*, 52, 065014.

Kaur, T., Blair, D., Stannard, W., Treagust, D., Venville, G., Zadnik, M., Mathews, W., Perks, D. (2018). "Determining the intelligibility of Einsteinian concepts with middle school students." *Research in Science Education*, 50, 2505–2532.

Kaur, T., Blair, D., Choudhary, R. K., Dua, Y. S., Foppoli, A., Treagust, D., Zadnik, M. (2020). "Gender response to Einsteinian physics interventions in school." *Physics Education*, 55, 035029.

Pitts, M., Venville, G., Blair, D., Zadnik, M. (2014). "An exploratory study to investigate the impact of an enrichment program on aspects of Einsteinian physics on year 6 students." *Research in Science Education*, 44, 363–388.

# 22 Introduction of Einsteinian physics to the upper secondary school physics curriculum in Scotland

## *Experiences and observations*

*Stuart Farmer*

**WHAT?**

This chapter provides an overview of the recent changes in the Scottish physics curriculum that have placed greater emphasis on topics of Einsteinian physics. Stuart Farmer takes general relativity as an example to discuss the preparedness of teachers to teach challenging topics and possibilities for successful professional learning.

**FOR WHOM?**

This chapter is intended for readers who are interested in curriculum development in Einsteinian physics, and also, for those who wish to further their pedagogical content knowledge in general relativity.

## INTRODUCTION

In recent years, a new curriculum has been introduced across Scotland which includes some Einsteinian physics. This has resulted in the need for effective professional learning activities for teachers as "*curriculum development must rest on teacher development.*" (Stenhouse, 1975, p. 24).

The Curriculum for Excellence reforms in Scotland resulted in a new curriculum for school pupils aged 3 to 18 years. New National Qualification courses taught to pupils aged 15 to 18 were implemented between 2013 and 2016. It is common for pupils progressing to university education to study six or seven subjects leading to National examinations at age 16, five subjects for the Higher examinations at age 17, and three subjects for the Advanced Higher examinations in their final year of school at age 18.

The previous curriculum included a small section on special relativity in the Advanced Higher Physics course. This only involved the pupils knowing that no object can travel faster than the speed of light in a vacuum and carrying out calculations involving rest and relativistic mass, an approach to special relativity not receiving universal support.

The revision of the Higher and Advanced Higher Physics courses placed a greater emphasis on modern physics topics. Special relativity, cosmology, and particle physics are now in the Higher course and general relativity, stellar physics, and the uncertainty principle in the Advanced Higher course. These topics are in addition to classical physics topics on mechanics, electricity, and waves, and more modern topics such as the photoelectric effect, spectra, and semiconductors which have been taught for some time. The course specifications and past examination papers are available on the Scottish Qualifications Authority physics homepage (Scottish Qualifications Authority, n.d.). These changes meant that many Scottish physics teachers required professional learning on Einsteinian physics topics.

## THE SCOTTISH CONTEXT

Scotland has a population of over 5 million people. It has a diverse geography including densely populated urban centres in the "Central Belt" but many relatively remote and sparsely populated rural areas to the north and south. Scotland has around 360 state-funded secondary schools and a small number of independent fee-paying schools. Almost all secondary schools offer Higher physics, but not all offer Advanced Higher physics as an option for pupils to choose. Each year typically 8500 people study the Higher physics course and 1600 the Advanced Higher physics course. All, but a very small number of small remote schools, are likely to have one or more qualified physics teachers. There are around 800 active physics teachers across the country. To teach a subject to age 15 to 18 pupils, teachers must have a degree containing at least 80 credits in the subject (typically two years of study). In addition, teachers require a teaching qualification, usually a Post-graduate Diploma in Education, and to be registered with the General Teaching Council for Scotland in that subject. Many of the physics teachers in Scotland have degrees, such as in engineering, which did not include the study of relativity, or studied it some time ago, or lack the Pedagogical Content Knowledge (Kind, 2009) required to teach these topics at school level well. There was, therefore, a significant need for professional learning for physics teachers on these new topics. In Scotland, teachers have an entitlement to 35 hours of professional learning per year, plus there are five "*in-service*" days when pupils are not in school. Some of these in-service days are likely to be organised on a whole-school basis and others on a local-authority (district) basis when physics teachers from across a local area may meet together.

## SUPPORT FOR PHYSICS TEACHER

If the system-wide introduction of a topic such as general relativity is to be implemented effectively, it is, therefore, essential to provide teacher professional learning to go along with the curriculum developments. There is a reasonable consensus around what makes effective professional learning for teachers (Cordingley et al., 2015; Timperley, 2008). Professional learning should be sustained, embedded in the context of the teacher, integrate new and worthwhile

Subject Content Knowledge with Pedagogical Content Knowledge, be truly collaborative, include challenge to the status quo, and be focussed on pupil outcomes. The CUREE (2018) review of subject-specific professional learning states that "*An important feature of effective continuing professional development is that it is either focussed directly on developing knowledge or practice in a subject area or focussed on developing an aspect of teaching and learning in ways that are contextualised for specific subjects*" (p. 5). It also cautions that "*There is also a need to ensure that research- and evidence-informed practice does not become distorted, and distract from a focus on subject-specific approaches and implications*" (p. 7).

There is evidence that most of the formally organised professional learning activities arranged by schools and local authorities for teachers in Scotland are not effective (Gilchrist, 2018, p. 69). Some professional learning is operational in nature, such as on the use of a new Information Management System; important but not intended to improve teaching and learning. Much is also generic in nature and not subject-specific. When professional learning is too generic, and teachers do not have time to translate the new generic knowledge gained into something meaningful in their own context, it is likely to have very limited impact on either teacher practice or pupil outcomes. A requirement to participate in such generic professional learning also reduces the time available for more impactful subject-specific professional learning which teachers require when introducing new and complex topics such as general relativity.

It is essential to maximise the quality of professional learning support for teachers and to try and do as much as possible in the limited time available. The professional learning has to address issues of lack of teacher Subject Content Knowledge in general relativity but also highlight common pupil misconceptions (which might also be shared by many of the teachers) and provide good pedagogical strategies to address these and improve the teachers' Pedagogical Content Knowledge. My aim is to improve all aspects of the Knowledge Quartet, as described by Rowland (2013). The Knowledge Quartet describes the knowledge needed to be an expert teacher as:

- Foundational knowledge (teachers' Subject Content Knowledge)
- Transformational knowledge (teachers' Pedagogical Content Knowledge)
- Connection knowledge (how a topic is connected to pupils' prior knowledge, other curriculum areas, and the wider environment)
- Contingency knowledge (teachers' ability to cope confidently and knowledgeably with unexpected events during teaching)

Contingency knowledge is especially important when teaching a complex and challenging topic such as general relativity.

Professional learning which addresses all four parts of the Knowledge Quartet in an area as complex as the teaching of general relativity is itself complex, and something that benefits from opportunities for collaboration and social interaction. This can be done when the professional learning delivered as part of a networked learning community (Stoll, Bolam, McMahon, Wallace, & Thomas, 2006). Professional learning delivered through a networked learning community allows for both the input of expertise from knowledgeable others in terms of Subject Content Knowledge, Pedagogical Content Knowledge and facilitation of the networked learning community, but also teachers learning from their peers. The delivery of professional learning in this way allows teachers, following initial input, to experiment with new knowledge and strategies before reflecting on this and sharing their experiences with their peers. As a result, the group can support each other through enaction and reflection cycles (Clarke & Hollingsworth, 2002), resulting in impactful professional growth.

This form of collaborative professional learning has been shown to be very effective and results in significant changes in culture that are not possible with one-off professional learning sessions. When teachers hear about successful developments from their peers, this local evidence, situated in their own context, provides greater motivation for change than even the most compelling evidence from an external expert or a research paper (Cain, 2019). Cordingley et al. (2015) state in their review of effective professional learning that "*The only common finding across all reviews was that peer support was a common feature in effective continuing professional development, with all participants having an opportunity to work together to try out and refine new approaches*" (p. 7).

Unfortunately, in Scotland, due to restrictions such as time, funding, and geography, the opportunities to operate in this way has been limited, and most of the general relativity professional learning has been delivered as one-off workshop sessions, often as a session at a larger event such as a conference. Drawing primarily on resources from the Perimeter Institute (2010; 2013a; 2013b), a two-hour workshop covering the basic general relativity Subject Content Knowledge required for advanced higher physics along with teaching strategies and resources has been developed. When the workshop is delivered, appropriate teaching strategies are modelled with the presenter discussing aspects of these with the delegates so that this contributes to the development of the delegates' Pedagogical Content Knowledge for the topic.

## THE AIMS OF THE WORKSHOP

The key learning outcomes for the workshop are as follows

1. Develop teachers' Subject Content Knowledge of qualitative concepts of general relativity, in particular, the equivalence principle and spacetime diagrams, as experience has shown that these are the two topics with which teachers require most support.
2. Develop teachers' Pedagogical Content Knowledge in relation to teaching general relativity in a conceptual, non-mathematical manner.

3. Introduce practical and video resources which teachers can use to illustrate and model aspects of general relativity.
4. Emphasise the role of modelling in science, the interplay between theory and experimental and observational evidence, and the importance of these when teaching about the nature of science.

## THE STRUCTURE OF THE WORKSHOP

Section 1– Introduction and modelling

The workshop begins, as for many Perimeter Institute resource-based workshops, with delegates observing a "*puzzle*" situation and are asked, in small groups, to devise a model which they think explains these observations. These observations then lead to a discussion of the role of models in science, how they are used to provide explanations and make predictions but then tested against observation and changed or abandoned as a result. Addressing the role of models provides for some explicit teaching about the nature of science and provides a context for subsequent discussion.

Section 2 – Force and acceleration models of gravity

Section 1 naturally leads to a comparison of Newton's "force model" and Einstein's "acceleration model". The "force model" of gravity is introduced using simple props such as masses on a bendy rod. This is a concept with which delegates should be very familiar and provides a secure foundation on which to build. Evidence is presented, which shows this model makes the wrong predictions, experimentally in terms of the orbit of Mercury and theoretically in violating the speed of light limit of the special relativity. The alternative "acceleration model" is then introduced. The comparison of the models is further illustrated by the use of the Alice and Bob video "*What keeps us stuck to Earth?*" (Figure 22.1) (Perimeter Institute, 2013b). This leads on to curved spacetime and what Einstein described as his "*happiest thought*", that "*An acceleration in one direction is identical to the result of a force in the other direction*", the equivalence principle.

Section 3 – Spacetime diagrams

Spacetime diagrams are introduced as a means of describing a situation described in the Alice and Bob video, but in a way which builds on the use of simple descriptions and calculations using the equations and graphs for uniform accelerations in free-fall. This builds on classical physics knowledge of delegates. Masking tape is used to trace out the paths of objects on flat and curved surfaces (Figure 22.2). These activities emphasise that the "acceleration model" of gravity is consistent with curved spacetime and with observations.

Section 4 – Clocks in spacecraft and gravitational fields

The equivalence principle in relation to a spacecraft in deep space and gravitational fields is then described. Clocks at the front and back of accelerating

22.1
(a) and (b) Stills from What keeps us stuck on Earth? (Perimeter Institute, 2013b) exploring the equivalence of force and acceleration. (*Source: Copyright by Perimeter Institute*)

spacecraft are then discussed, and the link is made that clocks, therefore, run at different rates at different altitudes in gravitational fields.

Section 5 – Modelling curved spacetime

The "acceleration model" is summarised and delegates then have the opportunity to use Lycra sheets, marbles, and masking tape to model four-dimensional curved spacetime in three dimensions. Delegates are asked to evaluate the pros and cons of this model and how it relates to a previous activity with a beach ball, for example, the comparison of positive and negative curvatures can be made and how this relates to the relative time dilation at different points on the curved surfaces.

Section 6 – Tests for general relativity

The workshop is completed by discussing tests for general relativity and the supporting evidence. Hands-on teaching strategies which can be used to visualise and illustrate some of these are introduced. These include wine glasses

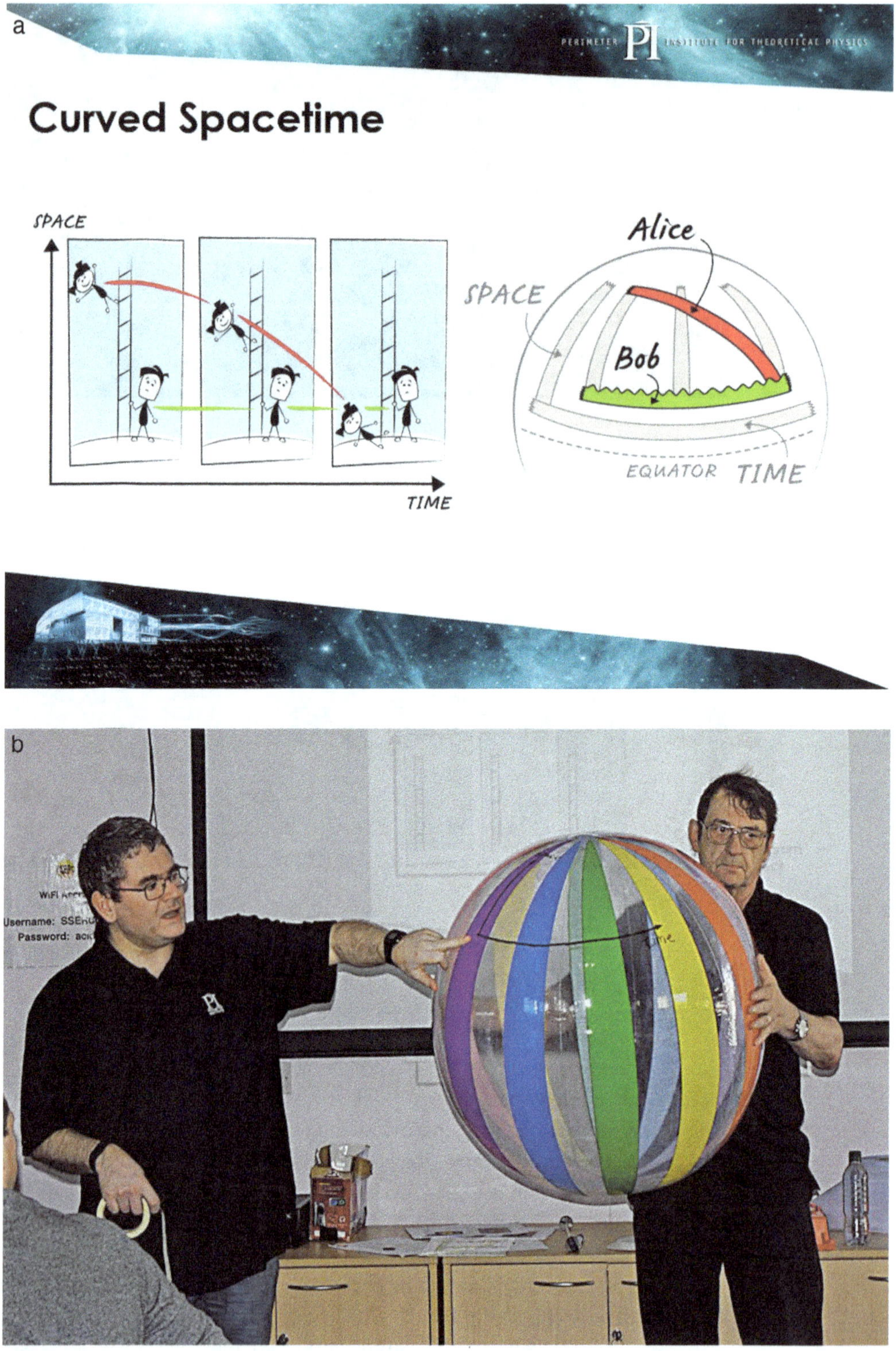


22.2
(a) and (b) Demonstrating accelerated and non-accelerated motion in flat and curved spacetime. (*Source: Copyright by Perimeter Institute*)

to model gravitational lensing and another masking tape activity for gravitational lensing which links back to and reinforces representations of constant velocities and accelerations in the flat and curved spacetime activities from section 3. The video "*The Global Positioning System and Relativity*" (Perimeter Institute, 2010) on the corrections for both special and general relativity on the clocks on GPS satellites is used to illustrate that both theories of relativity are not just abstract concepts but things that impact on our daily lives. The workshop concludes with the discovery of gravitational waves and the observation of the black hole in M87 and that they are as predicted by Einstein's "acceleration model" of gravity.

## EVALUATION OF THE WORKSHOP

Having delivered the professional learning workshop several times, as well as teaching the topic to several classes of pupils, it has become clear that the explanation of why clocks tick differently at different altitudes on Earth or at the front and rear of accelerating spacecraft is a particularly tricky issue for teachers. My approach has gradually evolved, breaking this down into more and more simple steps and improved graphics (Figure 22.3). This approach is consistent with the advice on effective teaching of Kirschner, Sweller, and Clark (2006) and Rosenshine (2012).

22.3
(a) and (b) Some of the slides showing how the explanation of the equivalence principle was broken down into progressively revealed small steps. (*Source: Copyright by Perimeter Institute*)

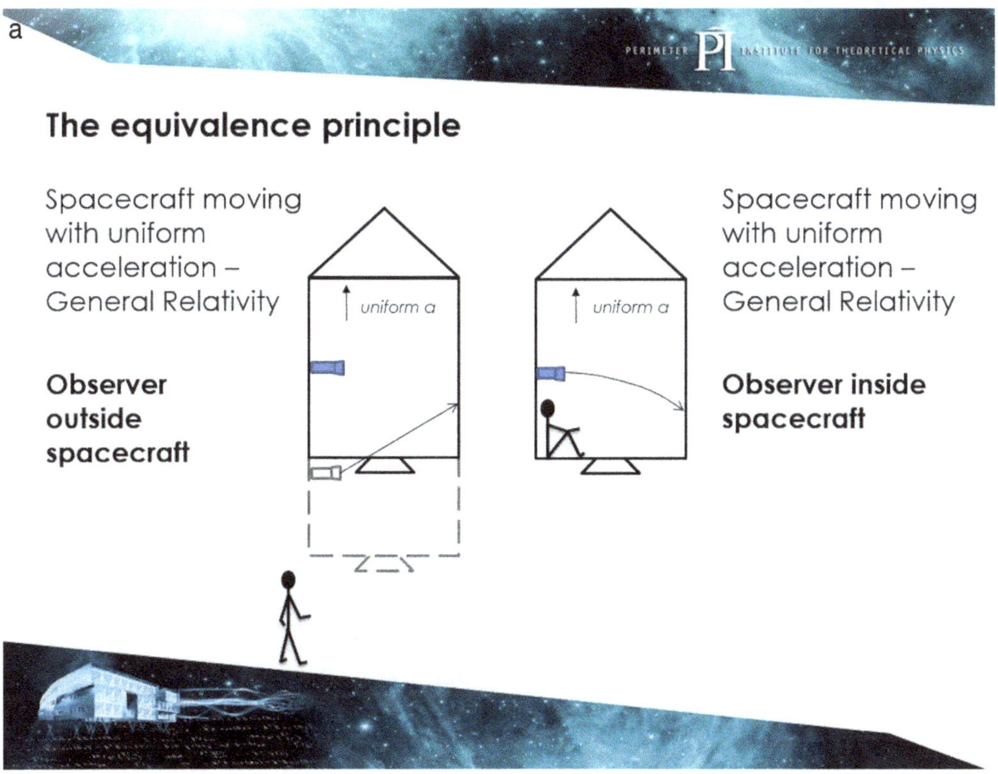

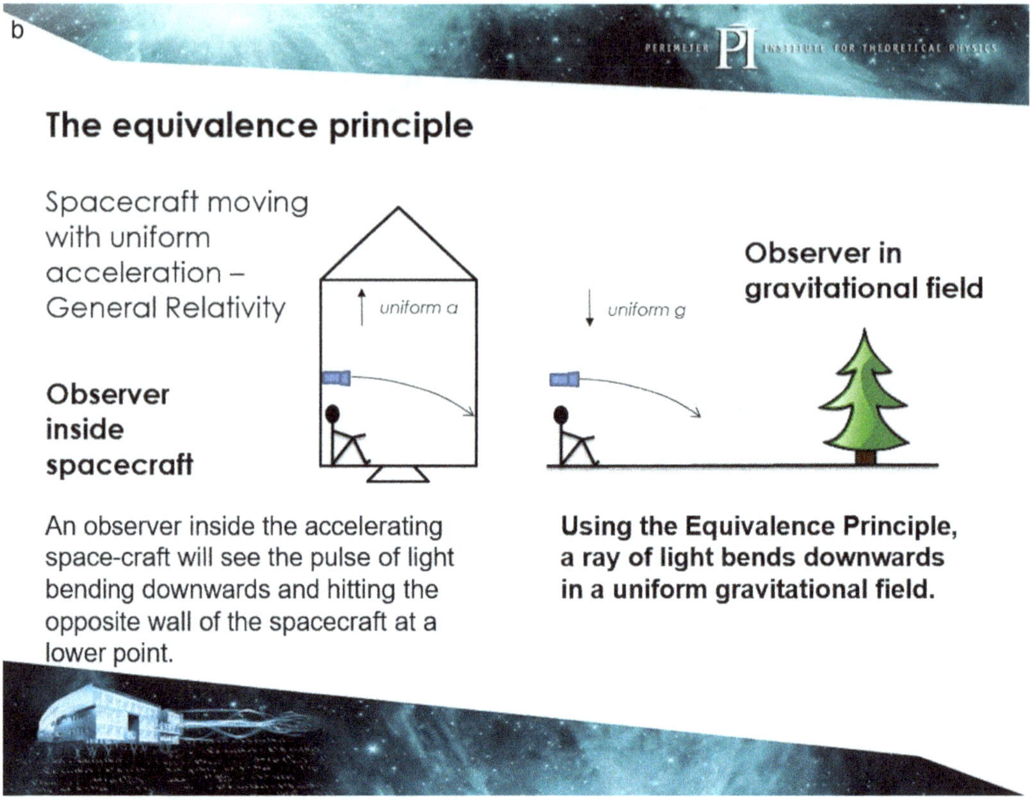


Teachers appreciate the modelling of a good explanation as well as having a bank of hands-on modelling activities to help teach what otherwise could be a very abstract and theoretical topic. A few teachers have returned to the one-off workshop for a second time openly admitting that this was not because the session had not been effective the first time around, but the opportunity to reflect and experiment between the sessions, an opportunity to work through the workshop again, and to ask further questions was valuable and improved their learning and confidence. This supports the literature indicating that revisiting topics during a series of networked learning community meetings is vital for embedding real change and increasing teachers' Subject Content Knowledge and Pedagogical Content Knowledge to a point where they exhibit good contingency knowledge.

22.3 *(continued)*

## OTHER RESOURCES

A webinar (SSERC, 2017) covering the required Subject Content Knowledge for the general relativity topic in advanced higher physics was given by Martin Hendry of the University of Glasgow and delivered through SSERC, the national support service for science and technology education in Scotland (SSERC, n.d.).

This has the benefit of being readily accessible to all teachers, including those in relatively remote and rural areas where establishing and sustaining effective face-to-face networked learning communities is challenging.

The Institute of Physics also manages an email forum only open to all teachers of physics in Scotland. This forum facilitates professional dialogue, including discussion of best teaching practices and the sharing of resources. Video conferencing and webinars are being used increasingly to connect teachers in remote areas, with the 2020 coronavirus pandemic increasing the frequency of use more generally.

## CONCLUSION (LESSONS LEARNED)

The introduction of general relativity into the Scottish curriculum has been in the most advanced part of the curriculum with a relatively expert physics teaching workforce. This introduction has, nevertheless, been challenging. If Einsteinian physics is to be embedded in any curriculum, it is crucial that curriculum-makers at all levels be convinced that this is worthwhile and an improvement on existing curriculum arrangements. This will require a paradigm shift from the current thinking of many, as described by David Treagust in Chapter 2. Such a change will likely involve Einsteinian physics being taught to a younger age range of pupils by mainly non-physics specialist teachers, unlike the Scottish example described above. For curriculum development to be successful, the professional learning needs of these teachers must be taken into consideration from the outset and teachers given appropriate support to enact curriculum guidelines in their context (Wallace & Priestley, 2017). Research has shown that "external experts" such as physicists creating and providing resources is not enough (Cain, 2019) and that while collaboration is necessary, it is not sufficient (Cordingley et al., 2015). The success of any curriculum development, especially one requiring a paradigm shift, can only be founded on effective teacher development (Stenhouse, 1975). This should include professional learning which: is embedded in the context of the teachers; focussed on the required Subject Content Knowledge and Pedagogical Content Knowledge; includes collaboration and peer support; includes input from appropriate knowledgeable others; provides opportunities for experimentation with, reflection on, and sharing of teaching experiences, and above all be focussed on improving the learning outcomes of the pupils involved.

## FURTHER READING

Chapter 2: Time for changing paradigms in science and in education
Chapter 8: Pushing the boundaries of Einsteinian physics education using virtual reality technology
Chapter 19: Designing learning resources and investigating student motivation and learning in general relativity and quantum physics in Norway

## REFERENCES

Cain, T. (2019). *Becoming a research-informed school why? What? How?* Abingdon: Routledge.

Clarke, D., & Hollingsworth, H. (2002). Elaborating a model of teacher professional growth. *Teaching and Teacher Education, 18*, 947–967.

Cordingley, P., Higgins, S., Greany, T., Buckler, N., Coles-Jordan, D., Crisp, B., & Coe, R. (2015). *Developing great teaching: Lessons from the international reviews into effective professional development.* London: Teacher Development Trust.

CUREE. (2018). *Developing great subject teaching: Rapid evidence review of subject-specific CPD in the UK.* London: Wellcome Trust.

Gilchrist, G. (2018). *Practitioner enquiry: Professional development with impact for teachers, schools and systems.* Abingdon: Routledge.

Kind, V. (2009). Pedagogical content knowledge in science education: perspectives and potential for progress. *Studies in Science Education, 45*(2), 169–204.

Kirschner, P. A., Sweller, J., & Clark, R. E. (2006). Why minimal guidance during instruction does not work: An analysis of the failure of constructivist, discovery, problem-based, experiential, and inquiry-based teaching. *Educational Psychologist, 41*(2), 75–86.

Perimeter Institute. (2010). GPS and Relativity – Perimeter Institute. Retrieved December 16, 2017, from https://resources.perimeterinstitute.ca/collections/individual-lessons/products/gps-and-relativity?variant=36262301510.

Perimeter Institute. (2013a). Revolutions in Science. Retrieved December 28, 2017, from https://resources.perimeterinstitute.ca/collections/lesson-compilations/products/revolutions-in-science?variant=17163026694.

Perimeter Institute. (2013b). What Keeps Us Stuck to Earth? – Perimeter Institute. Retrieved December 16, 2017, from https://resources.perimeterinstitute.ca/products/what-keeps-us-stuck-to-earth-1-day?variant=36262298566.

Rosenshine, B. (2012). Principles of instruction research-based strategies that all teachers should know. *American Educator, 36*, 12–19.

Rowland, T. (2013). The knowledge quartet: The genesis and application of a framework for analysing mathematics teaching and deepening teachers' mathematics knowledge. *Journal of Education, 1*(3), 15–43.

Scottish Qualifications Authority. (n.d.). Physics Homepage. Retrieved April 18, 2019, from https://www.sqa.org.uk/sqa/45729.html.

SSERC. (n.d.). Scottish Schools Education Research Centre. Retrieved May 13, 2019, from https://www.sserc.org.uk/.

SSERC. (2017). General Relativity ssercmeet. Retrieved May 15, 2019, from https://www.youtube.com/watch?v=YVjRPTprluQ.

Stenhouse, L. (1975). *An introduction to curriculum research and development.* Oxford: Heinemann.

Stoll, L., Bolam, R., McMahon, A., Wallace, M., & Thomas, S. (2006). Professional learning communities: A review of the literature. *Journal of Educational Change, 7*, 221–258.

Timperley, H. (2008). *Teacher professional learning and development.* Brussels: IBE.

Wallace, C. S., & Priestley, M. R. (2017). Secondary science teachers as curriculum makers: Mapping and designing Scotland's new curriculum for excellence. *Journal of Research in Science Teaching, 54*(3), 324–349.

# 23 Towards a study module on general relativity in the Netherlands

*Stanley Delhaye, Lesley de Putter, and Birgit Pepin*

**WHAT?**

This chapter reports on research activities that aimed to develop and evaluate curriculum materials to teach general relativity to upper secondary students in the Netherlands. On the basis of design-based research methods, the chapter puts particular focus on the design of the learning materials as well as students' responses to them.

**FOR WHOM?**

This chapter is intended for readers who are interested in curriculum design and development in Einsteinian physics, and also, for those who wish to teach general relativity at the upper secondary school level.

## INTRODUCTION

> After months of hard work, the task was finally finished; I had all of my lessons planned. Eager to teach my year 12 students about Einstein's general relativity, I did not expect what happened next. During the first lesson, I realised that my students were not really as motivated as I thought they would be. What was going on? I decided that I needed to address the proverbial elephant in the room. During the conversation that followed, I discovered the problem. Most of my students were already preparing for their final exams, and for them, the module on general relativity was a distraction. I had forgotten one of the most important of issues when preparing these lessons, and as a result, the students did not perceive the relevance of this module. This chapter has been written with this experience in mind.
>
> **S. Delhaye**

More than a century old, general relativity is more relevant than ever. People are interested in Einstein's general theory of relativity, which is supported by the fact that concepts like black holes, wormholes and time travel have been popularised in the media. Moreover, the public response to the recent discoveries like the detection of gravitational waves and the image of a black hole are examples of the enthusiasm surrounding general relativity. Unfortunately, general relativity is not taught at the secondary level in the Netherlands, although it could have been introduced when the Dutch physics curriculum was revised in 2013.

To modernise the curriculum, topics like quantum physics and relativity were incorporated in the revised curriculum. In practice, however, general relativity is still not being taught at secondary schools in the Netherlands. The guideline for relativity in the revised physics curriculum is that "*the candidate should be able to use thought-experiments and examples to explain the phenomena of time-dilation and length-contraction based on concepts such as the speed of light, simultaneity and frame of reference*". This guideline only references special relativity and not general relativity.

The reason for not teaching general relativity might stem from a widely held conviction that Einstein's general theory of relativity is "prohibitively difficult". Consequently, the curriculum renewal committee might have felt that general relativity was simply too difficult for upper secondary students. Nonetheless, recent research has shown that it is possible to teach the basics of general relativity to secondary students (Kaur, Blair, Moschilla, Stannard, & Zadnik, 2017; Kersting, Henriksen, Bøe, & Angell, 2018).

Our project aimed to develop and evaluate curriculum materials (in the form of a study module) to teach general relativity to upper secondary students in the Netherlands. In this chapter, we report on how we designed the materials and how didactic experts (teachers and teacher educators) and students responded to it. Subsequently, we discuss how general relativity can become a feasible topic for Dutch upper secondary students.

## DESIGN OF THE PROJECT

We used a design-based research (DBR) approach (Nieveen & Folmer, 2013) to design a study module on general relativity for upper secondary students in the Netherlands, which provides a systematic way to develop and evaluate interventions with, for example, new teaching materials. One of the key features of DBR is that it actively involves practitioners in the design and evaluation process, which increases the likelihood that the intervention will be relevant and valid. Another characteristic of DBR is that it consists of iterative cycles.

The intervention we designed and evaluated addressed the fact that general relativity was not taught in upper secondary classes in the Netherlands. Knowledge of general relativity would be of great benefit for students to better understand some of its cosmological implications (for example, the existence of gravitational waves and black holes). Firstly, we started with a review of the literature for the initial design principles for the intervention. The initial design principles were refined based on 1) feedback from didactic knowledge experts, 2) feedback from content knowledge experts, and 3) a textbook analysis of university textbooks and experimental materials previously used in secondary education and university textbooks. The resulting nine design principles formed the basis for the module. Secondly, we used the initial design principles to create curriculum materials: text, complimentary pictures, assignments, and so on. The materials are referred to as a study module on general relativity. Thirdly, we tested the module on general relativity in one of the classes the first author taught (final year of secondary education, physics) and evaluated it based on students' feedback. The evaluation consisted of a classroom discussion and an anonymous open-ended questionnaire.

### *Results of the first iteration*

#### *Initial design principals*

The literature review resulted in nine initial design principles shown in Table 23.1. These design principles were refined by using the three additional sources: 1) didactic knowledge experts, 2) a content knowledge expert, and 3) a textbook analysis.

Source 1. Didactic knowledge experts

During the Dutch annual physics teachers conference, we presented the initial design principles to a group of 24 didactic knowledge experts consisting of experienced physics teachers and physics teacher educators. Our goal was to consult didactic knowledge experts. The didactic knowledge experts agreed that it should be possible to teach general relativity to upper secondary students in the Netherlands if performed in an age-appropriate manner. This is in agreement with other studies (Kaur, Blair, Moschilla, Stannard, & Zadnik, 2017; Kersting, Henriksen, Bøe, & Angell, 2018).

One of the challenges when teaching general relativity is the complexity of the mathematics. The experts agreed that the level of mathematics should be similar to the level expected by the curriculum. Furthermore, one of the questions raised by the experts was whether teachers would have enough time in their busy schedules. We believe that embedding the module in the curriculum makes it easier for teachers to include general relativity in their lessons. Additionally, this also allows students to use their prior knowledge for learning general relativity. This increases the need for this module for students, as described at the beginning of the chapter. Some remarks by the didactic knowledge experts we considered after this phase were:

*Didactic expert 1:* "To ensure that students [learn], you have to make sure that this module is in the zone of proximal development (Figure 23.1)".

*Didactic expert 2:* "[Using] an interesting context can be helpful when teaching abstract concepts".

Source 2. Content knowledge expert

During the early design process, we consulted an associate professor who teaches general relativity to undergraduate students at the Eindhoven University

*Table 23.1* **Validation of each design principle based on a literature study as well as using one (or more) of the following sources: 1) didactic knowledge experts, 2) a content knowledge expert, and 3) a textbook analysis on academic textbooks on general relativity**

| *Design principles and rationale* | *Source* |
|---|---|
| **1. Curriculum materials should be presented in an age-appropriate way.**<br>• Students should be able to comprehend the curriculum materials.<br>• Exercises should approximately be of the same level as the national physics exams. | Didactic experts |
| **2. The level of mathematics within the module should be close to the level of mathematics that can reasonably be expected of an upper secondary pre-academic student.**<br>• A "math-first" approach is not feasible for upper secondary pre-academic students.<br>• The exercises should approximately be of the same level as the national mathematics exams. | Didactic experts |
| **3. The module should be embedded within the current physics curriculum.**<br>• This decreases the amount of time spent on "extra-curricular" activities, therefore it will better fit within the time-constraints that teachers and students experience in upper secondary classes in the Netherlands. | Didactic experts |
| **4. The module enables students to study the materials independently from the teacher.**<br>• Instructional scaffolding helps students achieve a high-level understanding of general relativity. | Didactic experts |
| **5. The module contains interesting contexts.**<br>• Interesting contexts foster student engagement. | Content expert and didactic experts |
| **6. The module should be based on four key concepts of general relativity.**<br>• The module covers as much of GR as possible. This ensures that students get the opportunity to get a good understanding what general relativity is about. | Content expert and textbook analysis |

*Table 23.1* **(*Continued*)**

| *Design principles and rationale* | *Source* |
|---|---|
| **7. There is enough visual support in the curriculum materials.**<br>• Visuals such as pictures, animated images, and videos enable students to grasp abstract concepts. | Textbook analysis |
| **8. The module contains hands-on experiences like exercises, worked examples, and student activities.**<br>• The engagement of the students is increased which stimulates learning. | Textbook analysis |
| **9. The module is appealing and encourages the student to read and study the curriculum materials.**<br>• The structure and layout of the materials is clear for the students and encourages learning. | Textbook analysis |

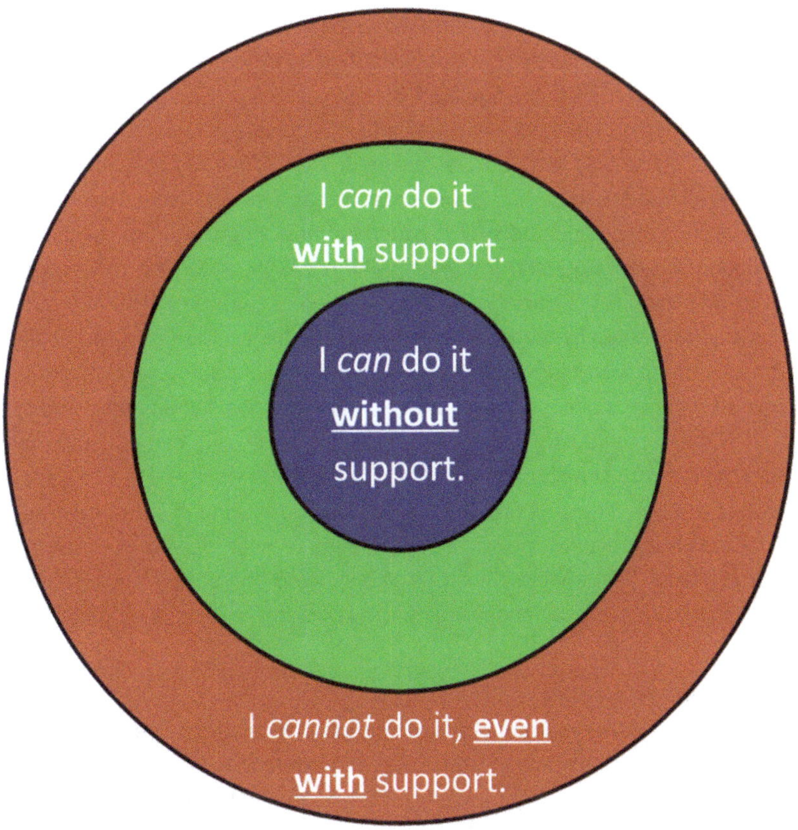


23.1
Zone of proximal development. The first (blue) circle is where the learner is right now. The third (red) circle represents tasks that are too difficult for the learner to complete. The second (green) circle is where learning is stimulated – this is the zone of proximal development. (*Source: Copyright by Stanley Delhaye*)

of Technology. He provided feedback on the intended contents and the structure of the module. We wanted to use interesting contexts to increase student engagement. The **content** knowledge expert agreed that when you teach general relativity to upper secondary students, you could do this by starting with an interesting context, something the didactic **knowledge** experts concurred with:

*Content expert:* "[...] you would have to focus more on the phenomenological aspects [of general relativity] and illustrate, based on examples, that [general relativity] is important to navigational systems. [...] For [upper secondary students] [...] it would be better to [start with an interesting context]".

Source 3. Textbook analysis

A textbook analysis of nine undergraduate and graduate textbooks was conducted. Their content was categorised based on the following topics that were present in the textbooks: special theory of relativity, (tensor) calculus, curvature, field equations, and applications of general relativity. Based on this cursory content analysis, we identified recurring themes that we consider to be close to upper secondary students' content knowledge. This resulted in the following four key concepts: 1) the principle of equivalence, 2) the principle of relativity, 3) the behaviour of light, and 4) the curvature of spacetime. The key concepts that we identified were similar to key concepts identified by other authors (Hartle, 2006; Kersting, Henriksen, Bøe, & Angell, 2018).

We decided to start with the principle of equivalence, which states that gravitational mass and inertial mass are equivalent. As a result, you can conclude that all free-falling objects experience the same acceleration. The principle of equivalence also connects well to students' prior knowledge of mechanics. Subsequently, we introduced the principle of relativity, according to which the laws of physics are the same in all inertial reference frames. It is, therefore, impossible to distinguish between free-falling objects on the Earth and objects that are in outer space far away from any gravitational source. And finally, we explained that gravity causes light to follow curved paths. When one combines this with the fact that light travels along straight lines, one can conclude that spacetime itself must be curved. After having been introduced to the concept of curved spacetime, students are likely to be able to understand better relativistic phenomena like gravitational waves, black holes or even gravitational redshift, and time dilation. The textbook analysis also showed that strong visual support, hands-on activities, and an appealing design could improve student learning.

### *Prototype study module*

We created our prototype study module on general relativity. The physics content was based on our design principles. The content expert considered the representation of this content as good. The module consisted of a booklet, a separate exercise booklet, and hands-on activities. The booklet contained two chapters. The first chapter started with an interesting context of the discovery of gravitational waves. The principles of equivalence and relativity were explained in the second chapter. Finally, geodesics (the straightest possible path between two points in curved spacetime) and curvature were explained with a hands-on activity based on sector models (Zahn & Kraus, 2014). The exercises in the exercise booklet addressed gravitational waves, the LIGO detectors, the solar eclipse of 1919, Mercury's perihelion precession, the Hulse-Taylor binary, and the Pound-Rebka experiment.

### *Class evaluation of the study module*

The module on general relativity was used to teach 28 upper secondary students. Twenty students provided feedback in open classroom discussions and anonymous open-ended questionnaires on the lessons they followed. Concerning age-appropriateness, we made sure that the language and the level of abstraction were aimed at year 12 students. Students had no difficulties understanding the text in the booklet. The exercises in the module were regarded as more difficult compared to exercises the students are familiar with. However, students felt that the difficulty of the exercises in the module was comparable to that of the final exam exercises.

We had opted for an approach that was aimed at conceptual understanding of general relativity instead of using complex mathematics. The students reported that the level of mathematics was fine, and they indicated that they were able to understand and work with the mathematical formulas. We tried to connect the exercises as much as possible to the mandatory physics curriculum. The idea was to create an exercise that would be in the zone of proximal development, for example, the Pound-Rebka experiment (Figure 23.2). This exercise was about gravitational redshift, and also included mandatory topics like nuclear physics, radiation, quantum physics, and oscillations and waves. Unfortunately, students did not recognise the connection between the exercise and the physics topics they had already studied in the previous chapters. They regarded it as "new" and "alien". After a conversation with the class, students realised that the module was tightly connected to the mandatory curriculum and even to the final exam. According to one student:

*Student 1:* "After you explained [the module was connected to the curriculum] I noticed that a lot of [mandatory topics] were present in the module. I started thinking that the module was actually useful. [...] I thought that the exercise on the iron-57 isotope, that Pound-Rebka experiment, was useful since it was clear that [the exercise] contained a number of exam topics".

We included worked examples and thought experiments as well as a glossary. This enabled students to achieve a high understanding of general relativity. Students appreciated the worked examples and the glossary:

**Exercise 3: The Pound-Rebka experiment**

The figure below shows the Jefferson laboratory of the Harvard building. Robert Pound and Glen Rebka conducted their experiment in this building with which they showed that gravitational redshift was real.

**IMAGE 1 REMOVED BY THE AUTHOR**

During the experiment Pound and Rebka used the unstable cobalt-57 isotope. A cobalt-57 isotope decays to an iron-57 isotope. During this process a co-called K-capture happens. An electron from the K-shell is captured by the nucleus and as a result a proton changes into a neutron.

**1. Write down the decay equation of the process in which cobalt-57 decays to iron-57.**

On average the newly formed iron-57 isotope will emit a photon after 141 nanoseconds.
**2. Explain clearly why the iron-57 isotope will emit a photon.**

The photon energy of the emitted photons is equal to 122 keV.
**3. Show with a calculation whether or not the emitted photons are gamma-photons.**

Pound and Rebka used two samples containing iron-57 isotopes to demonstrate the existence of gravitational redshift. One sample containing iron-57 isotopes was mounted on a loudspeaker cone which was located in the basement of the Harvard building. The other sample, also containing iron-57 isotopes, was placed near the top of the building.

**IMAGE 2 REMOVED BY THE AUTHOR**

Due to the movement of the loudspeaker cone a Doppler shift can be created. The idea was to create a Doppler shift that exactly cancelled the gravitational redshift. The explanation is that the change in wavelength due to the Doppler shift is equal (but opposite) to the change in wavelength due to gravity.

The following formula describes the relation between the gravitational redshift (left-hand side) and the Doppler shift (right-hand side):

$$\frac{g \cdot h}{c^2} = \frac{v}{c}$$

where *g* is the gravitational constant ($m\ s^{-2}$), *h* is the distance between the two samples (m), *c* is the speed of light ($m\ s^{-1}$) and *v* is the velocity of the loudspeaker cone ($m\ s^{-1}$).

Pound and Rebka were able to determine that gravitational redshift was cancelled out by the Doppler shift when the velocity of the loudspeaker cone was equal to 0.75 $\mu m\ s^{-1}$. The distance between the two samples was 22.5 meters.

**4. Show with a calculation that under these circumstances the gravitational redshift is cancelled out by the Doppler shift.**

23.2 This example exercise gives students the opportunity to learn more about gravitational redshift. (*Source: Copyright by Stanley Delhaye*)

*Student 2:* "[I liked] the worked examples. [You] could read the text, and if you didn't understand it entirely, you could look at the worked example. It helps to get the big picture".

*Student 1:* "It's nice that you're able to find the literal meaning of that word [in the glossary] instead of having to conjure up the meaning of it by reading the text".

To increase student engagement, we started the first lesson with the discovery of gravitational waves as an exciting context:

*Student 2:* "... I had never heard of it. [...] But I was curious. [...] I was interested. And obviously, afterwards, I thought 'I have learned something new'."

The structure of the module was based on the key concepts, and the students thought that this structure of the module was clear and consistent.

Additionally, students thought the visual support was helpful. The sector model activity and the exercises enabled students to gain a better understanding of certain concepts. Through the use of coloured textboxes (Figures 23.3 and 23.4), the module was appealing to the students, and it was inviting to read:

*Student 3:* "I thought [the textbook] was pleasant to read. I really like paragraphs instead of a wall of text. And there were also blue and yellow text[boxes]. [Concerning the] language, I just thought [these were] normal sentences. It was just like our normal schoolbooks."

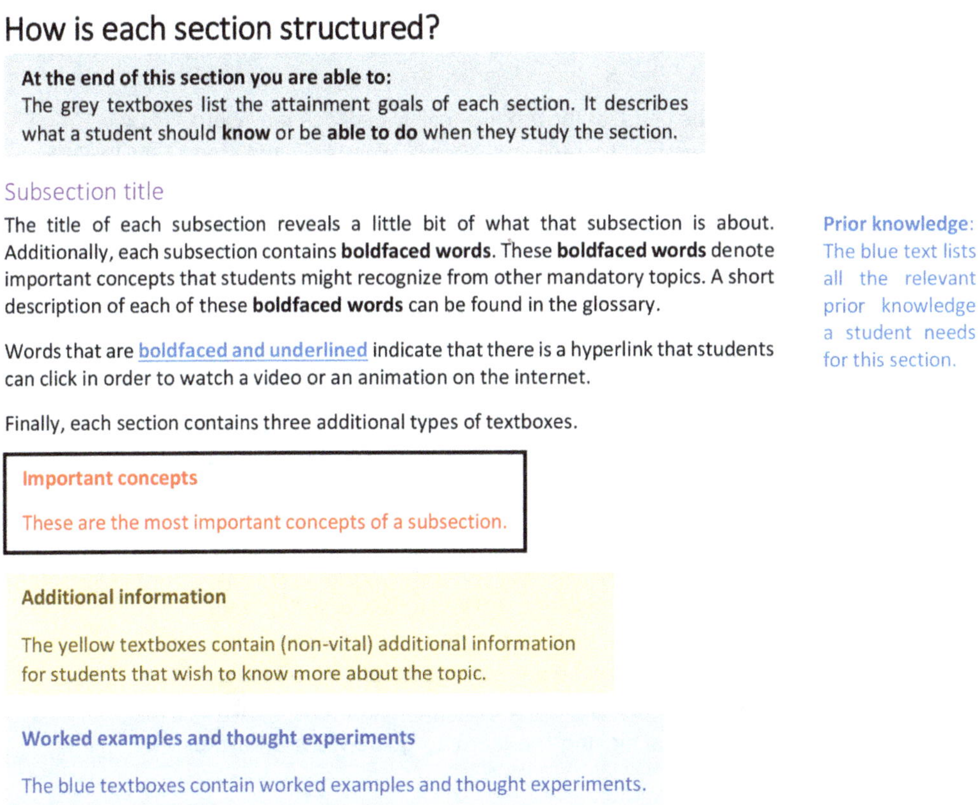

## How is each section structured?

**At the end of this section you are able to:**
The grey textboxes list the attainment goals of each section. It describes what a student should **know** or be **able to do** when they study the section.

### Subsection title

The title of each subsection reveals a little bit of what that subsection is about. Additionally, each subsection contains **boldfaced words**. These **boldfaced words** denote important concepts that students might recognize from other mandatory topics. A short description of each of these **boldfaced words** can be found in the glossary.

Words that are **boldfaced and underlined** indicate that there is a hyperlink that students can click in order to watch a video or an animation on the internet.

Finally, each section contains three additional types of textboxes.

**Prior knowledge**: The blue text lists all the relevant prior knowledge a student needs for this section.

**Important concepts**

These are the most important concepts of a subsection.

**Additional information**

The yellow textboxes contain (non-vital) additional information for students that wish to know more about the topic.

**Worked examples and thought experiments**

The blue textboxes contain worked examples and thought experiments.

23.3
Each section of the textbook contains elements that increase the readability. (*Source: Copyright by Stanley Delhaye*)

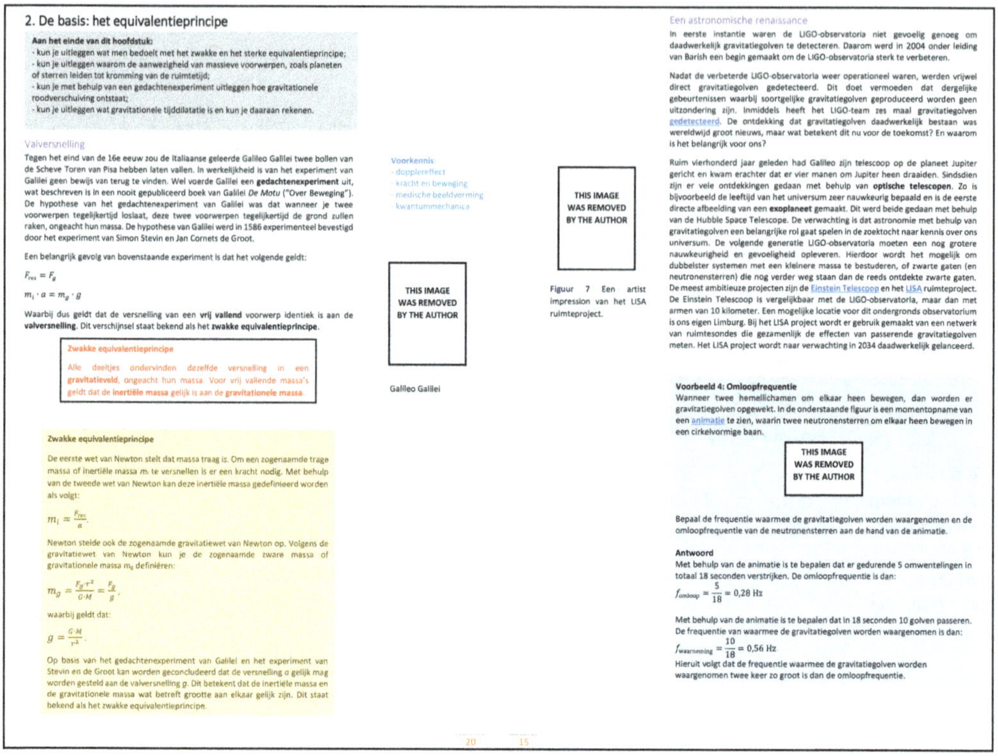

2. De basis: het equivalentieprincipe

**Aan het einde van dit hoofdstuk:**
- kun je uitleggen wat men bedoelt met het zwakke en het sterke equivalentieprincipe;
- kun je uitleggen waarom de aanwezigheid van massieve voorwerpen, zoals planeten of sterren leiden tot kromming van de ruimtetijd;
- kun je met behulp van een gedachtenexperiment uitleggen hoe gravitationele roodverschuiving ontstaat;
- kun je uitleggen wat gravitationele tijddilatatie is en kun je daaraan rekenen.

Valversnelling

Tegen het eind van de 16e eeuw zou de Italiaanse geleerde Galileo Galilei twee bollen van de Scheve Toren van Pisa hebben laten vallen. In werkelijkheid is van het experiment van Galilei geen bewijs van terug te vinden. Wel voerde Galilei een **gedachtenexperiment** uit, wat beschreven is in een nooit gepubliceerd boek van Galilei *De Motu* ("Over Beweging"). De hypothese van het gedachtenexperiment van Galilei was dat wanneer je twee voorwerpen tegelijkertijd loslaat, deze twee voorwerpen tegelijkertijd de grond zullen raken, ongeacht hun massa. De hypothese van Galilei werd in 1586 experimenteel bevestigd door het experiment van Simon Stevin en Jan Cornets de Groot.

Een belangrijk gevolg van bovenstaande experiment is dat het volgende geldt:

$F_{res} = F_g$

$m_i \cdot a = m_g \cdot g$

Waarbij dus geldt dat de versnelling van een **vrij vallend** voorwerp identiek is aan de **valversnelling**. Dit verschijnsel staat bekend als het **zwakke equivalentieprincipe**.

**Zwakke equivalentieprincipe**

Alle deeltjes ondervinden dezelfde versnelling in een **gravitatieveld**, ongeacht hun massa. Voor vrij vallende massa's geldt dat de **inertiële massa** gelijk is aan de **gravitationele massa**.

**Zwakke equivalentieprincipe**

De eerste wet van Newton stelt dat massa traag is. Om een zogenaamde trage massa of inertiële massa $m_i$ te versnellen is er een kracht nodig. Met behulp van de tweede wet van Newton kan deze inertiële massa gedefinieerd worden als volgt:

$m_i = \frac{F_{res}}{a}$.

Newton stelde ook de zogenaamde gravitatiewet van Newton op. Volgens de gravitatiewet van Newton kun je de zogenaamde zware massa of gravitationele massa $m_g$ definiëren:

$m_g = \frac{F_g \cdot r^2}{G \cdot M} = \frac{F_g}{g}$,

waarbij geldt dat:

$g = \frac{G \cdot M}{r^2}$.

Op basis van het gedachtenexperiment van Galilei en het experiment van Stevin en de Groot kan worden geconcludeerd dat de versnelling *a* gelijk mag worden gesteld aan de valversnelling *g*. Dit betekent dat de inertiële massa en de gravitationele massa wat betreft grootte aan elkaar gelijk zijn. Dit staat bekend als het zwakke equivalentieprincipe.

Voorkennis
- dopplereffect
- kracht en beweging
- medische beeldvorming
- kwantummechanica

THIS IMAGE WAS REMOVED BY THE AUTHOR

Galileo Galilei

THIS IMAGE WAS REMOVED BY THE AUTHOR

Figuur 7 Een artist impression van het LISA ruimteproject.

Een astronomische renaissance

In eerste instantie waren de LIGO-observatoria niet gevoelig genoeg om daadwerkelijk gravitatiegolven te detecteren. Daarom werd in 2004 onder leiding van Barish een begin gemaakt om de LIGO-observatoria sterk te verbeteren.

Nadat de verbeterde LIGO-observatoria weer operationeel waren, werden vrijwel direct gravitatiegolven gedetecteerd. Dit doet vermoeden dat dergelijke gebeurtenissen waarbij soortgelijke gravitatiegolven geproduceerd worden geen uitzondering zijn. Inmiddels heeft het LIGO-team zes maal gravitatiegolven gedetecteerd. De ontdekking dat gravitatiegolven daadwerkelijk bestaan was wereldwijd groot nieuws, maar wat betekent dit nu voor de toekomst? En waarom is het belangrijk voor ons?

Ruim vierhonderd jaar geleden had Galileo zijn telescoop op de planeet Jupiter gericht en kwam erachter dat er vier manen om Jupiter heen draaiden. Sindsdien zijn er vele ontdekkingen gedaan met behulp van **optische telescopen**. Zo is bijvoorbeeld de leeftijd van het universum zeer nauwkeurig bepaald en is de eerste directe afbeelding van een **exoplaneet** gemaakt. Dit werd beide gedaan met behulp van de Hubble Space Telescope. De verwachting is dat astronomie met behulp van gravitatiegolven een belangrijke rol gaat spelen in de zoektocht naar kennis over ons universum. De volgende generatie LIGO-observatoria moeten een nog grotere nauwkeurigheid en gevoeligheid opleveren. Hierdoor wordt het mogelijk om dubbelster systemen met een kleinere massa te bestuderen, of zwarte gaten (en neutronensterren) die nog verder weg staan dan de reeds ontdekte zwarte gaten. De meest ambitieuze projecten zijn de Einstein Telescoop en het LISA ruimteproject. De Einstein Telescoop is vergelijkbaar met de LIGO-observatoria, maar dan met armen van 10 kilometer. Een mogelijke locatie voor dit ondergronds observatorium is ons eigen Limburg. Bij het LISA project wordt er gebruik gemaakt van een netwerk van ruimtesondes die gezamenlijk de effecten van passerende gravitatiegolven meten. Het LISA project wordt naar verwachting in 2034 daadwerkelijk gelanceerd.

**Voorbeeld 4: Omloopfrequentie**

Wanneer twee hemellichamen om elkaar heen bewegen, dan worden er gravitatiegolven opgewekt. In de onderstaande figuur is een momentopname van een animatie te zien, waarin twee neutronensterren om elkaar heen bewegen in een cirkelvormige baan.

THIS IMAGE WAS REMOVED BY THE AUTHOR

Bepaal de frequentie waarmee de gravitatiegolven worden waargenomen en de omloopfrequentie van de neutronensterren aan de hand van de animatie.

**Antwoord**

Met behulp van de animatie is te bepalen dat er gedurende 5 omwentelingen in totaal 18 seconden verstrijken. De omloopfrequentie is dan:

$f_{omloop} = \frac{5}{18} = 0{,}28\ \text{Hz}$

Met behulp van de animatie is te bepalen dat in 18 seconden 10 golven passeren. De frequentie van waarmee de gravitatiegolven worden waargenomen is dan:

$f_{waarneming} = \frac{10}{18} = 0{,}56\ \text{Hz}$

Hieruit volgt dat de frequentie waarmee de gravitatiegolven worden waargenomen twee keer zo groot is dan de omloopfrequentie.

20 15

23.4 Two example pages that show the design of the textbook. (*Source: Copyright by Stanley Delhaye*)

Initially, students did not see the need for the module mainly because they were focussed on the approaching exams. Moreover, it was not clear from the start how the module was connected to the mandatory curriculum, and as a result, students were less motivated. One student suggested that it mainly had to do with the timing:

*Student 1:* "I thought [the module] was interesting, even fun, but I think it mainly had to do with the timing that I thought "do we have to do this now and couldn't we be doing something more useful?" If we would have done this halfway through the year, I think I would have thought very differently [about the module]. Then I would have thought "great, we're going to do something very different", and now I was constantly thinking about the final exams".

Student learning, as such, was not evaluated with a summative assignment. The feedback the students provided indicated that they were learning about the topic. Although the students found the topic a bit different from previous physics topics, their work on the exercises and their questions and feedback indicated they were capable of achieving the learning goals we set for the module.

## CONCLUSION

We contend that our first iteration of designing an intervention to teach general relativity to upper secondary students was successful. We have shown that one can teach general relativity to upper secondary students in the Dutch context. This can be achieved by applying the design principles from the analysis phase to the intervention. A lesson (series) on general relativity should:

1. use language and abstractions aimed at your students.
2. avoid complex mathematics and adopt a more conceptual approach instead.
3. connect to the mandatory curriculum.
4. connect to students' prior knowledge.
5. contain interesting contexts (e.g., gravitational waves or black holes).
6. be based on key concepts of general relativity.
7. have enough visual support.
8. contain hands-on experiences.
9. have an appealing design.

The formative evaluation phase revealed that it would be important to ensure that students perceive the need to learn about general relativity and its relevance. Hence, the lesson (series) on general relativity should be connected to 1) recent scientific developments, 2) practical applications, 3) the mandatory curriculum, and 4) students' prior knowledge (Figure 23.5). Following design-based research principles, we intend to improve the module in a second and third cycle, to include colleague physics teachers in the re-design. These cycles will involve teacher workshops and teacher development teams, in which teachers will further develop and appropriate the curriculum materials for their own use in the classroom.

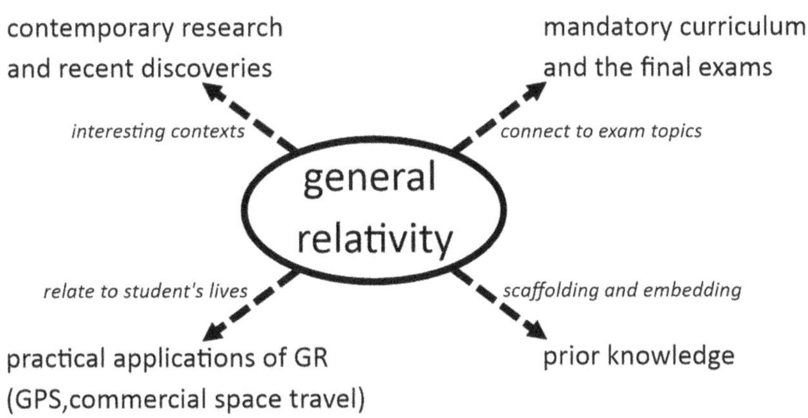


23.5
Four ways in which our module connects to the needs to learn from the students' perspective.
(*Source: Copyright by Stanley Delhaye*)

### *Epilogue*

Fast forward one year. Having learned from my initial mistake that I had not made clear why these lessons were relevant to them, this time I immediately explained to my new group of students how this module on general relativity connects to the mandatory curriculum, and also to their prior knowledge. I also started with a different context; the first "photo" of a supermassive black hole. This new group of students liked that topic a lot and responded that this context increased their motivation to learn more about general relativity. All in all, students were much more positive about the module this time around.

**S. Delhaye**

## FURTHER READING

The entire module can be downloaded here: https://bit.ly/3hpEySF
Chapter 19: Designing learning resources and investigating student motivation and learning in general relativity and quantum physics in Norway
Chapter 20: Towards a comprehensive general relativity course for secondary school

## REFERENCES

Hartle, J. B. (2006). General relativity in the undergraduate physics curriculum. *American Journal of Physics, 74*(1), 14–21.

Kaur, T., Blair, D., Moschilla, J., Stannard, W., & Zadnik, M. (2017). Teaching Einsteinian physics at schools: Part 3, review of research outcomes. *Physics Education, 52*(6), 065014.

Kersting, M., Henriksen, E. K., Bøe, M. V., & Angell, C. (2018). General relativity in upper secondary school: Design and evaluation of an online learning environment using the model of educational reconstruction. *Physical Review Physics Education Research, 14.*

Nieveen, N., & Folmer, E. (2013). "Formative evaluation in educational design research." In T. Plomp & N. Nieveen (Eds.), *Educational design research - Part A: An introduction* ( 206). Enschede: SLO.

Zahn, C., & Kraus, U. (2014). Sector models – A toolkit for teaching general relativity. Part 1: Curved spaces and spacetimes. *European Journal of Physics, 35*(5), 055020.

# 24 Introducing general relativity without special relativity

## *Classroom experience from the Czech Republic*

*Matej Ryston*

**WHAT?**

Without a prior discussion of special relativity, introducing general relativity can be challenging. Matej Ryston outlines a one-time extracurricular seminar about general relativity that does not require previous knowledge of special relativity.

**FOR WHOM?**

This chapter is intended for readers who wish to teach general relativity at the upper secondary school level.

### INTRODUCTION

Imagine you are a teacher who wants to introduce their students to general relativity. As such, you need to figure out two underlying problems. First, *how* you introduce general relativity – a question that is undoubtedly dealt with in many chapters of this book. Second, *when* to introduce general relativity. After all, you need to find time to teach all that you are supposed to teach *and* on top of that, some other topics you *would like* to teach such as general relativity.

One possible way is to offer students a one-time extracurricular activity; sort of a combination of a lecture and a workshop. The seminar, which I describe in this chapter, is based on two principles:

1. **Hands-on relativity**: General relativity is abstract enough as it is. With secondary school students (or anyone outside of technical universities), you need to be as comprehensible as possible. Let us try to make relativity hands-on. That means concrete equipment, activities for the students, and real-life analogies.
2. **No prior knowledge of special relativity necessary**: Of course, general relativity comes directly from special relativity, and ideally, if one wishes to dive into general relativity, one should be familiar with its special predecessor. However, as special relativity is taught usually in the last year of secondary school in the Czech Republic (if it is taught at all), preparing the

seminar for final year students (around 18- to 19-year-olds) would severely limit the number of possible attendees. Therefore, we will try to avoid needing special relativity as much as possible.

## PART 1: SPATIAL CURVATURE

The seminar consists of a chain of conceptual steps, which we will follow as well and which can be seen in Figure 24.1.

To keep things as short and straightforward as possible, we will focus on understanding just one sentence that expresses the fundamental idea of general relativity: *"Gravity is the curvature of spacetime."* Spacetime is a concept developed already in special relativity. As we do not want to rely on our students' prior knowledge of relativity, we will try to build up the concept of spacetime naturally throughout the seminar. Therefore, we will start with curvature, though not with the curvature of spacetime right away; not yet. To make things more understandable, let us split this strange concept into two parts, curvature of space and "curvature" of time.

To be honest, hardly anyone can imagine a curved space. Let us make it a little simpler and talk about curved surfaces because that is something we can work with.

Most secondary school students are not familiar with the concept of curvature in the geometrical sense. To introduce curvature, we first remind them of some well-known aspects of geometry that we learn in school – the Euclidean geometry, according to which, for example:

- The shortest distance between two points is a straight line.
- The sum of internal angles of a triangle is 180°.
- Two parallel lines are separated by the same distance everywhere.
- The radius of a circle $r$ and its circumference $C$ are related by $C = 2\pi r$.

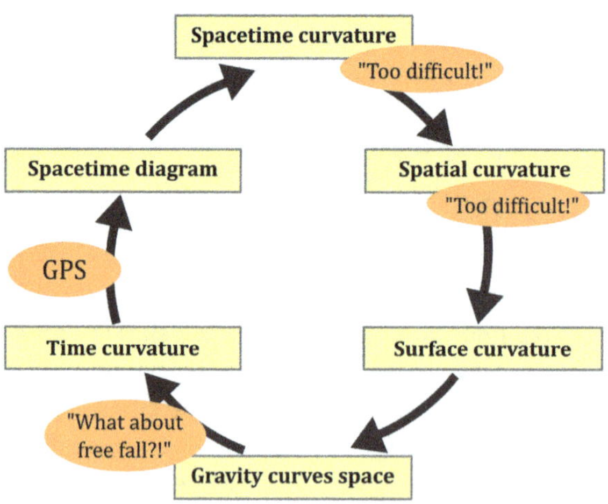


24.1 The overall structure of the seminar builds on small conceptual steps to illustrate the idea that gravity is curvature of spacetime. (*Source: Copyright by Matej Ryston*)

However, these statements do not hold on curved surfaces. The best example is the surface of a sphere. On a sphere, the closest distance between any two points is a part of a circle; all the triangles have the sum of internal angles bigger than 180° (depending on our triangle, it can be even 270°); two initially parallel lines get closer and eventually cross (think of two meridians starting at the equator on the globe); and any circle has a circumference smaller than its radius as measured on the surface times $2\pi$. We can easily illustrate all that using, for example, a beach ball (any ball would do, but beach ball is inflatable, therefore easily transportable and its visible seam is basically a meridian) or a globe. More importantly, the students can check everything themselves. All we need is a ball for every student or a small group. In this case, balloons are a great substitute for balls because they are cheap and easy to handle in larger numbers. It does not matter that balloons are not perfectly round; they are still a great example of a so-called **positively curved** surface.

To find the shortest distance between two points or create a triangle on a curved surface, we need to broaden our definition of a "straight" line. Of course, we will not be able to find any actual straight lines on curved surfaces, but we can define them as "following the same direction" while staying on the surface. Such curves that always go locally "straight" are called **geodesics**, and they play an essential part in geometry, and consequently in general relativity. We can approximate and visualise geodesics using long strips of paper. Put the strips on a flat surface, and they are perfectly straight, but they can bend and adhere to the curved surface, thus creating a path where every small step goes straight compared to the previous one, but altogether the path is curved. An example of such a geometrical demonstration can be seen in Figure 24.2.

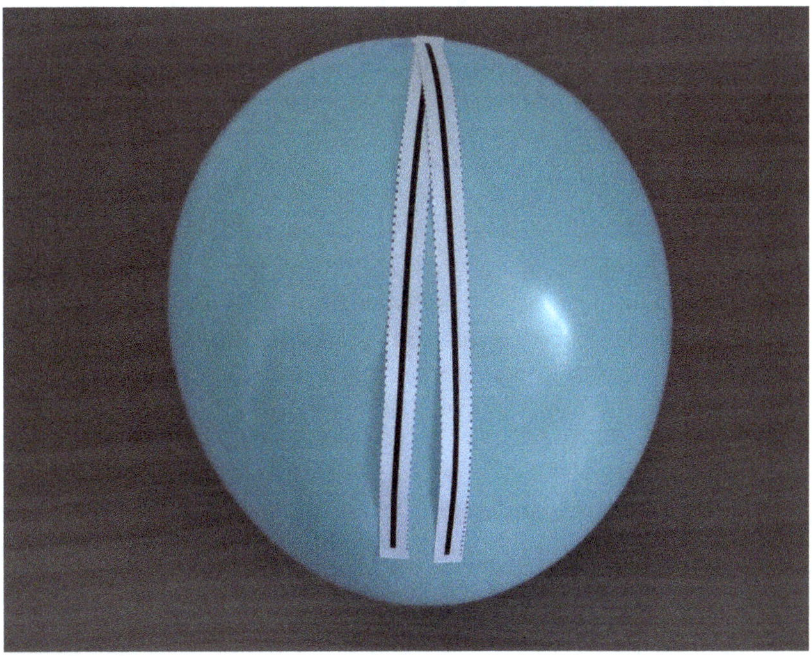

24.2
Two initially parallel straight lines cross on the surface of a balloon, which serves as an example of positive curvature. (*Source: Copyright by Matej Ryston*)

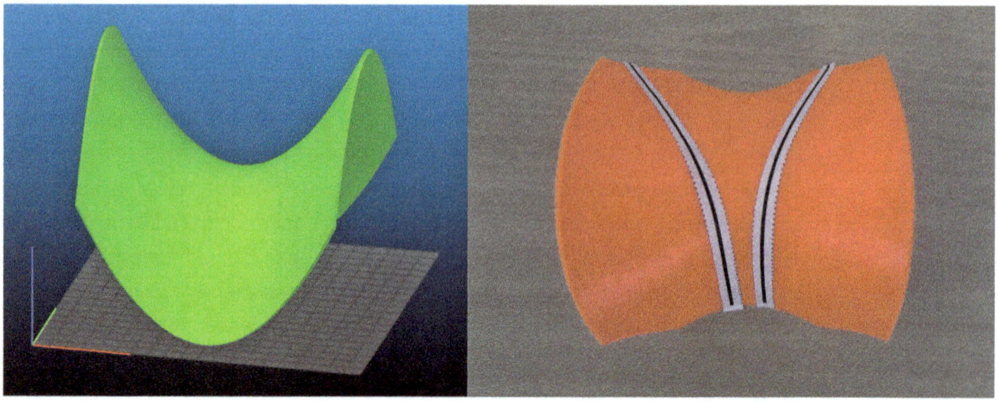

24.3
The hyperbolic paraboloid is an example of a negatively curved surface. A three-dimensional model of the hyperbolic paraboloid (left). A three-dimensional-printed model with the two initially parallel geodetics diverging (right). (*Source: Copyright by Matej Ryston*)

We can also show students a different kind of curvature, the so-called **negative curvature**. One example of a negatively curved surface is a hyperbolic paraboloid (see Figure 24.3, left), sometimes popularly called a "saddle surface", because it resembles a horse saddle. However, a horse saddle is not a particularly practical teaching aid, and also, virtually all my students live in the city. So, their real-life experience with horses and saddles is almost non-existent. Instead, I bring out a particular type of crisps (or chips for American-English speaking readers) of a well-known international brand, which shall remain unnamed, but starts with P. Suffice it to say, the crisps come in a can and their shape is quite characteristic. Students are usually familiar with this type of crisps, so the shape is not a complete surprise for them. A bowl of crisps is passed among the students, each gets a piece or two, and after they are done inspecting them, they can have a little snack – making this demonstration all the more popular.

Nevertheless, we want to again play with the geometry of the shape for which the crisps are not suitable. That is why we have three-dimensional-printed versions of the saddle surface ready that are big enough so that students can try all the previously mentioned geometrical activities. They can easily verify that initially parallel geodesics lead further apart from each other (see Figure 24.3, right); any triangle has the sum of internal angles smaller than 180°. Everything ends up being opposite to the positively curved case. Therefore, this is a different kind of geometry.

So, how can we imagine a curved space? The answer is we do not. Instead, we focus on the differences between curved and flat surfaces, and transfer that principle one dimension higher. In curved space, Euclidean geometry does not work. Geodesics are not straight lines; originally parallel lines diverge, and so on.

## PART 2: RELATIVISTIC CURVATURE

Let us now focus on an actual physical problem, a planet orbiting a star in a plane. General relativity tells us that the space around any massive body is curved in the geometrical sense, as we discussed above. We would now like to visualise the curvature. However, in order to visualise a two-dimensional curvature, we need to add a third dimension (just like on a piece of flat paper, there are only right-left and backwards-forwards, but if you crumble the paper it also goes up and down), thus creating a truly curved surface. We need to keep in mind, though, that the third dimension (height) is not a physical one, it is purely for the purposes of visualisation.

The actual shape of the surface must come from general relativity itself, of course. Luckily, even though general relativity is reasonably complicated, for the simplest case of a non-rotating, spherically symmetric black hole or the outside of a star (what we call the Schwarzschild solution of Einstein equations) it allows us to compute the shape of the curved surface (see Figure 24.4). You can find more information about the derivation of the shape and where it comes from in Ryston, 2019.

The shape we get is called Flamm's paraboloid. Flamm's paraboloid is the basis of both every image representation of general relativity and a possible inspiration for the often seen analogy of space around a gravitating object being like a flexible rubber sheet weighed down by a heavy object like a stone or a bowling ball in the middle. We will not focus on the rubber sheet analogy here, but one can find a thorough analysis in Kersting and Steier, 2018.

Using a 3D-printer, we can print any number of Flamm's paraboloids for the students to play with, and also, discover properties of the geometry of such curved spaces (Figure 24.5). We can again simulate geodesics on the surface using strips of paper and observe their properties. We can see that geodesics are (of course) curved and that initially, parallel geodesics diverge.

Now, we are ready to express another critical idea of general relativity: "Space tells matter how to move. Matter tells space how to curve."

According to Newton's first law of mechanics, any object when acted upon by no external force (or if all the acting forces are balanced) moves in a straight line. So is the case in the absence of gravity. If we "switch" gravity on, the trajectory of the free object becomes curved. Not because gravity is a force that acts on the moving object, but because it is the curvature of space (that is what

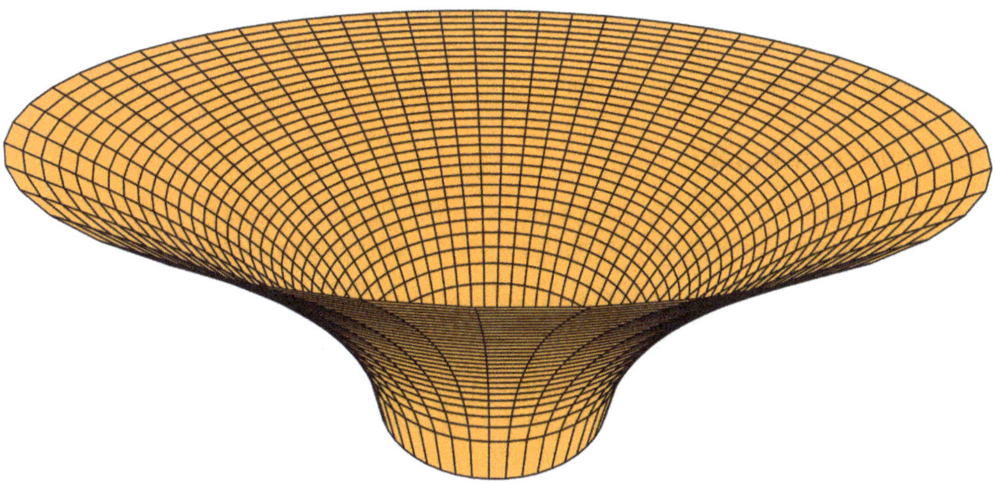

24.4
Flamm's paraboloid describes a surface where the curvature of the surface corresponds to real space curvature in a plane around a non-rotating spherical source of general relativity such as a star or a black hole. (*Source: Copyright by Matej Ryston*)

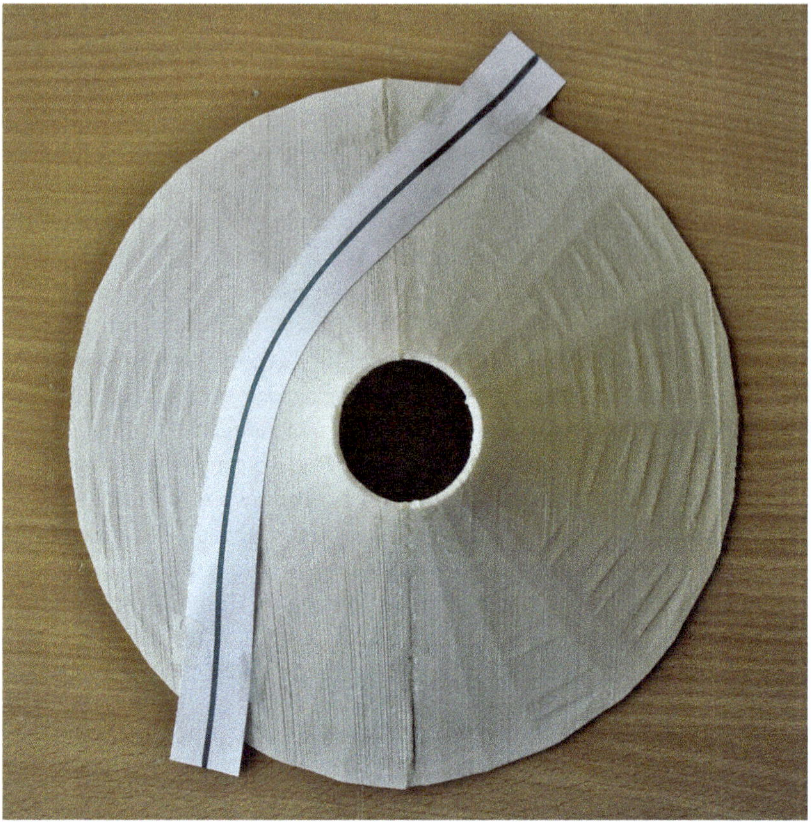

24.5
Curved geodesics on a Flamm's paraboloid illustrate the geometric view of gravity. The top view is important because we are modelling a two-dimensional movement. (*Source: Copyright by Matej Ryston*)

we tell the students now, remember, we are not at spacetime, yet!). Such is the viewpoint of general relativity. Gravity is not a real force, and therefore, there are still no external forces acting on the object, it is still free, and as such, it travels **locally** (meaning from point to point) straight. However, just like on a sphere or a hyperbolic paraboloid, even if you go straight from point to point, your overall trajectory is curved because of the curvature of the surface you are moving on – in the case of real bodies, curvature of space.

## PART 3: TIME "CURVATURE"

Now comes another important step in our seminar. Using only spatial curvature, we cannot explain the simplest of motions under the influence of gravity – free fall. Spatial curvature does not make you start moving; it only affects you if you are already in motion. This means that our description of gravity is not yet complete. Another clue to the incompleteness of our reasoning is that try as you may, you cannot create a closed orbit using geodesics on the Flamm's paraboloid. So, it cannot really explain planetary orbits. That is because we are missing an important part of relativity, the "curvature" of time.

First, let us explain what we mean by "curvature" of time because such a formulation must be understandably confusing. How can time be curved? By that, we mean that time does not "flow" the same way everywhere. Similar to space, time is curved or warped in the presence of gravity; in other words, it ticks differently at different places. More precisely, according to general relativity, time slows down in stronger gravitational fields, meaning at a higher altitude on the Earth, it should run faster.

We start with the famous Hafele-Keating experiment, where atomic clocks were flown on aeroplanes, both eastward and westward, and compared with atomic clocks left on the ground (Hafele and Keating, 1972). After describing the experiment to the students, we observe the results – taken from the paper itself, or using the dedicated Wikipedia page. What we see from the results is not only that there was indeed a discrepancy between the clocks but also that actually two relativistic effects were considered during the experiment. One effect was the height difference on the measurement of the clocks (so-called gravitational time dilation), which is the ominous time curvature we are talking about. The other was the effect of clocks on the ground and on the plane moving relative to each other (so-called kinematic or relativistic time dilation). This effect should be familiar to those students who already encountered special relativity. "Hold on"! I hear you say; this chapter is called "Introducing General Relativity without Special Relativity". And you are right. We do our best here not to depend on any knowledge of special relativity, but here we must cheat a little bit. And it is not really cheating. We simply state the fact that there is such an effect in special relativity called time dilation where observers moving relative to each other do not agree about the passage of time. We do not necessarily go into details (you could do this, of course, if you have time).

In my experience, students are mostly OK with this approach. Those who are already familiar with special relativity or at least have heard about some of its aspects nod in agreement, and those who have never heard of it just take it as another strange claim from a part of physics that they will get to later. In any case, to proceed correctly, we need to include the special relativity effect into our discussion because when testing gravitational time dilation on our planet, it is almost impossible to avoid it. Even if we consider the case of two clocks separated by height only, seemingly not moving, due to the rotation of the Earth, both clocks are moving on separate circles of different radii, thus moving with respect to each other.

We also mention some of the more recent experimental results. Thanks to the new generation of very precise atomic clocks, Chou et al. (2010) have been able to produce relativistic time dilation when two clocks were separated in elevation by only about 33 cm or moved with a relative speed of less than 10 $m \cdot s^{-1}$. Students are usually fascinated and amused by the fact that their head is older than their feet.

Now, we are finally able to answer the question regarding free fall. Because gravity is not just the curvature of space, nor is it just the curvature of time, it is both. Gravity is the curvature of spacetime.

## PART 4: CURVATURE OF SPACETIME

Let us consider the situation of a stone initially at rest at some altitude. Also, imagine for now that gravity is switched off and the stone is keeping still at rest. We can draw a graph depicting the dependence of the stone's height on time. Since the stone is not moving, we get a very simple constant function (Figure 24.6). Surprisingly for students, we have just drawn a **spacetime diagram**. That is a diagram where one axis represents time and the other represents space (in this case, simplified into one radial dimension, or simply the height). If this diagram represents spacetime, the curve of the graph (in this case, the line) represents the trajectory of the stone in spacetime – what we call its **worldline**. Even though the stone is not moving in space, just because it exists (it "experiences" passage of time), it moves through spacetime.

Now, we switch on the gravity. According to general relativity, spacetime gets curved, but this means that also the worldline of the stone must curve. We can illustrate this by drawing the diagram (two perpendicular lines) on a curved surface such as a ball or a balloon (see Figure 24.7). For the worldline, which is, in fact, a geodesic, we use the strip of paper as before. We immediately see the result. The worldline (originally going only in the time "direction") curves towards lower height. Just because it exists in curved spacetime, the stone must start moving in space; therefore, it starts to fall freely.

We have simultaneously introduced spacetime as the combination of space and time, and shown that in order for our geometrical description of gravity to work, we indeed need to treat it as the curvature of spacetime. That is the conclusion of our seminar.

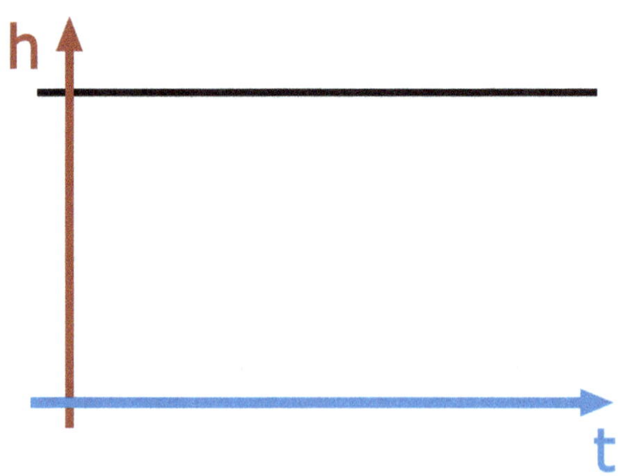


24.6
The graph of the stone's height as a function of time is also a spacetime diagram (even though that is traditionally drawn with a vertical time axis). (*Source: Copyright by Matej Ryston*)

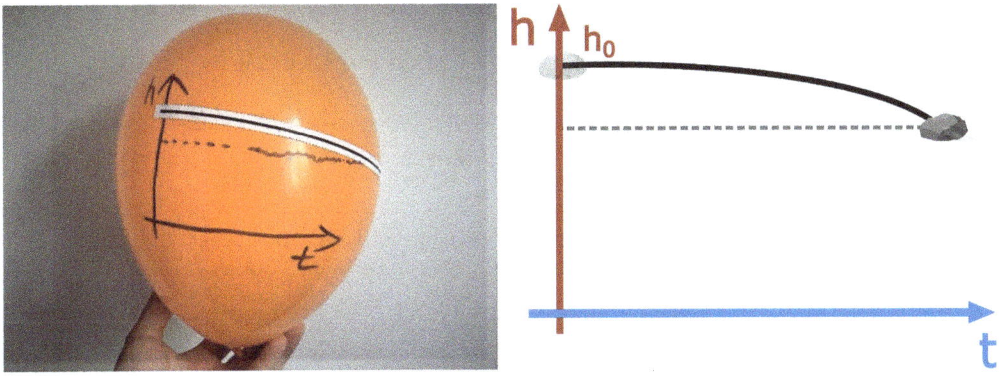


24.7
A makeshift spacetime diagram with the stone's worldline being a geodesic on a balloon surface representing curved spacetime (left). Mapping of the geodesic in the left picture back to the original spacetime diagram (right).
(*Source: Copyright by Matej Ryston*)

## CONCLUSION

The seminar, as described above, takes approximately three standard teaching lessons (45 minutes), especially because of the hands-on activities. Of course, all the mentioned activities can be worked with in more depth if time permits.

So far the seminar has taken place five times with students and also three times with teachers as part of Czech physics teacher conferences. After the end of the seminar, students fill out a quick feedback form.

The questionnaire included open items regarding the content of the seminar, asking for both positive (well-made points or interesting activities) and negative (missing or not fully explained) mentions. The most important finding came from the answers about positive features. Students mentioned many different aspects such as *three-dimensional-printed models*, *the curvature on the globe*, *visualising curvature*, *making paper models*, and many more. The point is that different students often picked different parts of the same activity (such as three-dimensional-printed models vs. paper models). The conclusion which I took from this feedback is that different people require and respond to different approaches; therefore, as teachers, we should try to offer more than one possible approach to a given concept or topic.

Hopefully, this outline of a one-time relativistic seminar has been beneficial to your plans of teaching general relativity. It is truly a fascinating subject, and I wish you good luck and a lot of fun teaching it.

## FURTHER READING

E-resources: Gravity versus geometry
Chapter 10: Models and analogies in general relativity
Chapter 14: Using the language of gravity to teach about space, time, and matter in general relativity

## REFERENCES

Chou, C., Hume, D., Rosenband, T. & Wineland, D. (2010). "Optical clocks and relativity." *Science*, 329(5999), 1630–1633.

Hafele, J. & Keating, R. (1972). "Around-the-world atomic clocks: Observed relativistic time gains." *Science*, 177(4044), 168–170.

Kersting, M. & Steier, R. (2018). "Understanding curved spacetime." *Science & Education*, 27(7–8), 593–623.

Ryston, M. (2019). "Interactive animations as a tool in teaching general relativity to upper secondary school students." *Journal of Physics: Conference Series*, 1286, 012049.

# 25 Towards understanding Einsteinian physics education

## *Relativity education as an exemplar*

*Hongbin Kim and Gyoungho Lee*

**WHAT?**

This chapter suggests a view towards physics education that reflects Einstein's perspective on the purpose, content, and methods of teaching physics. Hongbin Kim and Gyoungho Lee describe the South Korean context in which this view has been integrated into physics teacher education programmes.

**FOR WHOM?**

This chapter is intended for readers who wish to obtain historical and philosophical background knowledge in Einsteinian physics education, and also, for those who wish to further their cultural content knowledge in special and general relativity.

### INTRODUCTION

> So many people today—and even professional scientists—seem to me like somebody who has seen thousands of trees but has never seen a forest. A knowledge of the historic and philosophical background gives that kind of independence from prejudices of his generation from which most scientists are suffering. This independence created by philosophical insight is—in my opinion—the mark of distinction between a mere artisan or specialist and a real seeker after truth.
>
> A letter from Albert Einstein to Robert A. Thornton, 7 December 1944 (Howard, 2010)

Can *modern physics education* be carried out from the viewpoint of looking at the forest, as Einstein mentioned above, in order to help students have a bird's-eye view? This may seem a rhetorical question. One may say, "Yes, a bird's-eye view is important. Nevertheless, modern physics is so much more complicated than classical physics that for now, we need to concentrate on explaining individual facts". In many cases, there tends to be a bias toward individual phenomena and facts because there are so many extraordinary and fancy themes in modern physics (quantum phenomena, black holes, etc.). Particularly, this tendency to focus on individual facts can become even more prevalent in teaching modern physics since many students are just interested in these *facts* without understanding. Facing the complexity of modern physics, many educational researchers have developed various pedagogical methods for explaining such themes (e.g., Zollman, 2016). However, one critical appraisal could be raised: "Have these attempts helped students see not only the trees but also the forest of physics and led students to a deeper understanding of physics?" Unfortunately, we could not find positive responses from previous studies. In the face of this challenging situation, we as physics educators need to make diverse efforts to study how to teach modern physics.

This chapter introduces how we practice *Einsteinian physics education* in South Korea, which can help our students see the forest of physics and deeply understand the nature of physics. First, we explain the meaning of *Einsteinian physics education* in the Korean context and introduce our recent studies on it, focussing on relativity education.

## WHAT DOES EINSTEINIAN PHYSICS EDUCATION MEAN?

Interest in relativity is increasing gradually in science education all over the world. In South Korea, relativity has been included in the national curriculum since the 2009 reformed curriculum. The main goal of this change is to teach students how our understanding of space and time has been changed from a Newtonian view to the modern view. Although students are interested in this topic, we need more evidence-based research as to what students really learn and how we can help them deepen their understanding of nature and science. This kind of curriculum change can also be a serious challenge for teachers, and often, teachers have great difficulty in teaching relativity. For instance, teachers may wonder about the right level of exposition that is relevant for teaching relativity in their classes.

Since these are very complicated problems, we need to share diverse experiences and promote collaborative efforts to overcome these problems in modern physics education. In this regard, an international collaboration called Einsteinian Physics Education Research emerged from this need (Blair, Henriksen, & Hendry, 2016).

However, Einsteinian physics education research is just setting the preliminary stage. This means that we have much work to do, from basic to advanced (from conceptual issues to applications). For instance, because *Einsteinian physics education* is a term that only a few researchers have just begun to use, we need to pay more attention to elaborating on the definition of *Einsteinian physics education*, continuously studying and developing the definition and meaning of this term. We may immediately have questions about what *Einsteinian physics education* actually means? For instance, does it just mean teaching our new understanding of the universe, such as relativity and quantum theory before others? Or does it mean introducing these topics without teaching their philosophical and physical points of view? Is the term defined only by content knowledge that originated from Einstein? Or is the name "Einstein" just used as a symbol of 20th-century physics? Probably not.

As we know, Einstein's contribution to relativity and quantum theory is indisputable. However, in our understanding, it is not appropriate to see Einstein's works as limited to specific branches of physics. Instead, Einstein's work is characterised by drawing a consistent connection between old and new theories in physics. For example, Einstein's paper on the special theory of relativity shows that the theory emerged in the process of resolving the logical conflict between Newton's classical mechanics and Maxwell's field theory of electromagnetism (Einstein, 1905a). Moreover, Einstein's paper on the photoelectric effect shows that the quantisation of light is a concept that emerged from an attempt to understand radiation as something like molecules in Boltzmann's kinetic theory of gases (Einstein, 1905b). If we really want to use *Einsteinian physics education* as a term that shows appreciation of Einstein's genuine works, it should not be simply defined from physics content knowledge only. Rather, *Einsteinian physics education* would be a concept that encompasses the purpose of teaching, content knowledge, and teaching methods. In sum, we propose *Einsteinian physics education* as a new type of physics education that reflects Einstein's perspective on physics education and concerns the purpose (*why*), content (*what*), and methods (*how*) of teaching physics.

How then can we find Einstein's perspective on physics and teaching physics? There may be many ways to do this. Recently, we have noted that the book *The Evolution of Physics*, which was written by Einstein and his colleague, Infeld, reflects Einstein's perspective very well (Einstein & Infeld, 1938). Therefore, we decided to analyse *The Evolution of Physics* so that we could learn about Einstein's perspectives on the most important ideas for teaching physics. In the next sections, we present Einstein's perspective on special relativity and general relativity education as an exemplar of *Einsteinian physics education* as understood in the Korean context.

## WHAT ARE THE MOST IMPORTANT IDEAS IN TEACHING SPECIAL RELATIVITY?

We conducted this study by analysing *The Evolution of Physics* considering the following three questions (Kim & Lee, 2018): Why should we teach special relativity? What kinds of content knowledge should we teach in special relativity? How should we teach special relativity? To draw reasonable conclusions, we adopted a qualitative content analysis as an analysis method.

### *Why should we teach special relativity?*

This question is related to the purpose of special relativity education. This question examines what kind of *value* we intend to convey by teaching special relativity. Einstein wanted the readers to understand the intrinsic values of physics when readers learned special relativity. What then are the intrinsic values of physics? According to Einstein, they include (1) the attempt of the human mind to understand the *reality* of nature, (2) new physical and philosophical *viewpoints*, and (3) the characteristics of special relativity as a fundamental theory: *simplicity*, *harmony*, and *universality*. First, Einstein emphasised the field concept as a physical reality in contrast to the particle concept that is the main actor in the mechanical point of view. Although special relativity is not a theory which is based on field concepts (it is not a field theory), Einstein seems to have emphasised the need to modify Newtonian mechanics in terms of Maxwell's electromagnetism, which is a prototype of a field theory. Second, Einstein emphasised the changing of our viewpoint on nature through special relativity. This physical and philosophical viewpoint is mainly related to the spacetime continuum as an inseparable object. That is, spacetime is not just a name that can be used to refer to space and time combined at once, but is itself a single physical object. Third, Einstein wanted readers to understand the general characteristics of physical laws through special relativity. He emphasised the importance of *simplicity*[1] and *harmony*[2] as the main guidelines for special relativity and, in particular, the importance of the *universality* of the laws of physics by emphasising the term *invariant*, which means that the form of physical laws is the same for all inertial frames.

### *What should we teach in special relativity?*

This question deals mainly with the content knowledge of special relativity. We found that Einstein wanted the readers of *The Evolution of Physics* to achieve the purpose of teaching special relativity through learning integral knowledge that has the whole structure and overall flow (e.g. storyline) of the theory. An example of a type of integral knowledge would be cultural content knowledge. The cultural content knowledge is a theoretical framework of physics content knowledge that considers science a dynamic human activity of the construction of tentative – that is, not yet completed – but objective knowledge (Galili, 2012). The cultural content knowledge consists of three components: *nucleus*, *body knowledge*, and *periphery*. The *nucleus* contains a viewpoint of the physical world and fundamental principles. Individual phenomena and facts belong to *body knowledge*, which is important but not more important than the nucleus because *body knowledge* is induced from the nucleus. The *periphery*

1 This refers mainly to the simplicity of the assumptions.

2 This refers to harmony between various laws and principles. In particular, Einstein emphasised the harmony between Maxwell's field laws and the Galilean relativity principle in special relativity.

is what we cannot explain by the nucleus such as unresolved phenomena, and principles that conflict with the nucleus, which show that physics is still in the making.

From a cultural content knowledge perspective, the *nucleus* that Einstein emphasised is a new viewpoint of the physical world and fundamental principles. In this regard, Einstein devoted much of the book, *The Evolution of Physics* to explaining the two postulates of special relativity: the reason why the Galilean relativity principle (that is, the laws of mechanics are the same in all inertial frames) should be extended to *all* physical laws, and why the speed of light is constant. Individual phenomena and facts such as time dilation and length contraction belong to *body knowledge*, which is necessary but not more important than the *nucleus* because *body knowledge* is based on the principles contained in the nucleus. Einstein noted that while these individual facts of special relativity are worth teaching, it is more important for students to eventually learn the characteristics of physics such as the inner consistency of the physical theory. Finally, Einstein also emphasised the anomalous phenomena in the *periphery*. That is, Einstein seemed to think that we should also teach phenomena that we cannot explain using special relativity such as unresolved phenomena and principles that conflict with special relativity, which show that physics is still in the making.

***How should we teach special relativity?***

This question is about teaching methods. First, Einstein showed ways to lead readers to participate in the process of the unification and generalisation of the theory in the history of physics and within its overall structure. That is, Einstein wanted to give a vivid picture of how special relativity appeared in the history of physics by presenting dialogues between physicists, who represented old and new theories. In this manner, readers can see how physical laws are unified into a single principle and see the overall structure of physics in ways that are not just limited to special relativity. Second, Einstein instructed the readers to understand the nature of science, meaning that science is not complete but is still in the making. This is why Einstein tended to keep conclusions open, saying that "still more difficult problems arise which have not been definitely settled as yet" when he finished a section or a chapter.

Based on our analysis, we found Einstein's emphasis on special relativity education: (1) Are we pursuing the *intrinsic values* of physics as the purpose of teaching special relativity? (2) Are we teaching *integral knowledge* as content knowledge? and (3) Are we using a *hermeneutic approach* (that is an approach concerning interpretation) and a *communicative approach* as the teaching method? Einstein's emphases and their opposites are summarised in Figure 25.1.

Using Einstein's emphasis on special relativity education, we analysed research papers on special relativity education published in South Korea over the last five decades. We found that many studies adopted a perspective that

| | Einstein's Emphasis* | The Opposite of Einstein's Emphasis |
|---|---|---|
| Purpose (why) | Intrinsic value<br>• To figure out the attempt of the human mind to understand reality of nature<br>• To understand a new physical and philosophical viewpoint<br>• To understand the fundamental principles that represent *simplicity*, *universality*, and *harmony* | Instrumental value<br>• To learn new scientific concepts explaining physical phenomena<br>• To satisfy students curiosity<br>• To understand new technological applications in the modern era<br>• To improve students scientific literacy in order to be citizens of a democratic society by utilising scientific thinking |
| Content Knowledge (what) | Integral knowledge<br>Treatment of the whole structure and overall flow focussing on key words<br>• Viewpoint of the physical world/fundamental principles<br>• Body knowledge which represents inner consistency<br>• Knowledge in periphery | Fragmentary knowledge<br>Piecewise/fragmented propositions, concepts, and facts<br>• Relativity of simultaneity<br>• Time dilation<br>• Length contraction<br>• Lorentz transformation<br>• Equivalence of mass and energy |
| Teaching Approach (how) | Hermeneutic/communicative approach<br>• Ways to lead students to participate in the process of the unification/generalisation of the theory within the overall structure<br>• Ways to instruct students to understand the nature of science that is not completed but continuing, as an inquiry about the reality of nature | Partial/Functional approach<br>• Ways to make students become familiar with individual/partial concepts and content related to relativity without an overall context (ways of better understanding, teaching methods, *etc.*)<br>• Ways to instruct students to think that science is a set of completed knowledge |

25.1 What Einstein emphasised from *The Evolution of Physics* on special relativity education and its opposite. (Adapted from Kim & Lee (2018). *Emphasis from *The Evolution of Physics*.) (*Source: Reproduced with permission from the Journal of the Korean Physical Society*)

significantly differed from Einstein's perspective. Regardless of the researchers' intentions, their studies are likely to lead other research on relativity education and educational practices toward the opposite of Einstein's emphasis, such as toward instrumental purpose, fragmentary content knowledge, and a partial or functional approach.

## WHAT ARE THE KEYWORDS IN TEACHING GENERAL RELATIVITY?

Just as we identified Einstein's view of special relativity education, we analysed *The Evolution of Physics*, focussing on the keywords that Einstein emphasised in teaching general relativity. We identified seven keywords that can guide teaching and instruction in general relativity (Lee, Kim, & Lee, 2020a):

### *Abolition of inertial frame*

Einstein emphasised that the concept of inertial frames, which is assumed in Newtonian mechanics and even in special relativity, has been *abolished* in general relativity. This is comparable to the concepts of "ether" that appeared and then disappeared in the history of physics. In *The Evolution of Physics*, Einstein described the logical process of abandoning the concept of (global) inertial frames in the process of developing the general theory of relativity.

### *General covariance*

Einstein emphasised that the laws of physics are the same in all coordinate systems using the term *general covariance*. In particular, general covariance is the most important characteristic that a good physical law should have.

### *Equivalence principle*

The equivalence principle states that there is a complete physical equivalence between a gravitational field and a corresponding acceleration of the reference frame. While it may be easy to expect that Einstein would emphasise this fact of the equivalence principle, one remarkable feature of Einstein's explanation of this is that the coordinate transformation can eliminate or produce gravity. That is, the gravitational field depends on observers. After all, according to Einstein, the main implication of the equivalence principle is the *arbitrariness of gravity*.

### *Non-Euclidean geometry*

Even though *The Evolution of Physics* is a book for non-specialists, Einstein did include the explanation of non-Euclidean geometry in a detailed manner. For instance, Einstein explained that the ratio between the radius and the circumstance of a circle could be different according to different observers. Einstein considered geometry a branch of physics. In other words, geometry has an experimental character by which we can understand the geometrical character of our physical world, which Einstein referred to as *practical geometry* (Einstein, 1922).

### *Spacetime*

Einstein explained in detail the necessity of combining time and space, and treated *spacetime* as a single object, emphasising a change in our worldview.

### *Field (equation)*

Einstein explained that general relativity is a field theory, comparing it to Maxwell's electromagnetic field theory. In particular, Einstein contrasts the *mechanical view*, which describes phenomena based on the interaction between point particles, and the *field view*, which describes phenomena based on the dynamics of the field. In this way, Einstein presented that general relativity respects Maxwell's field theory. That is why a great deal of effort has been devoted to explaining the meaning of *the structure law*, which describes the dynamics of the field, in *The Evolution of Physics*.

### *Observational phenomena*

In *The Evolution of Physics*, Einstein also introduced representative phenomena (e.g. bending of light, the precession of the perihelion of Mercury) that general relativity predicts, and emphasised the importance of confirmation of the theory through observations. However, this is not emphasised as much as the previous six keywords.

Based on these seven keywords, we developed a framework for analysing textbooks that explain general relativity. We then analysed six "modern physics"

| Text book | Abolition of Inertial Frame | General Covariance | Equivalence Principle | Non-Euclidean Geometry | Spacetime | Field (equation) | Observational Phenomena |
|---|---|---|---|---|---|---|---|
| A | - | - | M | - | M | - | S |
| B | - | M | S | M | M | - | S |
| C | - | M | S | - | - | - | S |
| D | M | M | M | - | - | - | S |
| E | - | M | S | M | M | M | S |
| F | - | - | M | - | M | - | S |

A: Beiser (6th ed.)
B: Serway, Moses & Moyer (3rd ed.)
C: Tipler & Llewellyn (5th ed.)
D: Harris (2nd ed.)
E: Krane (3rd ed.)
F: Thornton & Rex (4th ed.)

S: satisfactory explanation / M: simple mention / N: no mention

25.2
Results of analysis of the general relativity chapters in "modern physics" textbooks. (Adapted from Lee, Kim, & Lee (2020a).) (*Source: Copyright by Hongbin Kim, Gyoungho Lee*)

textbooks that are widely used as main textbooks for college students all over the world. The results show a vast difference in emphasis between Einstein and these textbooks. These are the main characteristics of the description on general relativity in "modern physics" textbooks: (1) poor descriptions of the reality of physical objects (inertial frame, field, spacetime continuum), (2) poor descriptions of the geometrical properties of the physical world (non-Euclidean geometry), (3) poor descriptions of the universality of physical theories (general covariance), and (4) relatively rich descriptions of observational phenomena (see Figure 25.2).

Taking into account Einstein's emphasis, teachers could explain the above seven keywords in their classroom in appropriate ways. However, understanding a theory means understanding not only the keywords that constitute the theory but also its structure. One example illustrating the overall structure and flow (including storyline) as well as the above keywords is shown in Figure 25.3.

In South Korea, we implemented this framework of general relativity to explain not only the keywords, which Einstein emphasised but also the overall structure and meaning of general relativity. For instance, in the previous general relativity courses for physics teachers, we as instructors summarised the contents (including keywords) by drawing the framework and provided them as a big picture as shown in Figure 25.3. After exposure to the framework, we had a discussion time with physics teachers about what is new for them, and what remains unclear. At the end of the courses, we have found responding positively to the framework. For example, some teachers considered the role of the framework as a guide or a map.

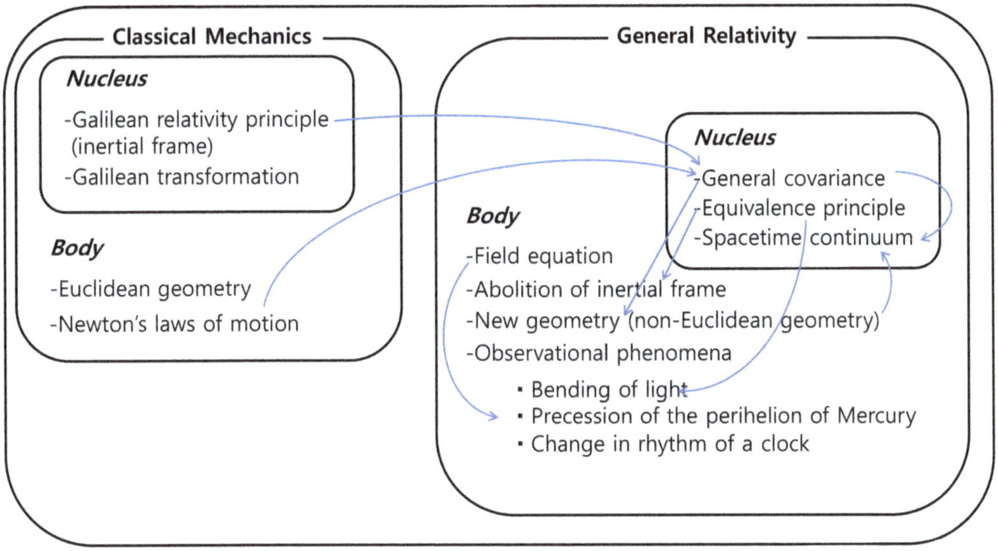


25.3
A framework of general relativity, which shows the relation among the seven keywords of the theory. (Adapted from Lee, Kim, & Lee (2020b).) (*Source: Copyright by Hongbin Kim, Gyoungho Lee*)

*Teacher a:* "Sometimes I got lost in learning relativity because I had no idea about which is central and which is peripheral. [...] It is possible to distinguish core idea from body knowledge via the framework of relativity so that I would not get lost".

*Teacher b:* "This [the framework] would play the role of a compass".

## CONCLUSION

In this chapter, we presented Einsteinian physics education as a new physics education area. Einsteinian physics education in the Korean context is a physics education that reflects Einstein's perspective on physics education, which concerns the purpose (*why*), the contents (*what*), and the methods (*how*) of teaching physics. We provided two examples of Einsteinian physics education, as understood in the Korean context. We propose a vision for reflecting on and providing an outlook on physics teacher education programmes, secondary school physics curricula, and related materials based on our perspective of Einsteinian physics education. Figure 25.4 presents our ongoing work and future directions.

The first step (Phase A) is to establish a conceptual foundation for Einsteinian physics education. To accomplish this goal, we have examined Einstein's perspective on physics and teaching physics by analysing his writings (e.g., books, letters, and academic papers). As an outcome, we have developed a framework of relativity and have developed criteria for evaluating

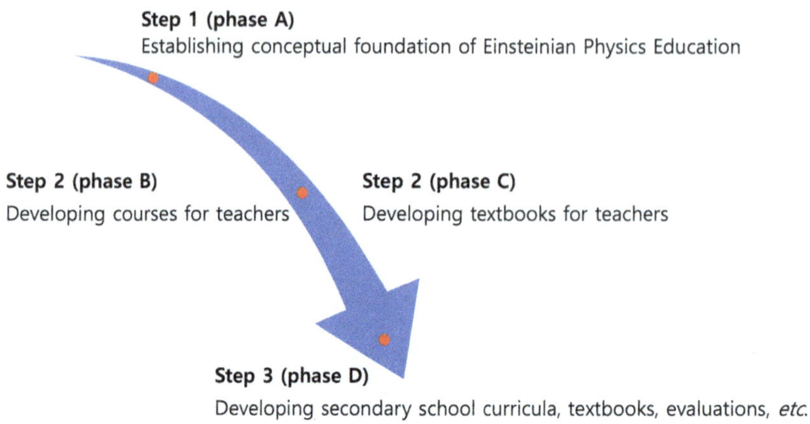


25.4 Research programme toward Einsteinian physics education in Korea. (*Source: Copyright by Hongbin Kim, Gyoungho Lee*)

textbooks and research papers on relativity (e.g., Kim & Lee, 2018). The second step is two-fold. First is course development for teachers in colleges of education (Phase B). From 2015 to 2019, we developed general relativity lectures for teachers. We have used diverse forms for those lectures: (1) using an undergraduate-level textbook, (2) using Einstein's 1916 original paper, and (3) using Einstein's *The Evolution of Physics* with the original paper as supplementary material. We are still working on the development of a general relativity course for teachers. This kind of work will be expanded to other branches of physics (electromagnetics, quantum mechanics, *etc.*). We are also working on general relativity textbook development for teachers (Phase C). Because there is an inseparable relation between course development and textbook development, this work will be conducted parallel to course development. Finally, the third step (Phase D) will be to develop secondary school curricula, textbooks, and evaluations based on the achievements of previous steps.

## FURTHER READING

Chapter 2: Times for changing paradigms in science and in education
Chapter 14: Using the language of gravity to teach about space, time, and matter in general relativity
Chapter 22: Introduction of Einsteinian physics to the upper secondary school physics curriculum in Scotland: Experiences and observations

## REFERENCES

Blair, D., Henriksen, E. & Hendry, M. (2016). "Why don't we teach Einstein's theories in school?" *The Conversation*, 69991.

Einstein, A. (1905a). "On the electrodynamics of moving bodies." *Annalen der Physik*, 17, 891.

Einstein, A. (1905b). "Concerning a heuristic point of view toward the emission and transformation of light." *Annalen der Physik*, 17, 132.

Einstein, A. (1922). *Sidelights of Relativity.* London: Methuen & Co Ltd.

Einstein, A. & Infeld, L. (1938). *The Evolution of Physics: From Early Concepts to Relativity and Quanta.* New York: Simon and Schuster.

Galili, I. (2012). "Promotion of cultural content knowledge through the use of the history and philosophy of science." *Science and Education,* 21(9), 1283.

Howard, D. (2010). "Einstein's philosophy of science," The Stanford Encyclopedia of Philosophy, Edward N. Zalta (Ed.).

Kim, H. & Lee, G. (2018). "Reflection and outlook on special relativity education from the perspective of Einstein: Focusing on research papers published in Korea." *Journal of the Korean Physical Society,* 73(4), 422.

Lee, K., Kim, H. & Lee, G. (2020a). "An analysis of modern physics textbooks from Einstein's perspective: Focusing on core concepts in general relativity." To be submitted.

Lee, K., Kim, H. & Lee, G. (2020b). "An analysis of modern physics textbooks from Einstein's perspective: Focusing on framework of general relativity." To be submitted.

Zollman, D. (2016). "Oersted lecture 2014: Physics education research and teaching modern physics." *American Journal of Physics,* 84(8), 573.

# Teaching Einsteinian Physics in Schools Glossary Terms

**acceleration model of gravity** The acceleration model of gravity is a teaching model in general relativity that serves as an intermediate step between rejecting Newton's force model and introducing Einstein's curved spacetime model of gravity. The model makes use of the principle of equivalence to explain gravity near the surface of a planet as an upwards acceleration. (see teaching model) (see principle of equivalence)

**activity-based learning** Activity-based learning is a process of learning characterised by doing activities as opposed to asking students to simply listen and take notes. It encourages students to actively participate in their own learning experience through practical activities such as problem-solving and independent investigation.

**ageing Earth model** The ageing Earth model is a teaching model in general relativity that illustrates gravitational time dilation. The model provides an example of how the difference in the gravitational potential of the Earth leads to time passing differently at the surface and the centre of the Earth. (see teaching model) (see time dilation)

**amplitude** Amplitude is a measure of the displacement of a wave, such as the height from a peak to a trough. The square of the amplitude gives the intensity of a wave. Amplitude can be measured as peak to peak amplitude or root mean square (RMS) amplitude. Mains electricity in many countries has an RMS amplitude of 220 Volts, but its peak to peak amplitude is more than 600 Volts. (see wave)

**analogy** Analogies in science education are instructional tools that can help communicate scientific concepts. Analogies are statements that characterise one thing in terms of another; they are relations of "is like" between some aspect of what is to be taught and an element of a teaching model. (see teaching model)

**annihilation** In particle physics, annihilation (from latin: nihil = nothing) describes the process that occurs when a particle collides with its antiparticle so that both particles disappear, releasing energy carried by other particles such as photons. (see antimatter)

**antimatter** Antimatter is matter made from antiparticles such as anti-protons, anti-neutrons and positrons. Each type of matter particle has a corresponding type of antiparticle, with the same mass, but opposite charges (such as opposite electric charge). Antimatter annihilates when it meets matter, thereby creating other particles, often photons. (see annihilation)

**atomic mass** The atomic mass of a single atom is its total mass. Protons and neutrons in the atomic nucleus make up the largest part of the total mass, with minor contributions coming from electrons and the binding energy that keeps the atom together. The atomic mass is typically expressed in atomic mass units, abbreviated amu: also called one Dalton. 1 amu = 1 Dalton= $1.66 \times 10^{-27}$ kg.

**ballistic light deflection** Ballistic light deflection (also known as Newtonian light deflection) is a method to calculate the deflection of light. The underlying idea is that light rays are composed of tiny particles that are influenced by the Newtonian gravitational attraction in the same way as planets are. The amount of light deflection, according to this method, is one half of the amount predicted by general relativity.

**behaviourism** Behaviourism is a theory of learning that emphasises observable behaviour. Based on scientific principles derived from learning tasks with animals and humans, the main idea behind behaviourism is that human behaviour is predictable and controllable. (see cognitivism) (see sociocultural theory)

**big bang** In many of the cosmological models describing an expanding universe, the cosmos has evolved from an early hot and dense state known as the big bang, or the big bang phase. In general relativity, the beginning of that phase is a spacetime singularity, an event where curvature and matter density take on infinite values. That singularity is also called the big bang. Our own universe is thought to have emerged from its big bang phase 13.8 billion years ago.

**blackbody radiation** Blackbody radiation is radiation emitted from something that perfectly absorbs incoming radiation, like a black stove. The spectrum of radiation from hot objects like a red-hot steel bar or a furnace or even a human body is a close approximation to black body radiation. The spectrum (range of wavelengths) depends on the temperature, hence expressions red hot and white hot that describe temperature in terms of colour.

**black hole** A black hole is a region of space in which the gravitational field is so strong that no light can escape. General relativity predicts the existence of black holes that often come into being as the stellar remnant of some types of massive stars. (see Schwarzschild solution)

**boson** Bosons make up one of the two classes of particles, the other being fermions, that are distinguished by whether or not their wavefunctions are symmetric or antisymmetric. Bosons have an integer spin (0, 1, 2, ….) which lead to a symmetric wavefunction. An unlimited number of bosons can exist in the same state, unlike fermions. Photons and gravitons are fundamental bosons with spin

1 and 2 respectively, while helium-4 and phonons are composite bosons. (see fermion) (see wave function)

**classical physics** Classical physics has two different meanings. First, classical physics can describe the Newtonian theory of mechanics and gravity (classical mechanics) as opposed to relativity and quantum theories. Second, classical physics can describe all theories (including the theory of relativity) in which particles and fields can be assigned a definite state at each moment in time, as opposed to quantum theories. In both cases, classical physics i) describes physical quantities as continuous (i.e. they can have any value along a continuum) rather than quantized (i.e. they can only have certain values) ii) assumes physical systems to behave deterministically iii) assigns a definite state to particles and fields at each moment in time.

**classical determinism** The view of classical determinism entails that we can predict the outcome of a single event if we know all the initial conditions. Classical physics is deterministic. Quantum physics breaks with classical determinism because it cannot predict the outcome of single events. What quantum physics can predict with great accuracy is the statistical distribution of many events. (see statistical description)

**cognitivism** Cognitivism is a theoretical framework that focuses on mental processes, including how people perceive, think, remember, learn, and solve problems. Cognitivism developed as a response to behaviourism which did not explain cognition. (see behaviourism) (see sociocultural theory)

**Compton scattering** The phenomenon of Compton scattering illustrates that photons have momentum. Compton scattering describes the elastic collision between an X-ray or gamma photon and a weakly bound electron in an atom. Energy and momentum are conserved in the collision. (see photon)

**constructivism** Constructivism is a theory of learning that emphasises students' knowledge and experiences prior to instruction. According to this view, students use and build on their pre-existing understanding of the world to actively construct new knowledge. (see sociocultural theory) (see behaviourism)

**constructive interference** Constructive interference describes two or more waves at a certain location combining to produce a single amplitude equal to the sum of the amplitudes of the individual waves. Mathematically, when phasors co-align with each other they create a constructive interference. (see amplitude) (see phasor) (see destructive interference)

**contingency knowledge** Contingency knowledge is the knowledge possessed by teachers giving them the ability to cope confidently and knowledgeably with unexpected events occurring during teaching.

**Copenhagen interpretation** The Copenhagen interpretation of quantum physics, associated with Niels Bohr, is a philosophical position assuming that a quantum system does not have a definite physical state prior to measurement; rather, the act of measuring forces the system into one out of several possible states. According to this view, quantum physics can only predict probabilities for the possible outcomes of a measurement. The question of how to interpret quantum physics is still an ongoing philosophical dispute. (see realist view)

**cosmology** Cosmology is the science of the origin, development, and overall shape of the universe. Modern cosmological models describe an expanding universe, which emerged from a hot and dense big bang phase 13.8 billion years ago.

**cosmic expansion** Cosmic expansion describes the systematic change in distances between distant galaxies. In cosmological models, all such distances change in direct proportion to a time-dependent, universal cosmic scale factor. The raisin cake model is a teaching model that illustrates the cosmic expansion via a raisin cake that rises in the oven. (see raisin cake model)

**cosmic microwave background** The cosmic microwave background is the remnant radiation from an early stage of the Universe. Being the oldest electromagnetic radiation in the Universe, the cosmic microwave background holds information about the large scale structure of the Universe.

**Cultural Content Knowledge** Cultural Content Knowledge is a theoretical framework in science education that describes science as a dynamic human activity of constructing tentative but objective knowledge. The framework consists of three components: nucleus, body knowledge, and periphery. The nucleus contains a viewpoint of the physical world and fundamental principles. Individual phenomena and facts belong to body knowledge. The periphery is what we cannot explain by the nucleus, such as unresolved phenomena, and principles that conflict with the nucleus, which shows that science is still a process in the making.

**curvature (positive, negative)** Curvature is a geometric concept that has several specific meanings depending on the mathematical context. Broadly, the notion of curvature captures how much curves deviate from being straight lines or how much surfaces deviate from being flat planes. On positively curved surfaces such as a sphere or an ellipsoid, the shortest path between two points is an arc, parallel lines converge, and the angles of triangles add up to more than 180°. On negatively curved surfaces such as a hyperboloid or a hyperbolic paraboloid, the shortest path between two points is a part of a hyperbola, parallel lines diverge, and the angles of triangles add up to less than 180°.

**Design-based Research** Design-based Research is a methodological framework in education research that aims to build a bridge between research and practical classroom applications. A characteristic of Design-based Research are repeated rounds of development and close collaboration with practicing teachers.

**destructive interference** Destructive interference describes two or more waves cancelling each other out, to produce zero amplitude, and in the case of probability waves, zero probability amplitude. Mathematically, when phasors anti-align with each other they create a destructive interference. (see constructive interference) (see phasor)

**differential geometry** Differential geometry is the branch of mathematics that describes surfaces, spaces, and more abstract generalisations of spaces in terms of their geometric properties. Curvature is one example of such a geometric property that can be measured locally at each point in space. General relativity is formulated in the language of differential geometry and describes gravity as the curvature of spacetime. (see Euclidean geometry) (see general relativity)

**diffraction** Diffraction occurs when the propagation of classical waves (like ocean waves) or probability waves gets interrupted by obstacles or a slit. As a consequence of diffraction, the wave spreads out or reaches regions of geometrical shadow. (see wave)

**Five-E Lesson Plan** The 5-E Lesson Plan is a constructivist instructional model in which the phases of learning are based on engaging, exploring, explaining, elaborating, and evaluating. (see constructivism)

**elastic sheet model** (see rubber sheet model)

**electric field** The electric field describes a force pattern experienced by one charged object near to another charged object. The strength of the electric force on a standard unit of charge is called the electric field strength. The field strength is measured in units of Newtons per Coulomb, where Coulomb is the unit of electric charge; this is identical to the unit of volts per meter.

**electron** Electrons are subatomic elementary particles with a single quantum unit of negative electric charge and a single unit of quantum spin. Electrons have no detected internal structure, so they appear to be pointlike. Their mass of $9 \times 10^{-31}$ kg is 1836 times less than the mass of a proton. Because of their tiny mass, the quantum wavelength of electrons that depends on their velocity usually dominates their behaviour. (see particle)

**electromagnetic spectrum** The electromagnetic spectrum describes the range of all types of electromagnetic radiation that exist from radio waves, millimetre and submillimetre radiation, infrared and visible light to ultraviolet, x-rays and gamma rays. One can classify the electromagnetic spectrum by its wavelength (measured in metres), or equivalently by frequency. Frequency is measured in Hertz (HZ) and counts the number of waves that pass by a given point per second. The wavelength is the distance from the peak of one wave to the peak of the next. The larger the frequency, the smaller the wavelength. (see light) (see waves)

**Einstein, Albert** Albert Einstein (1879-1955) was a theoretical physicist who greatly contributed to the development of modern physics. Einstein developed the special and general theory of relativity and discovered the law of the photoelectric effect which paved the way for the development of quantum physics. Einstein received the 1921 Nobel Prize in Physics for his services to theoretical physics and his name has become synonymous with genius.

**Einstein's field equations** Einstein's field equations lie at the heart of general relativity, Einstein's theory of gravity, time, and space. The field equations relate the geometry of spacetime to the mass/energy distribution within it. Einstein used the tools of differential geometry to formulate and publish his field equations in 1915 (see general relativity) (see differential geometry)

**Einstein-First Project** A group of physicists and physics educators in Australia formed the Einstein-First project to introduce modern concepts of physics by reforming the school science curriculum. The Einstein-First project has developed hands-on learning resources for introducing general relativity and quantum physics in schools.

**Einstein's first law** Einstein's first law is a generalisation of Newton's first law. It states that objects that are not influenced by forces move along geodesic curves in spacetime. (see geodesic curve) (see inertia) (see warped-time model).

**Einsteinian Physics** Einsteinian physics is a branch of modern physics that is based on two major theories of physics: the theory of special and general relativity that describes space, time, and gravity at the cosmic scale, and quantum physics that describes the interactions of matter and radiation at scales down to the smallest subatomic particles.

**Einstein ring** An Einstein ring is a ring-like circular image of a distant light source that is produced by a gravitational lens. The ring arises for reasons of symmetry if the observer, the gravitational lens, and the light source are perfectly aligned. (see gravitational lens) (see gravitational lens effect)

**entanglement, entangled photons** Entangled particles (including photons) must be described as different parts of a system that shares the same quantum state. Thus, they cannot be described as independent of each other. A measurement on one particle, therefore, measures the quantum state of both particles. Entanglement entails that a measurement made in one place can have immediate consequences in a completely different place. (see measurement) (see quantum physics) (see local realism)

**Einsteinian Physics Education Research (EPER) collaboration** The international Einsteinian Physics Education Research (EPER) collaboration aims to pool and combine innovative learning approaches in Einsteinian physics. Members of the EPER collaboration acknowledge the need to find innovative ways to conduct Einsteinian physics education research on a global scale by disseminating learning resources and research results across a range of countries. (see Einsteinian physics)

**equivalence principle** see principle of equivalence

**Ether** The ether is a substance filling the whole universe, and in particular "empty space". Physicists have discussed the existence and nature of ether for centuries, and the concept has created many controversies. In the 19th century, physicists used the ether to explain the propagation of light waves. At the end of the 19th century, the Michelson-Morley experiment attempted to detect the existence of such a carrier of light waves by comparing the speed of light in perpendicular directions. While the results did not disprove the existence of ether, the experiment showed that there is a terminal value for the speed of light. This experiment, therefore, represents one of the fundamental tests of special relativity. In quantum physics, the modern concept of the vacuum of space resembles the concept of ether. (see special relativity) (see light postulate)

**Euclidean geometry** Euclidean geometry follows the postulates laid down by Euclid around 300 BC, in which lines that are parallel to each other remain parallel, and never intersect. In Euclidean geometry, the sum of the angles of a triangle is 180 degrees, and the ratio of circumference and diameter of a circle is equal to $\pi$. (see curvature) (see non-Euclidean geometry)

**event** An event (often also called spacetime event) is the spacetime analogue of an idealised point in space. an occurrence that is specified by a single time value and a single spatial location.

**event diagram** The event diagram is a graphical representation of spacetime that shows the position of events for subsequent moments in time. The event diagram can be used as a reasoning tool to perform thought experiments. (see event)

**event horizon** (see Schwarzschild solution)

**expanding space interpretation (of cosmic expansion)** The expanding space interpretation of cosmic expansion interprets the distance changes between Hubble-flow galaxies as distinct from those galaxies' motions through space, namely as expansion of the space between the galaxies (while the galaxies themselves remain at rest). (see Hubble flow) (see cosmic expansion)

**Fermat's principle** Fermat's principle is a fundamental principle in the optics of light rays in isotropic media. Its original formulation in 1657 (reaching back to Fermat) reads: *Light propagates along that path connecting two given points for which the travel time adopts a minimum.*

A more recent, generalised version (the principle of stationary optical path) reads: *Among all conceivable paths light might traverse between two given points, the one realised in nature is characterized by its optical path length – the product of refractive index and geometrical path length – adopting an extremal value.*

**fermion** Fermions are particles that have half a quantum spin unit. Electrons, protons and neutrons are examples of fermions. Fermions are unable to share the same quantum state. (see particle) (see boson)

**field** (see gravitational field)

**fission** Fission is the name given to a certain type of nuclear reaction in which one nucleus splits into two or more smaller nuclei. (see fusion)

**Flamm's paraboloid** Flamm's paraboloid is a funnel shape that serves as the basis of many visualisations of general relativity. It is a rotationally symmetrical parabolic surface which can be obtained by embedding the equatorial plane of the Schwarzschild spacetime for constant time in three-dimensional Euclidean space. (see curvature)

**force of gravity** A force is a physical quantity that measures an interaction between objects that leads to acceleration of these objects. In Newton's theory, the force of gravity is an attractive force between massive objects that causes the acceleration of falling bodies. In Einstein's theory, there is no force of gravity and objects in free fall are not accelerating. (see gravity)

**free fall** An object that is only influenced by gravity is in free fall. In classical mechanics, a freely falling object moves only under the influence of the force of gravity. In general relativity, there is no force of gravity. Freely falling bodies follow geodesic curves through spacetime. In both cases, freely falling bodies experiences weightlessness. (see geodesics) (see gravity)

**frequency** Frequency is the number of occurrences of an event per unit time. For example, an adult human heart beats at an average of 72 times in a minute which means its frequency is 72. Frequency also equals the number of peaks or troughs in a wave per unit time. Frequency is the inverse of wavelength. (see wavelength)

**fusion** Fusion describes two nuclei fusing into one usually with the release of energy. Fusion is the opposite of fission. An example is protons and neutrons combining to make helium in stars. (see fission)

**general covariance** General covariance is an important characteristic of physical laws that assumes that the laws of physics are the same in all coordinate systems.

**general relativity** General relativity is Albert Einstein's theory of space, time, and matter and our current best understanding of gravity. General relativity generalises special relativity and replaces Newton's force model of gravity by the spacetime model of gravity. According to general relativity, gravity is a manifestation of the curvature of spacetime. Matter tells spacetime how to curve, while curved spacetime tells matter how to move. This dynamic interplay between matter and the geometry of spacetime is captured by Einstein's field equations. (see Einstein's field equations) (see gravity) (see special relativity)

**geodesic (curve)** Geodesics or geodesic curves generalise the notion of "straight lines" to make the concept meaningful in curved spaces. A geodesic is the straightest possible path between two points in a (curved) space, or on a (curved) surface. In general relativity, freely moving particles and light move along geodesic curves. (see Einstein's first law)

**geometry** (see differential geometry)

**gluon** In the language of modern physics, one can say that the electromagnetic force is mediated by photons, which are the quanta of the electromagnetic field. Likewise, the strong nuclear force that holds atomic nuclei together is mediated by particles called gluons. They act like "glue" that hold quark together forming hadrons such as protons and neutrons. (see particle)

**GPS positioning** GPS positioning is a navigational system based on satellites: an observer receives a time signal and orbital data from at least four satellites and, knowing that light propagates at constant speed, uses that information to determine their own position. GPS satellite clocks need to be corrected for relativistic time dilation. If relativistic effects were neglected, there would be measurable satellite clock effects. (see time dilation) (see warped time)

**gravity** In Newton's theory, gravity is explained by an attractive force between massive objects that gives weight to physical objects. In Einstein's theory, gravity is not a force but the consequence of the curved geometry of space and time. (see general relativity)

**gravitational field** The gravitational field is a physical quantity that has a value for each point in spacetime and that can change with time. The gravitational field of an object encodes all the possible gravitational forces that the object can exert on a small test particle. By multiplying the gravitational field at a specific location with the mass of the test particle you obtain the gravitational force that the object exerts on the test particle at that location.

**gravitational frequency shift** When light (and other electromagnetic radiation) propagates away from a massive object, the wavelength of the light will increase. If light moves towards a massive object, its wavelength will decrease. This phenomenon is called gravitational frequency shift and it is closely linked to gravitational time dilation. (see time dilation)

**gravitational lens** A gravitational lens is a celestial body with a strong gravitational field that deflects light emitted by a distant source. Examples of gravitational lenses are black holes, galaxies and clusters of galaxies. (see gravitational lens effect)

**gravitational lens effect** The gravitational lens effect describes the deflection of light of a distant source such as a star or a galaxy under the influence of the gravitational field of another celestial body such as a star, black hole, galaxy, or cluster of galaxies that is located somewhere between the light source and the observer. The effect results in multiple or distorted images of the light source and in an amplification of its brightness. (see gravitational lens)

**gravitational waves** Gravitational waves are ripples or disturbances in the fabric of spacetime. Accelerating masses create gravitational waves that propagate from the source at the speed of light. Einstein predicted gravitational waves in 1916 and the detectors of the Laser Interferometer Gravitational-Wave Observatory (LIGO) first observed gravitational waves on Earth in 2015.

**gravitational time dilation** (see time dilation)

**Gullstrand-Painlevé spacetime** The Gullstrand-Painlevé spacetime is a solution of Einstein's field equations that describes a spherically symmetric black hole. The coordinates of this spacetime are adapted to a freely falling observer and are coordinate transformations of the usual Schwarzschild solution. (see Schwarzschild metric) (see Einstein's field equations)

**heuristics** In physics, heuristics is a tool of intuition to approximate complex physical systems by a few simple characteristics that might be found by studying particularly simple examples. Heuristics is a powerful driver of scientific investigation. (see intuition)

**Hubble flow** The Hubble flow describes the motion of astronomical objects due to expansion of the universe and serves as evidence for the big bang model. In an expanding universe, galaxy motion is a combination of distance changes due to cosmic expansion and motion due to gravitational influences in galaxy clusters (or simply due to initial conditions in the formation of a specific galaxy). Idealized galaxies without these additional velocity components are said to be "in the Hubble flow". (see cosmic expansion) (see big bang)

**hyperbolic paraboloid** A hyperbolic paraboloid is a type of surface that is purely negatively curved. Its shape reminds one of a horse saddle, therefore it is also known as a saddle surface. (see curvature)

**inertia** Inertia is the property of physical objects to resist changes in their motion. Newton described the principle of inertia in his first law of motion: an object not subject to any net external force moves at a constant velocity. (see Einstein's first law)

**inertial frame, inertial observer** An inertial frame or inertial frame of reference is a coordinate system in which an object with zero net force acting on it stays at rest or moves at a constant velocity. In general relativity, there are only local inertial frames which are characterised by observers in free fall, for whom the laws of physics are those of special relativity. (see principle of equivalence) (see inertia)

**interference** Interference is a phenomenon in which waves add or cancel each other to produce a resultant wave. Interference occurs whether or not the wave is a classical wave or a probability wave. In the case of quantum objects, interference is sometimes called quantum interference. (see constructive and destructive interference).

**intuition** In physics, intuition is a basic understanding of how things work and fit together that requires little or no mathematics. Intuition is a tool for deepening understanding that works in parallel with any mathematical development of a physics subject.

In physics education, the development of physical intuition is a crucial goal. New concepts should be introduced using both intuition and mathematics, at whatever level is appropriate for the student. (see heuristics)

**isotope** Isotopes are atoms of one element with the same number of protons, but different numbers of neutrons.

**The Karlsruhe Physics Course** The Karlsruhe Physics Course is a teaching concept for the lower and upper secondary school. The course is available in the form of textbooks for the students and teaching aids for the teachers. It was developed at the University of Karlsruhe and has been tested in numerous schools in Germany and other countries. The focus of the course is on the "substance-like" quantities energy, momentum, entropy, electric charge and amount of substance. The main feature of the course is that it draws analogies between the classical fields of science: mechanics, thermodynamics, electrodynamics and chemistry. This makes the lessons both compact and easy to understand.

**Kilonova** A kilonova is a bright explosion of nuclear matter caused by the collision of two neutron stars. (see stellar evolution)

**knowledgeable other** A knowledgeable other is a person bringing expertise (subject knowledge, pedagogical knowledge or facilitation skills) to a professional learning situation. (see pedagogical content knowledge)

**length contraction** Length contraction is a phenomenon of special relativity that occurs when objects move at speeds close to the speed of light relative to an observer. The observer measures the length of the moving object to be shorter than the length measured in the rest frame of the object. (see special relativistic phenomenon) (see observer)

**LHC** The LHC is the Large Hadron Collider, the world's largest particle accelerator spread over a circumference of 27 km beneath the Swiss-French border near Geneva. The LHC is used to study subatomic particles by colliding protons or ions at velocities so close to the speed of light that the mass-energy is of the order of $10^4$ times the proton rest mass energy.

**light** In classical physics and relativity, light is the visible part of the electromagnetic spectrum, i.e. the part that can be perceived by the human eye. The wavelength of light lies between 400-700 nanometers. In empty space, light always propagates at a definite velocity of 299,792,458 meters per second which is one of the postulates of special relativity.

In quantum physics, the particles of light, the photons, are quantum objects which have both wave and particle properties. Quantum mechanics provides mathematical models that allow us to make predictions about nature at the smallest scales. Depending on the experiment, physicists treat light either as waves or particles. (see quantum physics) (see electromagnetic spectrum) (see special relativity) (see photon) (see wave-particle duality)

**light cone (future light cone, past light cone)** In special and general relativity, a light cone is the set of all spacetime paths (worldlines) of light signals that can be emitted or received at a single event. The future light cone is the set of all spacetime paths of light-signals that can be emitted at an event; the past light cone is the set of all spacetime paths of light-signals that can be received at an event. The light cone is the union of the future and past light cones. (see spacetime perpendicularity) (see light clock diamonds)

**light clock** A light clock is a common device in special relativity to illustrate relativistic phenomena such as relativistic time dilation. The light clock is designed via a light-signal that reflects periodically between two mirrors kept at fixed separation. Counting how many signals occur along the way allows for measuring time. (see special relativity) (see time dilation)

**light clock diamonds** A light clock diamond is a figure on a spacetime diagram produced by the spacetime paths (worldlines) of light-signals that are reflected in a light clock. Light clock diamonds can be interpreted as the intersection of the future light cone of the emission event and the past light cone of the reception event. In the literature, other names for light clock diamonds include "casual diamond", "optical parallelogram" and "light rectangles". (see light clock) (see worldline) (see light cone)

**light postulate** Historically, the light postulate is one of two postulates of special relativity. In empty space, light always propagates at a definite velocity of 299,792,458 meters per second, also called c, independent of the state of motion of the emitting body. Combined with the principle of relativity, this means that light has the same finite speed relative to all inertial observers or in all inertial frames. This observation leads to the notion of the absolute speed of light. (see principle of relativity) (see special relativity) (see light)

**local realism (locality principle)** Local realism describes a principle in physics that assumes, first, that objects have physical properties (such as velocity or position) independently of whether or not a measurement is performed and second, that objects can only be directly influenced by their immediate surroundings. Quantum physics breaks with local realism since the position and velocity of quantum objects is not defined before measurement is done, and because the objects may in certain cases influence each other without local contact. (see entanglement) (see nonlocality) (see Copenhagen interpretation) (see realist view)

**luminosity** Luminosity is the amount of electromagnetic energy emitted per unit time with the radiation of a star, galaxy or other radiating astronomical object. Luminosity has units of joules per seconds or watts.

**matter** Matter is any substance that has mass. (see mass)

**mass-energy equivalence** The mass-energy equivalence states that the two fundamental quantities of mass and energy are directly related: mass is a form of energy; the mass of any object represents a certain amount of energy and vice versa. This equivalence is captured in Einstein's most famous equation: $E = mc^2$ – where $E$ is the energy of an object, $m$ its mass, and $c$ the speed of light. (see rest mass) (see mass)

**mass** Mass describes the inertial property of matter: the tendency to keep moving and to resist acceleration if a force is applied. The mass of an object also determines the strength of the gravitational attraction to other objects. Mass has units of kilogram (kg). In special relativity, mass has two meanings: rest mass is the same for all observers in all reference frames; relativistic mass increases with velocity and is observer dependent. General relativity describes how mass connects matter to space. (see matter) (see mass-energy equivalence) (see rest mass) (see relativistic mass)

**measurement** In quantum physics, measurements must be seen as an interaction between the measuring device and the system being measured. Rather than simply detecting the state of a quantum system, measurements will cause the system to go into another state. (see "Copenhagen interpretation" and "superposition")

In special relativity, a measurement usually refers to an inertial observer assigning a time and place to an event. (see inertial observer, observer)

**mechanical world view** The mechanical world view was a prevalent view among physicists in the 19th century that emphasised the reality of atoms and their interaction. The mechanical world view assumed that natural phenomena could be explained by the action of forces representing either attraction or repulsion, depending only upon distance and acting between unchangeable particles. In the 20th century, the theory of relativity and quantum physics replaced the mechanical world view by the nonmechanical view captured in Einstein's field equations and in quantum physics with its indeterminacy. (see Einsteinian physics) (see statistical description) (see Einstein's field equations)

**micron** 1 micron is 1-millionth of a metre commonly known as a micrometre.

**Minkowski space** Minkowski space is the mathematical description of Einstein's theory of relativity. Minkowski space combines three-dimensional Euclidean space and time into a four-dimensional structure called spacetime. (see spacetime) (see Euclidean space)

**Mission Gravity** Mission Gravity is a virtual reality programme developed by the Education and Outreach Team at the Australian Research Council Centre of Excellence for Gravitational Wave Discovery (OzGrav). Mission Gravity was designed with a focus on the holistic incorporation of curriculum, pedagogy, content, and technology to highlight how virtual reality technology can be used to effectively teach contemporary physics concepts.

**model** (see scientific model) (see teaching model)

**momentum** Momentum is a physical quantity that corresponds to the ability of moving objects to push other things. In a closed system, the total amount of momentum remains unchanged and is conserved. Newton's second law of motion states that the rate of change of momentum of an object is equal to the net force acting on it. For massive objects, momentum is the product of mass and velocity. For photons, momentum is energy divided by the speed of light. Momentum has units of kilogram metre per second. (see mass) (see inertia) (see radiation pressure)

**monochromatic** Monochromatic describes electromagnetic radiation consisting of one wavelength or frequency. This term is mostly used to express the single wavelength and frequency of light and sound. For example, light emitted from a laser is highly directional and has a very narrow frequency range, so it is almost monochromatic and can be regarded as having only one frequency. (see light)

**multimessenger astronomy** Multimessenger astronomy combines the observation of different forms of radiation or other messengers from the Universe to study astrophysical objects. These observations rely on the electromagnetic spectrum that ranges from radio waves to gamma rays, cosmic particles such as neutrinos, and gravitational waves.

**nature of science** Many science curricula around the world ask students to require both knowledge in science and knowledge about science. Knowledge about science concerns the nature of science. Such knowledge includes understanding of the scientific method and the acknowledgement that science is a human endeavour. In their attempt to explain natural phenomena, scientists constantly refine and re-examine scientific knowledge which is, therefore, always subject to change. (see scientific model)

**neutron star** A neutron star is the remnant of the core of a massive star, which is left after the star has exploded in a supernova explosion. Neutron stars consist mostly of neutrons, and their density is comparable to that of an atomic nucleus. Neutron stars have typical masses between one and two times the mass of the Sun, but at a diameter of a mere 10 to 20 km. (see stellar evolution)

**neutrino** Neutrinos are quantum particles with no electrical charge, which only interact with other particles via the weak nuclear force or via gravity. They are typically produced when neutrons or protons transmute into each other via radioactive beta decay. Neutrinos are created in nuclear reactions in the sun, and pass through matter very freely, with hardly any absorption. They have extremely low rest mass, about $10^6$ times less than an electron. (see particle)

**networked learning community** A networked learning community is a collaborative group of teachers, usually teaching the same subject, from across a number of schools.

**Newton, Isaac** Sir Isaac Newton (1642–1727) was a natural philosopher widely recognised as one of the most influential scientists of all time. Newton developed the mathematical formalism of calculus to formulate the laws of nature. His book *Mathematical Principles of Natural Philosophy* laid the foundations of classical mechanics by introducing the three basic laws of motion and the force model of gravity. (see mechanical view)

**Newtonian limit** On Earth, we mostly experience the laws of physics in accordance with Newton's theory of gravity. Physicists call gravitational conditions that reduce to this classical description the Newtonian limit. In this limit, particles move slowly compared to the speed of light and gravitational fields are weak and not changing in time. (see gravity)

**noise (in gravitational wave detectors)** Gravitational wave detectors have to filter the weak gravitational wave signals out of the noise of the measured data. The main noise sources are:

- Seismic noise is the ever-present internal motion of the Earth and its effect on all components of the detector.
- Thermal noise is the effect that temperature has on the detector. The thermal motion of the atoms cause noise in detector signals.
- Quantum noise is a collective term for all noises associated with the quantum-nature of light, i.e. the fact that detector light beams are made up of photons.

→ (see gravitational wave) (see light)

**non-Euclidean geometry** In mathematics, non-Euclidean geometry describes geometries in which the parallel postulate of Euclidean geometry does no longer hold. Two straight parallel lines do not remain at constant distance from each other but either curve away or curve toward each other. (see Euclidean geometry) (see curvature)

**nonlocality** In physics, nonlocality is a principle that entails that quantum objects that share the same quantum state may in certain cases influence each other at a distance without local contact. Nonlocality is opposed to the principle of locality that assumes that objects can only be directly influenced by their immediate surroundings. Nonlocality also describes the aspect of the quantum world that makes objects indeterminate until the time of measurement. (see entanglement) (see local realism)

**observer** The notion of an observer has different meanings in different theories in physics. Often, these meanings differ from the ordinary English meaning of observer.

In the theory of relativity, an observer usually stands for a specific reference frame from which events and phenomena are being measured. Such an observer could be a global reference frame which assigns time and location coordinates to events. Alternatively, an observer could be an inertial reference frame where time is measured on a local clock. (see inertial frame, inertial observer)

In quantum mechanics, observers are measurement devices where measurements are understood as interactions between the observer and the system being measured. (see measurement)

**observation** Observations are crucial components of all physical experiments. In the theory of relativity, observations are understood as assignments of time and location coordinates to events. In quantum physics, observations are understood as quantum measurements, i.e. interactions between the measuring device and the system being measured. Rather than simply detecting the state of a quantum system, observations cause the system to go into one of its several possible states. (see observer) (see event)

**OzGrav** OzGrav is the acronym of the Australian Research Council Centre of Excellence for Gravitational Wave Discovery. The mission of OzGrav is to use the historic first detections of gravitational waves to understand the extreme physics of black holes and warped spacetime and to inspire the next generation of Australian scientists and engineers through this new window on the universe. (see Mission Gravity) (see EPER collaboration)

**paradigm** A paradigm is a shared world view of theoretical assumptions, methodological principles and norms of practice of a discipline. Paradigms provide the frameworks that underlie a way of thinking about phenomena; if you change paradigms, you change the way you think about something. The shift from classical to Einsteinian physics is one example of a paradigm change in physics. (see Einsteinian physics) (see zeitgeist)

**particle** In physics, a particle is an object or entity that is delimited to a small area in space. The size of particles varies greatly from subatomic particles like electrons, to microscopic particles like atoms, to macroscopic particles like granular materials. (see wave nature of particles)

**pedagogical content knowledge** Pedagogical content knowledge describes the knowledge possessed by teachers about how to make a topic accessible to learners.

**perihelion precession, anomalous** By Kepler's first law, an object with a small mass orbiting a body with a great mass will follow an elliptical orbit, with the heavy central body in one of the focus points. Einstein showed that, in general relativity, the ellipse as a whole would slowly rotate, the orbital point closest to the central body precessing slowly in a circle. For the planet Mercury, such perihelion precession was found in the late 19$^{th}$ century. Most of the precession is due to the gravitational pull of the other planets in the solar system. The magnitude of this effect can be predicted using Newtonian mechanics. But a small part could not be accounted for using Newtonian mechanics. Albert Einstein was able to explain that this discrepancy was caused by curvature of spacetime. (see general relativity)

**phasor** A phasor is a vector that describes the phase of a wave at a particular point. Phasors rotate in an anticlockwise direction as time passes or as distances along a wave increase. The length of the phasor vector represents the wave amplitude. Phasors allow the addition of waves of the same frequency to be reduced to simple addition of vectors. (see wave) (see amplitude)

**phonon** A phonon is a quantum or wave packet of the collective vibrations of groups of atoms. Phonons are the basic constituents of both sound and heat. Phonons are sometimes called quasiparticles. Phonons are the quantised vibrations of the relative position of atoms, that have energy given by $E = hf$, the same as photons.

**photon** A photon is the particle of light and the fundamental constituent of electromagnetic waves. Photons have energy and momentum, but zero rest mass. Each photon has the energy: $E = hf$ where $f$ is the frequency and h is Planck's constant with $h = 6.62607015 \times 10^{-34}$ J·s. (see light) (see electromagnetic spectrum) (see particle)

**photoelectron** A photoelectron is an electron emitted from an atom or a metal when the absorption of a photon causes it to be ejected. (see photoelectric effect)

**photoelectric effect** The photoelectric effect illustrates the particle-nature of light. In the experimental setup of the photoelectric effect, electrons are emitted from a metal surface only when light of sufficiently high frequency hits the surface. The wave-model of light cannot explain the photoelectric effect because a low-frequency beam of light at a high intensity cannot build up the energy required to produce the effect. Einstein used the concept of photons (particles of light) to explain that one photon transfers all its energy to one electron. (see photon) (see light)

**physics education research** Physics education research (PER) is an interdisciplinary form of research that investigates learning and teaching in physics with the aim of improving instructional practices. Physics educators comprise a diverse group of people: physicists, educational researchers, physics teachers, science communicators, public outreach staff, and even educational policy makers often work in the service of physics education research.

**professional learning** Professional learning describes activities supporting the professional growth or development of teachers. (see pedagogical content knowledge)

**probability wave** (see wave function)

**principle of equivalence** The principle of equivalence is a fundamental principle of general relativity. Einstein called the principle of equivalence the happiest thought of his life. In its original form, the principle states that locally, it is not possible to distinguish between a gravitational field and an accelerated system.

An observer who is falling freely in a gravitational field measures no local effects of gravity at all; she is weightless. (see weak equivalence principle)

**principle of relativity** The principle of relativity describes one of three key principles in physics:

1) principle of relativity (classical mechanics)

The principle of relativity in the context of classical mechanics goes back to Galileo Galilei and states that the laws of mechanics are the same in all inertial frames of reference that move with constant velocity with respect to each other.

2) principle of relativity (special relativity)

The principle of relativity in the context of special relativity states that the laws of physics are the same for all inertial frames. Einstein extended the classical version of the principle of relativity to replace 'the laws of mechanics' by 'the laws of physics' including the laws of electromagnetism. (see inertial frame) (see special relativity)

3) principle of relativity (general relativity)

The principle of relativity in the context of general relativity states that the laws of physics are the same for all frames of reference. Einstein extended the principle of relativity in special relativity to include all reference frames instead of just inertial frames. That is, all coordinate systems (even accelerating frames) are treated equally in general relativity. (see principle of equivalence)

**proper time** Proper time is a concept in special relativity that describes the time associated with an observer, as measured by her own wristwatch following the worldline of the observer. "Proper" refers to "property" or "ownership", not correctness. Proper time is independent of the choice of coordinates. (see wristwatch) (see worldline) (see special relativity)

**pulsar** A pulsar is a highly magnetised rotating neutron star that emits beams of electromagnetic radiation from its poles. As it rotates, the beams act as a "lighthouse" and this effect causes a pulsed appearance of the emission. (see neutron star) (see stellar evolution)

**qualitative content analysis** Qualitative content analysis is a method of qualitative data analysis that aims to identify meanings of content within a given text through systematic reading and coding of the text. Qualitative content analysis is a common method in physics education research. (see physics education research)

**quantum physics** Quantum physics is a fundamental theory of theoretical physics that describes nature on an atomic and sub-atomic scale. If physicists apply the mathematical model of quantum physics to macroscopic objects, they obtain the same results as in classical mechanics. Nevertheless, quantum physics breaks with three central concepts of classical physics: continuity, determinism, and local realism. Quantum physics breaks with continuity because energy and certain other quantities are not continuous but quantized. Quantum physics breaks with determinism because it cannot predict the outcome of single events. What it can predict with great accuracy is the statistical distribution of many events. Quantum physics breaks with local realism because quantum objects do not have a real position or velocity before a measurement is taken and because quantum objects may in certain cases influence each other without local contact. (see local realism) (see Copenhagen interpretation) (see statistical description)

**quantum scattering** Quantum scattering is a process in which particles collide and exchange energy and momentum with other particles, causing new particles to emerge with different energy and momentum. (see Compton scattering) (see photoelectric effect)

**quantum state** A quantum state describes the state of a quantum system, i.e. a part of the physical world that is to be studied in an experiment. Quantum states assign probabilities of the occurrence of different outcomes in the experiment for each physical quantity that can be measured. Knowledge of the quantum state allows physicists to make predictions about the future development of the system. (see quantum physics) (see statistical description)

**quantum vacuum fluctuations** Quantum vacuum fluctuations describe the temporary appearance of energetic particles out of empty space. These temporary changes of the amount of energy at a given point in space are a consequence of the uncertainty principle in quantum physics. Quantum fluctuations allow for the creation of virtual particle – antiparticle pairs. (see antimatter) (see uncertainty principle)

**radiation pressure** Radiation pressure is the momentum per square metre transferred to an object when photons are either absorbed or reflected. If photon(s) impact on a surface they create radiation pressure. Although the force exerted by photons is miniscule in the context of day-to-day life, radiation pressure becomes important in sensitive instruments such as gravitational wave detectors and it also has significant effects on spacecraft trajectories.

**raisin cake model (of cosmic expansion)** The raising cake model of cosmic expansion is a teaching model of general relativity that uses a thought experiment to demonstrate the cosmic expansion. The pair-wise distances between raisins in a cake that rises in the oven change in a similar way as the distances between (Hubble-flow) galaxies during cosmic expansion. (see Hubble flow) (see cosmic expansion) (see thought experiment) (see teaching model)

**realist view (on the description of physical phenomena)** A realist view is a philosophical position that assumes that quantum objects have actual, definite properties independently of our description of them and regardless of whether or not a measurement is performed on the objects. This is in contrast to the Copenhagen interpretation of quantum physics. The question of how to interpret quantum physics is still an ongoing philosophical dispute. (see Copenhagen interpretation) (see quantum physics) (see local realism)

**reference frame** A reference frame is a coordinate system with a set of physical reference points that allows describing the world through standardised measurements. Reference frames specify the relationship between observers and observed phenomena. (see inertial frame)

**relativistic mass** Relativistic mass is a term for the velocity-dependent notion of mass. In contrast to the concept of rest mass that is the same for all observers in all reference frames, relativistic mass is observer dependent. Many physicists criticize the usage of the term relativistic mass because two different meanings of mass can create conceptual confusions. Therefore, many prefer to use the term mass to mean rest mass and to not use the term relativistic mass at all. (see mass) (see rest mass) (see mass-energy equivalence)

**relativistic time dilation** (see time dilation)

**relativistic length contraction** (see length contraction)

**relativity principle** (see principle of relativity)

**relativity of simultaneity** Simultaneity describes the relation between two spacetime events that happen at the same time in a given reference frame. Special relativity predicts that simultaneity of events is not absolute but depends on the observer and her reference frame. Two events that are simultaneous but separated in space according to one observer need not be simultaneous according to another observer in relative motion to the first observer. (see event) (see special relativity) (see reference frame)

**ReleQuant** ReleQuant is a physics education project in Norway that has developed digital, research-based learning resources in general relativity and quantum physics for upper secondary physics students. ReleQuant has combined the development of digital learning resources and research on students' learning processes and motivation in physics. (see EPER collaboration)

**rest mass** According to special relativity and due to the mass-energy equivalence, the relativistic mass of a body is dependent on its velocity, temperature and other variables. The rest mass of a body is the value that the mass assumes when the centre of mass of the body is at rest. The rest mass is a quantity which is the same for all observers in all reference frames. (see mass-energy equivalence) (see mass) (see special relativity) (see relativistic mass)

**river model (of black holes)** The river model is a teaching model of general relativity that represents spacetime as a river with variable flow speed, and test particles as fish with a maximum swimming speed. The model can be used to explain the concept of a black hole horizon. (see black hole) (see teaching model)

**rubber sheet model (of curved spacetime)** The rubber sheet model (often called rubber sheet analogy) is a teaching model of general relativity that illustrates the basic mechanism of gravity. An elastic sheet, weighted down in the middle by a heavy sphere, represents curved spacetime. The way the sheet's curvature influences the motion of smaller spheres rolling on the sheet represents the way distorted spacetime influences the motion of test particles. (see teaching model) (see analogy) (see gravity)

**rubber balloon model (of cosmic expansion)** The rubber balloon model is a teaching model of general relativity that relies on a rubber balloon whose surface represents space, with stickers affixed to the surface representing galaxies (in the Hubble flow). The model can be used as a physical model of the systematic changes in distances between galaxies in an expanding universe. (see cosmic expansion) (see teaching model)

**rubber band model (of cosmic expansion)** The rubber band model is a teaching model of general relativity that relies on a one-dimensional representation of the expanding universe. The rubber band model is well-suited for physical measurements. Marks on the (linear) rubber band represent galaxies; a stretching of the rubber band simulates cosmic expansion. (see teaching model) (see cosmic expansion)

**saddle surface** (See hyperbolic paraboloid)

**SAMR Model** The SAMR Model is an educational framework to assist in the evaluation of the impact of technology in the classroom, focusing on four categories for varying degrees of technological integration: Substitution, Augmentation, Modification, and Redefinition (see TPACK framework)

**Schrödinger's cat** Schrödinger's cat is a thought experiment proposed by Erwin Schrödinger to show that the idea of superposition of quantum states has seemingly absurd consequences. In the thought experiment, a cat is enclosed in a box containing a radioactive source and a flask of poison. If the radioactive source decays (and emits an energetic particle or a photon), it will release the poison and the cat will die. But a radioactive nucleus is a quantum system that can only be described by a probability for the nucleus to emit radiation during a certain period of time. We cannot, therefore, know exactly when the radiation will be emitted: the nucleus is in a superposition of the two states "emitted radiation" and "not emitted radiation". An observation of the system will cause it to go into one specific state, for instance "emitted radiation". Since the cat's destiny depends on whether the nucleus decays or not, can we then say that the cat is in a superposition of the states of being dead and being alive? (see superposition) (see thought experiment) (see quantum physics) (see observation)

**Schwarzschild solution (Schwarzschild metric)** In general relativity, the Schwarzschild solution (also known as Schwarzschild metric) is the solution to Einstein's field equation that describes the spacetime geometry around a spherically symmetric static object such as a star or a black hole. The Schwarzschild solution predicts the existence of a singularity. Surrounding the singularity is a surface we call the horizon, which is a one-way membrane: things from outside can cross to inside, but not the other way around. Physicists are troubled by the singularity because the laws of physics as we currently understand them do not tell us what happens when an infalling object reaches the singularity. (see black hole) (see Einstein's field equations)

**Schwarzschild radius** The Schwarzschild radius is the physical parameter $R_S = 2GM/c^2$ that appears in the Schwarzschild solution of Einstein's field equations. This parameter describes the event horizon of a spherically symmetric, non-rotating and electrically neutral black hole with mass *M*; here, *G* is the gravitational constant and *c* the speed of light. The event horizon marks the region of space in which the gravitational field is so strong that no light can escape. The ratio of the true size of an object and its Schwarzschild radius determines how important general-relativistic effects are. (see Schwarzschild solution) (see black hole) (see river model)

**scientific inquiry** Scientific inquiry is the process of discovery in science that involves asking questions, making predictions, testing hypotheses, and studying the natural world. As part of that process, scientists build scientific models to explain natural phenomena. (see scientific model) (see nature of science)

**scientific model** Scientists build scientific models to explain natural phenomena; models build the basis of our physical intuition. There can be different models of the same physical phenomenon. Both Newton's force model and Einstein's spacetime model provide explanations of gravity. Einstein's model has a greater explanatory scope than Newton's model because it predicts phenomena (such as gravitational time dilation) that the force model cannot explain. (see scientific inquiry) (see nature of science) (see intuition)

**sector model** Sector models are geometric models that represent curved surfaces, spaces or spacetimes true to scale. They have been developed for teaching general relativity without using advanced mathematics. (see curvature)

**simultaneous** Two events are simultaneous when they have the same time coordinate in a given reference frame. (ree relativity of simultaneity) (see reference frame)

**singularity** (see Schwarzschild solution) (see big bang)

**sinusoidal wave** A sinusoidal wave describes a smooth periodic oscillation. Sinusoidal waves can be generated by the trigonometric sine function related to the rotation of the wheel.

**screencast** Screencast describes the transmission of video data to an external screen. In this book, the video data comes from a virtual reality headset and the screencast allows students working in groups to communicate and collaborate. (see virtual reality)

**sociocultural theory** Sociocultural theory is a theory of learning that understands learning as a social process. According to this view, collaboration, social interaction, and cultural tools (such as language) mediate learning and development. (see constructivism) (see cognitivism) (see talking physics)

**space** Space is a three-dimensional medium which encompasses the entire universe and in which objects and events have relative position and direction. Space has a shape which is determined by the influence of masses. Even what we consider empty space is filled with dark matter and dark energy whose nature is still unresolved. (see spacetime)

**spacetime** Spacetime is the four-dimensional continuum defined by the three dimensions of space and one dimension of time. Neither space nor time are independent entities; rather, they are inseparably linked. Spacetime is often conceptualised as a smooth fabric to help visualise the underlying structure of the Universe. In special relativity, the observed rate at which time passes for an object depends on the relative velocity of the object to the observer. In general relativity, spacetime interacts dynamically with massive objects in the Universe leading to gravitational phenomena. (see spacetime diagram) (see Minkowski space)

**spacetime diagram** A spacetime diagram is a graphical tool that illustrates movement in spacetime and helps building conceptual understanding and intuitions of relativistic phenomena such as relativistic time dilation or length contraction. Spacetime diagrams are two-dimensional coordinate systems with one space and one time dimension. The units of measurement are chosen in such a way that light moves along 45° trajectories. (see special relativity) (see Minkowski diagram) (see special relativistic phenomenon)

**spacetime event** (see event)

**spacetime perpendicularity** In relativity, an observer's sense of space is "perpendicular" to her sense of time. Spacetime perpendicularity describes the time-axis of an observer as being perpendicular to the observer's space-axis (which represents an instant of time for that observer). The diagonals of light clock diamonds of an observer encode what it means to be perpendicular in the geometry of spacetime. (see light clock diamonds) (see spacetime diagram)

**Spacetime-Travel Project** The physics education group at Hildesheim University, Germany, runs the project Spacetime-Travel to facilitate the teaching of special and general relativity at school and university. The group develops novel teaching materials, in particular, visualisations and models; the group also develops teaching modules for school and university education. (see sector model)

**special relativity** Special relativity is Einstein's theory of space and time in the absence of gravity. Special relativity differs from classical mechanics in that inertial observers come to different conclusions regarding simultaneity, lengths, and the passage of time, but find the same constant value for the speed of light. Einstein based special relativity on two postulates: the light postulate and the principle of relativity. (see light postulate) (see principle of relativity)

**special relativistic phenomenon** A special relativistic phenomenon is a phenomenon that can be observed and is explained in the context of special relativity. Examples of such phenomena are the relativity of simultaneity, relativistic time dilation, relativistic length contraction, and the energy-mass equivalence. (see special relativity) (see time dilation) (see length contraction)

**speed of light principle** (see light postulate)

**spontaneous models of light propagation** Spontaneous models of light propagation are ways of reasoning that students exhibit spontaneously and unprompted while performing thought experiments in special relativity. (see constructivism) (see thought experiment) (see light postulate)

**Standard Model of particle physics** If we go down to the very fundamental components of matter, we reach the very end, popularly known as the Standard Model of particle physics. Broadly, the Standard Model describes fermions (quarks and leptons) as the basic building blocks of matter and bosons (gluons, photons, W, Z-bosons and Higgs) as the mediators of force carriers.

**standing wave** A standing wave is a wave that oscillates in time while its peak amplitude at each point does not move in space. Standing waves are patterns of vibration that occur when the medium is moving in the opposite direction to the wave, or when two waves moving in opposite directions interfere. (see wave)

**statistical description (of quantum physics)** In quantum physics, the outcome of measurements must be described in terms of probabilities. Quantum physics can predict with great accuracy the statistical distribution of the outcomes of many measurements, but not the outcome of single events. Quantum physics thus breaks with classical determinism. (see classical determinism) (see quantum physics) (see measurement)

**stellar evolution** The stellar evolution describes the life cycle of a star defined by the various physical parameters and changes over a star's lifetime. (see stellar remnant) (see neutron star) (see black hole)

**stellar remnant** Stellar remnants are dense objects that remain when a star reaches the end of its life: white dwarfs for low-mass progenitor stars, neutron stars or even black holes for high-mass stars. (see stellar evolution) (see black hole)

**STEAM** STEAM stands for Science, Technology, Engineering, Arts and Mathematics. The term adds the arts to the term STEM in an attempt to integrate all these subjects to teach more holistically across disciplines. (see STEM) (see physics education research)

**STEM** STEM stands for Science, Technology, Engineering and Mathematics. The term is usually used in the context of education, education policy, and curriculum design. (see STEAM) (see physics education research)

**supernova** A supernova is a stellar explosion triggered by the gravitational collapse of a high mass star's core. (see stellar remnant) (see stellar evolution)

**superposition of quantum states** A superposition of quantum states describes a quantum system that exists in several different quantum states at the same time. A measurement on the system will cause it to fall into one of its possible states. (see quantum states) (see measurement) (see Copenhagen interpretation)

**student-centred education** Student-centred education focusses on students' learning for meaningful understanding. Students actively construct their own ideas and knowledge rather than being dependent on the teacher or lecturer to present these ideas and knowledge to them. In this book, we use student-centred education as one example of a paradigm in education.

**talking physics** Talking physics is an instructional approach that invites students to use oral language to promote conceptual understanding. For example, open discussion tasks give students the opportunity to reason with their peers about concepts of Einsteinian physics. (see sociocultural theory) (see ReleQuant)

**teacher-centred education** In teacher-centred education, the teacher or lecturer presents ideas and knowledge to the students in a structured and well-organised manner, leaving limited opportunity for students to express their ideas. In this book, we use teacher-centred education as one example of a paradigm in education. (see student-centred education) (see paradigm)

**teaching model (of general relativity)** A teaching model of general relativity is a concrete or abstract entity that represents an aspect of general relativity, or structures related to general relativity's concepts and applications, in a simplified way, for the purpose of teaching about general relativity and its applications. (see rubber sheet model)

**thought experiment** A thought experiments is an essential tool of reasoning in science and science education. Common to thought experiments is a structured way of thinking with the goal of exploring consequences, predictions, or explanations of phenomena. This structured way of thinking consists of several elements: posing a question, drawing on theory, imagining an experimental setup, obtaining the outcome through deductive reasoning and arriving at an overall conclusion. (see Schrödinger's cat)

**time** Time is the indefinite continued progress of existence and events that occur in an apparently irreversible succession from the past, through the present, into the future. We cannot travel into the future, but we can change the rate of travelling into the future. Time appears as a relative quantity that can vary for different observers, depending on their speed through space. Theoretically and experimentally, objects moving faster than us will age less rapidly. Time will also slow down in response to gravity. Matter curves space and warps time. (see spacetime)

**time dilation** Einstein's theory of relativity predicts two different types of time dilation: relativistic and gravitational time dilation. Relativistic time dilation is a phenomenon of special relativity: For an observer in an inertial frame, a clock that is moving relative to her is measured to tick more slowly than a clock that is at rest in her inertial frame. Gravitational time dilation is a phenomenon of general relativity: Time passes more slowly in the direct neighbourhood of a massive object than it does further away. Gravitational time dilation is closely linked to the phenomenon of gravitational frequency shift. This link can be illustrated by light clocks in a gravitational field that measure the passing of time via light signals. (see inertial observer) (see gravitational frequency shift) (see light clocks) (see warped time) (see warped-time model) (see ageing Earth model)

**TPACK framework** The TPACK framework is a framework developed to highlight the Technological Pedagogical Content Knowledge teachers need to effectively teach with technology. (see pedagogical content knowledge) (see SAMR model)

**twin paradox** The twin paradox is a false paradox in special relativity arising from the misconception that a frame of reference that is "at rest" is equivalent to a frame that is inertial. The starting point of the twin paradox is a thought experiment involving two identical twins on Earth one of whom leaves the planet in a rocket that travels at a high speed close to the speed of light. Since each twin sees the other twin as moving, a naïve application of relativistic time dilation leads to the conclusion that each twin will find the other one to have aged less. However, the situation between the two twins is not symmetrical. The journey of the traveling twin involves accelerating frames; any frame of reference can be at rest—including accelerating frames. However, accelerating frames cannot be inertial frames. (see special relativistic phenomenon) (see time dilation)

**Ultra-relativistic particle** An ultra-relativistic particle is a massive particle whose total energy, consisting of rest-mass energy and kinetic energy, is almost totally due to kinetic energy. These particles move almost at the speed of light. An example is the neutrino. Particles for which the rest-mass energy dominates are called non-relativistic and their speed is much less than that of light. (see rest mass) (see mass-energy equivalence) (see neutrino)

**uncertainty principle** In quantum physics, the uncertainty principle (often called Heisenberg's uncertainty principle after German physicist Werner Heisenberg) describes a fundamental property of quantum systems inherent to the wave-like nature of these systems. The principle states that there are pairs of quantities in nature that cannot have well-defined values at the same time. For example, the uncertainty principle tells us that the more well-defined a particle's position, the less well-defined its momentum - and vice versa. A related, yet distinct, effect is the observer effect that captures the fact that no measurement of quantum systems can be made without affecting the system. (see quantum physics) (see measurement) (see statistical description) (see wave-particle duality)

**vector** A vector is a mathematical quantity described by an arrow. Vectors quantify the magnitude and direction of an object in a system. For example, force, velocity, acceleration, displacement, and momentum are vector quantities. Phasors are examples of vectors. (see phasor)

**virtual reality** Virtual reality describes the use of computer technology to create a simulated, often three-dimensional, environment in which the user experiences to be immersed in the environment. Users usually experience virtual reality by use of a wearable headset. (see Mission Gravity) (see virtual reality headsets)

**virtual reality headset** Virtual reality headsets (or virtual reality goggles) are head-wearable devices that completely cover the eyes to allow the user to experience immersive computer-simulated environments. (see virtual reality)

**virtual photons** Virtual photons are imaginary entities which do not have any physical existence until measured. Virtual photons do not have precise energy and momentum like real photons. Their mechanics is best described by the path integral formulation. The probability amplitude of a photon travelling between two points can be predicted by taking the vector sum of all the phasors attributed to multiple virtual photons.

**warped time** Warped time describes the phenomenon of general relativity that time passes more slowly in the direct neighbourhood of a massive object than it does further away. The rate at which time passes depends on the clock's position relative to sources of gravity. (see time dilation) (see ageing Earth model) (see warped-time model)

**warped-time model** The warped-time model is a teaching model of general relativity that compares Einstein and Newton's theories of gravity. The model draws on an interactive height-time diagram in which the time axis can be warped (i.e. stretched out of shape). The model allows linking free fall movement to a geometric model of gravity via "Einstein's first law". (see warped time) (see teaching model) (see Einstein's first law)

**wave** A wave is a disturbance that travels through a medium and transports energy from one location to another.

**wave function** The wave function (also called probability wave) is a variable quantity in quantum physics that mathematically describes the wave characteristics of a particle or quantum state. As a mathematical entity, the wave function has properties very similar to that of a classical wave, such as superposition and interference. The square of the amplitude of the probability wave at any given point determines the probability of finding a quantum particle at that point. (see wave nature of particles)

**wavelength** The wavelength of a wave describes the distance between successive crests, or alternatively successive troughs, of a wave. The unit of wavelength is meter. (see wave)

**wave nature of particles** In quantum physics, particles exhibit wave nature when they display properties typical of waves. For example, electrons passing through a double slit form an interference pattern on a screen in a similar way as light or water waves would. Quantum physics tells us that quantum objects including electrons (which would classically be seen as particles) may be ascribed a wavelength and a frequency and can be described mathematically by wave functions. (see wave-particle duality) (see wave function)

**wave-particle duality** In quantum physics, the wave-particle duality describes the idea that quantum objects such as photons or electrons behave as waves in some experiments and as particles in others. However, these objects can never be observed as both wave and particle at the same time. (see quantum physics) (see wave nature of particles) (see light)

**wave train** A wave train describes a group of waves of equal or similar wavelengths travelling in the same direction. (see wave)

**weak equivalence principle** The weak equivalence principle states that the acceleration at which a test body falls does not depend on the test body's internal structure or its composition. At a given point in a gravitational field, all test bodies experience the same acceleration which is called the universality of free fall. (see gravity) (see free fall) (see principle of equivalence)

**weight** In classical physics, weight is the magnitude of the force exerted by the Earth on a body to keep it in place: the weight of a body is the quantity measured by a spring scale. In general relativity, weight is the magnitude of the force that deviates objects from their straight path through curved spacetime. Objects in free fall are weightless. (see mass) (see gravity) (see free fall)

**weight** In classical physics, weight is the magnitude of the force exerted on a body to keep it in place (balancing the pull from the force of gravity): When you stand on a spring scale, you exert a force on the scale equal to the pull of the Earth's gravitational force on your body. The spring scale exerts an equal and oppositely directed force on you, keeping you in place; this is your weight. In general relativity, weight is the magnitude of the force that deviates objects from their straight path through curved spacetime. Objects in free fall are weightless. (see mass) (see gravity) (see free fall)

**worldline** A worldline is a one-dimensional curve in spacetime that corresponds to the movement of a particle. Usually, worldlines are illustrated as curves in spacetime diagrams. (see spacetime diagram) (see Minkowski space)

**wristwatch** In special relativity, a wristwatch is the clock belonging to an observer that displays the observer's proper time. (see proper time) (see inertial observer)

**X-ray imaging** X-ray imagining describes the use of x-rays – photons with high energy and short wavelength – to form images of the interior of a human body or some other system. A stream of photons produced in an X-ray tube is directed at a patient, and a detector is placed behind the patient. On their way through the body, some photons are absorbed or scattered, whereas other photons pass straight through the patient and hit the detector, where they are registered. Different structures in the body absorb or scatter X-ray photons to different extents, giving lighter or darker areas on the detector screen and thus forming the image. (see electromagnetic spectrum)

**zeitgeist** Zeitgeist is the spirit or mood of a particular period of history as shown by the predominant ideas, ways of thinking, and beliefs of the time. (see paradigm)

**zone of proximal development** In educational research and practice, the zone of proximal development describes learning that occurs when a student interacts with a more knowledgeable other such as a peer or a teacher to learn skills that are close to being mastered.

# Index

www.ingramcontent.com/pod-product-compliance
Lightning Source LLC
LaVergne TN
LVHW080804011026
849160LV00015B/1526
*9781760877712*